CE MARKING

Of Electrical and Electronic Products

By
Chetan Kathalay
B.E. (Electronics)

FIRST EDITION

Pune, India.

CE MARKING –Of Electrical and Electronic Products
-Chetan Kathalay

First Print: 2020

Set in Garamond.
Copyright © Chetan Kathalay 2019.

ISBN: *9798643675440*

9 798643 675440

To my wife Madhuri and daughter Aditi

And

To my parents who always encouraged me to learn

CONTENTS

CHAPTER 3: DESIGN FOR SAFETY

CHAPTER 4: THE EMC DIRECTIVE

CHAPTER 5: EMC FUNDAMENTALS, STANDARDS AND TESTING

CHAPTER 6: EMC DESIGN METHODOLOGIES

CHAPTER 7: OTHER DIRECTIVES

CHAPTER 8: TECHNICAL DOCUMENTATION/FILE

PREFACE

CE marking has become THE most sought after marking for manufacturers since it not only introduces them to new avenues for expanding their business to the European Union (EU) but also enhances product acceptance in their domestic market. Although the CE marking simply means that the product complies with applicable European laws, such has been its impact that many have started considering it as an ultimate symbol of quality (which it isn't) and that gives a big boost to product sales and gives them universal acceptability.

As compared to other conformity assessment processes, CE marking remains by far the simplest since it allows *self-declaration* on part of the manufacturer.

People who are new to the CE marking process and who have gone through a rather painful process of other product and system certifications, the concept of self-declaration is out of this world. In fact, the process for majority of the products (which are in the *low risk* category) is so simple that many find it hard to believe. The plight of these people is like an Australian who is visiting Africa for the first time and is shown a Giraffe. The Australian exclaims that *such an animal does not exist* although the animal is right there in front of him! The reason why people don't believe in the simplicity of the process is because they try to draw parallels between CE marking and other product certifications (like UL, FCC, SABS, C-tick etc) or even between system certifications like ISO 9000. This is the reason that misconception, misinformation and confusion rule the roost and that there is no clarity. The confusion is also partly created because many CE marking consultants also happen to be ISO 9000 consultants and who fail to drive home the point that CE marking is a product certification as opposed to ISO 9000 which is a system certification. Also, these consultants in an attempt to play safe, rely on over-compliance insisting on certain steps which may be unwarranted and which, often than not, become the norm that is then followed blindly.

It should be noted that although the CE marking process is simple, it requires compliance of the product with certain technical requirements. It is here that the real challenge lies especially to those manufacturers who are new to the process. This book, while providing a step-by-step approach to CE marking of electrical and electronic equipment also seeks to provide insights into product design and test methodologies so that the product meets the technical requirements. It also seeks to clarify the many doubts and misconceptions about CE marking. Contrary to other certification processes, CE marking has a two-pronged approach with the conformity process being split into two –the *legal* process and the *technical* process which is one of its unique features. The legal process requires that the manufacturer himself/herself declare conformity to certain *essential* requirements given by a legal document called as the *directive*. The basis of this declaration could be a simple technical justification or a laboratory test report that demonstrates conformity to a *standard* (which translates the essential requirements into technical tests). This standard is prepared by a standardisation body that has been authorised by the very legal entity that makes the directive, so that the process comes a full circle.

The book begins with a chapter that introduces the reader to the nuances of the CE marking process, the ways or methods to approach conformity and to compile supporting documents that illustrate the process. The book is restricted to five directives which more or less cover a majority of the electronic and electrical equipment range viz. low voltage directive (LVD), electromagnetic compatibility (EMC) directive, medical devices directive (MDD), radio equipment directive (RED) and the directive on restriction of use of certain hazardous substances (RoHS). Although the directives are five, they essentially contain EMC and electrical safety requirements (barring RoHS) and as such, EMC and safety requirements are dealt with in detail. The chapter on safety describes the principles of safety as found in the international IEC and European harmonized standards. It provides ways and means to improve product design so as to ensure reasonable compliance when a product is subject to safety evaluation by a test laboratory. The information is generally applicable to most product types such as information technology equipment (ITE), test and measurement devices, appliances, machinery, and other similar equipment. The design tips concentrate on problematic areas that manufacturers most often encounter during their first safety assessment. Then there are two chapters dedicated to EMC. One explains the fundamentals, standards and the test methodology while the other deals with EMC design. The design chapter contains ways and means to incorporate EMC measures at the design stage so that the product can comply with the EMC tests with a minimum of iterations. The design means discussed are very practical in nature and are given in such a way that the design engineer can immediately incorporate them without worrying too much about theory.

As is the case with a majority of international and European standards, one of the newest requirements of all the directives is the emphasis on risk analysis and risk reduction. All the directives now-a-days require a detailed risk assessment to be carried out in addition to testing as per standards. Thereafter the risk assessment needs to be documented so as to demonstrate how the risks have been reduced/eliminated. The book deals with this aspect in detail for all the directives under consideration.

And last but not the least, the CE marking procedure is not complete unless the entire process is documented through the so-called *technical file* or *technical documentation.* This document resides inside the European Union and is kept with the manufacturer (if European) or the importer/representative (if the manufacturer is non-European) to be produced if demanded by the European surveillance authorities. The chapter on technical documentation explains how this documentation is to be compiled as required by the directives.

It is important to note that a book of this nature can, at best, only draw out the meaning, significance and practical consequences of the directives to which it refers. It can give a simpler explanation of the legal text but cannot replace it or change what the law-making agencies have decided. Hence it is always prudent on behalf of the reader to refer to the actual text of the directive or the standard wherever he/she thinks a conflict exists. Needless to say, the contents of the actual directive or standard take precedence over the contents of this book.

It is sincerely hoped that the book will prove to be an invaluable aid towards understanding and simplifying the process of CE marking and that the manufacturer can approach it with professionalism, zeal and confidence.

Chetan Kathalay
Pune, 2020.

1

CE MARKING

1.1   INTRODUCTION

CE is the French acronym of "Conformite` Europe`ene". In English it stands for European Conformity i.e. conformity of a product to European laws. It is a mandatory compliance marking printed on all products intended to be placed in the European Union (EU) or more precisely the European Economic Area (EEA) which consists of 27 nations of EU (at the time of writing this book) plus Iceland, Liechtenstein and Norway. The EEA is also referred to as "community market". It should be mentioned here that Turkey is presently not a part of the EU but has implemented all EU laws and as such CE marking is applicable here as well. Again, Switzerland is not a member of the EU, but for some products it accepts the CE marking based on mutual recognition agreement (MRA) (see section 1.6). In addition, the CE marking is applicable over some overseas colonies of the EU member countries like Guadeloupe, French Guyana, Martinique, Réunion, Saint-Barthélemy, Saint-Martin, the Azores, Madeira and the Canary Islands. The CE marking consists of the label "CE" placed on the product (or its nameplate) as a visible proof indicating that the product meets all essential requirements (concerning health and safety) as prescribed by the applicable EU *directives* (laws). Once a product bears the CE marking, it is free to enter the EU and to move freely within the EU and hence many people consider CE marking as a *passport* to the EU.

1.2   BACKGROUND OF THE EUROPEAN UNION

The second world war had caused enormous damage to life and property and had resulted in large scale destruction of most of Europe. After the war, the idea of a unified Europe began to develop, in order to maintain peace and tranquillity in the region and to avoid future wars. Furthermore, the US and the USSR had become superpowers both militarily and economically. There was also a thought that the European nations, if united, could more effectively face economic challenges posed by these superpowers. The seeds of the EU were sown when six countries namely Belgium, France, Italy, Luxemburg, Netherlands and West Germany met in Rome and signed the European Economic Community (EEC) treaty in 1957. Customs duty between these countries was abolished and common policies in trade and agriculture were put into place.

Further in 1960, four countries namely Iceland, Liechtenstein, Switzerland and Norway formed the European Free Trade Agreement (EFTA) to promote free trade and economic cooperation between these countries. Denmark, Ireland and the United Kingdom joined the EEC on 1st January 1973, raising the number of member states to nine. By 1986, Greece, Spain and Portugal also joined the EEC taking the membership to twelve and the single European act was signed in the same year. This treaty provided the basis for a six-year programme allowing free flow of goods across EU borders and the concept of *single market* was conceived. In a major upheaval, the Berlin wall was pulled down on 9 November 1989, and the border between East and

West Germany was opened for the first time in 28 years. This led to the reunification of Germany. Finally, in December 1991, the European Union treaty or the Maastricht treaty was signed between twelve nations and the EU finally came into being on 1st November 1993. In 1994, the EU along-with three EFTA members namely Iceland, Liechtenstein and Norway formed the European Economic Area (EEA). In 1995, Austria, Finland and Sweden joined the EU taking the membership to fifteen. In 2004, in one of its biggest expansions, as many as thirteen nations joined the EU. However, in 2020, the unification process received a major setback when the UK decided to leave the EU. It remains to be seen (at the time of writing this book) whether it still accepts the CE marking or replaces it with its own. Presently the EU consists of 27 nations namely Austria, Belgium, Bulgaria, Croatia, Cyprus, Czech Republic, Denmark, Estonia, Finland, France, Germany, Greece, Hungary, Ireland, Italy, Latvia, Lithuania, Luxembourg, Malta, Netherlands, Poland, Portugal, Romania, Slovenia, Slovakia, Spain and Sweden.

1.3   OBJECTIVES OF THE EU

The main objective of the EU is the creation of a common internal market without trade boundaries which allows free movement of goods, personnel, services and capital. These are sometimes referred to as the *four freedoms*. It is commonly observed that varying national legislations cause barriers in trade restricting free movement of goods. Hence it was essential to ensure that the same law is applicable over the entire EU. This process is called as *harmonisation* (i.e. one law applicable over the entire EU) and it is one of the basic features of the EU.

The other objective, which is not immediately obvious, is to bring about cost savings to manufacturers. Before harmonization, if a manufacturer desired to export his product to one country, he had to comply with the laws of that country. If, at a later stage, he had orders from another European country, then the process had to be repeated for that country as well. This not only involved additional cost and time delays but also resulted in unnecessary hassles for the manufacturers located both within the European region and outside of it. In the present scenario, the manufacturer has to comply with only one law and after that he is free to export his goods to any EU member state. This avoids duplication of efforts thus saving both time and money. Another objective is to ensure/enhance product safety. Since the manufacturer has the obligation of complying with EU laws including *directives*, the community market is reasonably protected from unsafe products. Further, as the surveillance mechanism is also harmonised, it provides the EU surveillance/inspection/customs authorities with a uniform procedure to ensure entry of safe products as well as identification of unsafe products, transmitting information about unsafe products to member states and removal of unsafe products.

1.4   CE MARKING AND OTHER MARKS

CE marking differs vastly from other safety/ compliance marks (see Fig. 1.1). It is important to note that CE marking is neither a voluntary marking nor a quality marking. It is not even a marking of guarantee. It is a compulsory marking having legal implications and is not an *option* for manufacturers.

Fig. 1.1: CE Marking and Other Marks

CE marking is also not an approval of any authority i.e. it is not approved by EU government. This may come as a surprise as there is often a misconception that the CE marking is authorized by EU government. This speculation is fuelled by requirement of other marks like the C-tick or the SABS marking for example, which require authorization from the Australian or the South African governments respectively and for which these governments charge a fee. CE marking is different from these marks in the sense that it is not owned by the EU and as such does not require authorization from the EU nor do the EU authorities charge a fee or royalty from manufacturers using the CE marking. However, products that bear the CE marking are definitely under surveillance of the EU authorities once they enter the community market and there exists a procedure to weed out products that may be wrongly marked as CE.

1.5   WHO GIVES CE MARKING?

There are also other questions and misconceptions prevalent as regards to CE marking. The first is —who *gives* the CE marking? Now the biggest myth about CE marking is that one has to *go* to a certifying agency or some authority that will *give* the *CE certificate*. The fact of the matter is that no one *gives* CE marking in normal sense of the word, not even EU authority or government. This comes as a surprise since it is normal to assume that CE marking is a European marking, so they must be naturally its *owners*. The truth is that the EU authorities are not the owners of CE marking so they cannot give it. The CE marking is not given by your consultant either (although they may coerce you to think that way!). It is also not given by any certifying agency not even European labs. And mind you, the term *CE certificate* does not even exist in the EU laws! Well, then who gives CE marking? The CE marking is given (or rather *affixed* on the product) by the manufacturer himself through a procedure called   *self-declaration* wherein the manufacturer himself ensures that the product meets essential requirements given by applicable directives, he himself declares conformity (through a document called as *Declaration Of Conformity* (D.O.C, see section 1.14)) and he himself affixes the CE marking on his product. The reader may now be somewhat confused here since self-declaration beats common sense. If the manufacturer does everything, the reader may argue, then he is free to do almost anything -even go to the extent of false or bogus CE marking. Although this is possible, the EU authorities have put certain measures in place to ensure that the manufacturers desist from such practices. This will be clear in section 1.18 which talks about market surveillance mechanism which has been put into place to identify risky or dangerous products that may somehow enter the community market. The measures include weeding out products falsely marked CE, putting heavy penalties, blacklisting and/or taking legal action against the manufacturer or if the manufacturer is non-European, against the importer/distributor/assembler who is based inside the EU (see section 1.17)).

Then there is the second question –who is responsible for CE marking? Well the answer is that the manufacturer is fully responsible for CE marking. But if the manufacturer is non-European then the importer/distributor/assembler is responsible for CE (see section 1.17). It must be clarified here that for certain category of high risk products (active implantable medical devices for example), one must as a rule, take services from European notified bodies (NBs) which are test laboratories based inside one of the EU member countries and who, after examining the product or its design, will give an *EC Type Examination* certificate or *Design Examination* certificate respectively (see section 1.12 CE marking procedure) for which the NB is responsible. Still, it is only the manufacturer who has to sign the D.O.C and affix the CE marking on the product.

1.6   ENTITIES IN THE EU LAW MAKING PROCESS

To understand the law-making process of the EU, it is worthwhile to take a look at the its institutional structure. As shown in the Fig. 1.2, at the apex sits the Court of Justice which has a function of interpreting the community laws and to monitor their implementation by all member *states* (the EU member countries are called member states). The court consists of judges –one for each member country, appointed for six-year renewable terms.

Then there is the European Commission (EC) simply referred to as the *Commission*, which consists of commissioners -one for each member states who are under oath to relinquish national interests. The commission is headed by a president and has six vice-presidents. The commission initiates all proposals for EU *directives* and has the responsibility of negotiating international agreements. One of the more significant of these is the mutual recognition agreements or the *MRAs* with non-member countries so as to mutually accept each other's reports, certificates and markings. The European Parliament or simply the *Parliament* is the EU's directly elected body with members elected to 5-year terms by member country citizens. Its function is to provide opinion on all draft laws (including the directives) by holding hearings and developing recommendations on proposed laws.

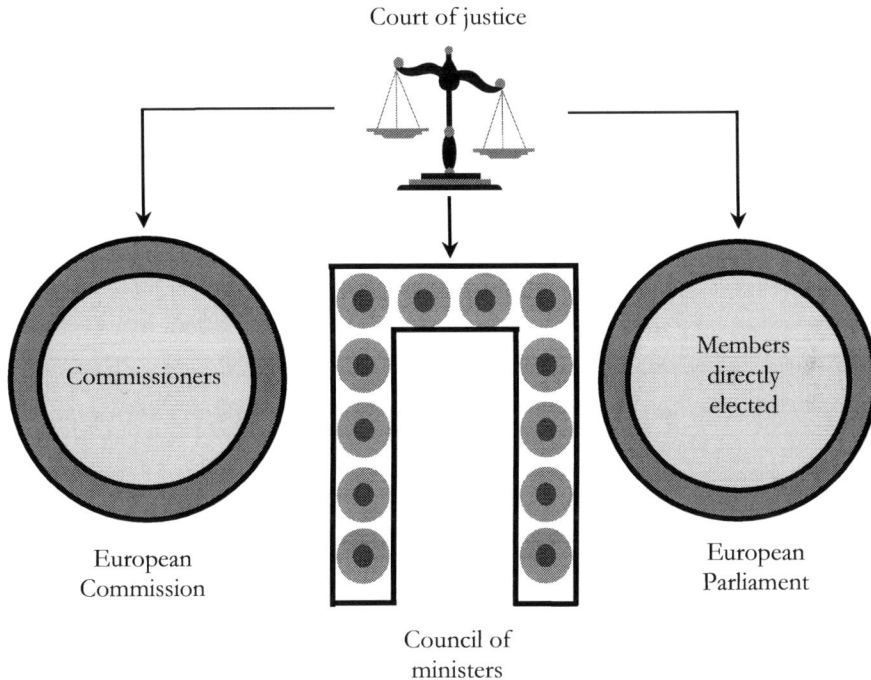

Fig. 1.2: EU Institutional Structure

The parliament's decisions are not binding, but amendments are often incorporated plus it has the power to dismiss the commission. Finally, there is the European council of ministers or simply the *Council* which has ministers in various policy making areas, appointed by national governments and who act on behalf of national interests. It is the EU's final decision-making body which takes decisions by a voting process where the votes are apportioned among member states and weighted according to population and economic output. Stronger economies like Germany, France and Italy have 29 votes each, Spain and Poland have 27 each, Romania has 14 while a small country like Malta has only 3. Out of a total of 345 votes, qualified majority needs 71% i.e. 255 votes. It is clear therefore that the stronger economies have more sway in the EU.

1.7   TYPE OF LAWS

As far as product conformity is concerned, the outcomes of the EU institutions are four legislative actions viz. regulations, directives, decisions and opinions/recommendations. The regulations are directly applicable as law over the entire EU and they do not require approval by individual member states. For example, regulation *756/2008 giving requirements for accreditation and market surveillance.* The directives, on the other hand, are broad legal guidelines prepared by specialists in the commission. All directives do not lead to CE marking. Some directives give *essential requirements* for product safety and conformity, the fulfilment of which leads to CE marking. They are not directly applicable as law over the EU. Member states must modify their laws so as to incorporate a particular directive in their law. This process is called *transposition.* Decisions can be issued by Council or Commission and are binding only on those parties to whom they are addressed to. Opinions/recommendations have no binding effect but can influence course of future legislation, decisions or judgments are given by the court of justice.

1.8   ESSENTIAL REQUIREMENTS

The aim of the directives is to limit legislation only to certain requirements that are essential to public interest. These are embodied in the directives as *essential requirements.* The essential requirements deal with the protection of health and safety of users (usually consumers and workers) and are designed to provide and ensure a high level of protection. They either arise from certain hazards associated with the product (for example electrical, electromagnetic, mechanical, flammability, chemical, biological, radioactivity etc.) or refer to the product or its performance (for example provisions regarding materials, design, construction, manufacturing process, instructions drawn up by the manufacturer etc.) or lay down the principal protection objective (for example by means of an illustrative list). Often, they are a combination of these. As a result, several directives may be applicable to a given product at the same time, since the essential requirements of different directives need to be applied simultaneously in order to cover all relevant public interests. The essential requirements are set out in relevant sections or annexures of the directives.

The necessary and sufficient requirement for manufacturers marking their products CE is to demonstrate that it complies with the essential requirements. He/she may follow different methods to meet these requirements (testing of products from a third-party laboratory to comply with European or "EN" standards being one of them). It is important to note that the manufacturer himself is responsible for designing and manufacturing the product in accordance with essential requirements laid down by the directives, he is also responsible for carrying out conformity assessment in accordance with the procedure(s) laid down by the directive(s) and therefore only he is solely responsible for CE marking. The manufacturer has to understand that if it is found that products bearing CE marking do not meet conformity requirement, he (the person signing the D.O.C) can be tried in the court of law and can be held liable for damages. In case the manufacturer in non-European then the importer/distributor/assembler/installer can be held liable for damages.

1.9   THE NEW APPROACH TO CONFORMITY

This section throws light on the approach or way(s) to achieve the CE marking. In what is called as the *new approach*, the compliance process is split into directives (legal compliance) and standards (technical compliance). The directives contain general guidelines that give only the *essential requirements* along with other requirements (to ensure free movements of goods). The technical requirements (i.e. tests and evaluations) to meet the essential requirements are given in the technical standards. All European standards begin with letters "*EN*" (Euro Norm) and are *harmonized* meaning that the same standard is applicable over the entire community market (EEA). Directives are prepared by specialists within the commission while the standards are prepared by the EU standardization bodies namely the CEN (Committee' Europe'enne de Normalisation), the CENELEC (Committee' Europe'enne de Normalisation Electrotechnique) and the ETSI (European Telecom Standards Institute). Once a product complies with a standard, it is *presumed* to comply with the essential requirements of a directive. This principle of *presumption of conformity* which links the directives (the legal process) with their corresponding standards (the technical process) is one of the most interesting features of the new approach. Another important feature of the new approach and most loved by manufacturers is that of self-declaration which was discussed earlier.

To further improve the internal market for goods and strengthen the conditions for placing a wide range of products on the EU market, the *new legislative framework* (NLF) was adopted in 2008. It is a package of measures that aims to improve market surveillance to protect both consumers and professionals from unsafe products, especially those imported from outside the EU. In particular, this applies to procedures for products which can pose danger to health or the environment. The NLF sets clear and transparent rules for the accreditation of conformity assessment bodies and at the same time, boosts the quality of and confidence in the conformity assessment of products through stronger and clearer rules on the requirements of notified bodies (especially regarding sub-contracting of jobs by the notified bodies to laboratories and subsidiaries outside the EU). It also clarifies the use of CE marking and creates a toolbox of measures for use in product legislation.

1.10   EU DIRECTIVES

As said earlier, directives are legislative acts prepared by specialists in the European Commission. It actually sets out certain *goals* that all EU countries must achieve. Once a directive is published, all member countries of the EU are directed to change their laws to include the directive in their law and repeal all conflicting national laws within a certain *transition period*. When this is done, the directive is said to be *transposed* into the law of all member states and becomes mandatory over the community market. A directive is designated by the number and the year in which it was published, for example the low voltage directive or LVD is designated by 2014/35/EU i.e. the directive was published in year 2014, 35 is the number of the directive and EU stands for European Union. The directives are published in the Official Journal of European Union (OJEU) which is a gazette of record for the EU. The OJEU comprises three series:

- The *L-series* contains EU legislation including regulations, directives, decisions, recommendations and opinions.

- The *C-series* contains EU information and notices including the judgments of the European courts, calls for expressions of interest for EU programmes and projects, public contracts for food aid etc.
- The supplementary *S-series* contain invitations to tender

All the directives are available for free. They can be downloaded from https://ec.europa.eu/growth/single-market/european-standards/harmonised-standards_en .

1.10.1   When Do Directives Apply To Products?

The legal obligations of the EU directives are applicable (i.e. conformity is to be demonstrated) to finished products (i.e. those that can be put to use with minimum connections) which are *placed in the market* or *put into service* by economic operators (i.e. manufacturers, importers, distributors, assemblers, employers etc.). Needless to say, the product must comply with that version of the directive that was in place at the time of its placing on the market. *Placing in the market* occurs when the product is *made available* (for distribution/consumption/use) *first time* in the community market and takes place when the product is transferred from the manufacturer (or importer) to the distributor. *Putting into service* takes place the moment the product is *first* used. The directives are also applicable to any subsequent action in between placing in the market and/or putting into use. Once the product reaches the *end user* (see section 1.17.7) and is put into service, it is no longer considered to be a new product and the directives cease to apply. However, the product should continue to work to the level of safety and protection in relation to the directives in its intended lifespan. The EU directives are also applicable to a product which is still in the distribution chain as long as it is a new product.

As said earlier, once the product is put into service, the directives cease to apply. However, after placing in the market or putting into service, if the product is subject to such changes that can modify its original performance, then the product is considered as new and must comply newly with the requirements of the directives. The directives also apply to used or second hand products imported from a non-European country for sale in the community market when such products enter the community market for the first time. The directives are not applicable if the product is not placed in the community market. This can happen in the following instances:

– When the product is transferred by the manufacturer to his authorised representative (see section 1.17.2) or to a European laboratory for testing.

– When the product is displayed in trade fairs under controlled conditions (like expert operators, restrictions to public contact, avoiding inappropriate interaction with other neighbouring products). The product should bear a sign that the product has not yet been placed in the market.

– When the product is in stock with the manufacturer or the importer.

– When the product is manufactured inside the EU, solely for export.

– Products for personal use bought by consumer from a non-European country while physically present in that country. Now, the end-user is not one of the economic operators who bear responsibilities under the directives and hence legal obligations do not apply to him/her. Also, the directives do not apply if the product is transferred from this person to another by the way of gifts.

As discussed before, the new legislative framework (NLF) was adopted in 2008 to further strengthen the directives. Under the NLF, many directives were *recast* i.e. the original directive and all its subsequent amendment were incorporated in a single new directive. All these directives after recast, have the same general structure. Clauses regarding placing in the market, free movement, roles of manufacturers/importers/distributors/assemblers/employers, presumption of conformity, declaration of conformity, affixing rules for CE marking, market surveillance, action on non-compliances, safeguard procedure, penalties, transposition and transitional provisions remain the same. These common clauses (called *articles*) have been explained in detail in the chapter on low voltage directive. For other directives, only references to these articles have been made to avoid repetition. Let's take a look at some of the frequently referred directives for electrical and electronic products that are being considered in this book. All the directives are discussed in detail in the latter part of the book.

1.10.2 The Low Voltage Directive (LVD) (2014/35/EU)

This directive was published on 26th February 2014 and is legally binding from 20/04/2016. It covers equipment designed for use between 50V ac to 1000 V ac and between 75V dc to 1500 V dc. It ensures primarily electrical safety and requires products to be manufactured such that it will not pose safety hazard so as to endanger safety of persons, domestic animals and property. The *essential requirement* is protection against electric shock through suitable insulation system, isolation barriers and/or earthing methods, protection against fire through minimizing risk of ignition, use of low flammability components and fire enclosures. It also covers other hazards like mechanical, chemical, radiation etc. applicable under certain circumstances.

1.10.3 The EMC Directive (2014/30/EU)

This directive covers electromagnetic compatibility (EMC). It is in force since 20th April 2016 and is applicable to electrical equipment. It aims to ensure EMC in community market and the *essential requirement* is that any equipment (under its normal operational environment) should not emit electro-magnetic disturbance above a certain limit and should have an adequate level of immunity to electro-magnetic disturbance expected in its normal operational environment.

1.10.4 The Medical Devices Directive (MDD) (93/42/EU)

It aims to regulate medical devices and their accessories by requiring them to provide an adequate level of safety to patients, users and third party. The original designation was 93/42/EC. However, it was amended in 2007 under the designation 2007/47/EC which combined amendments for two other directives namely active implantable medical devices (90/385/EC) and biocidal products (98/8/EC). In this book, we are going to concentrate on 93/42/EC along with amendments specific to medical devices in 2007/47/EC.

1.10.5 Radio Equipment Directive (RED) (2014/53/EU)

The directive is regarding placing radio equipment on the community market and is mandatory since 13th June 2016. It ensures a single market for radio equipment by setting *essential requirements* for safety, electromagnetic compatibility and the efficient use of the radio spectrum. It also provides the basis for further regulation governing some additional aspects like technical features for the protection of privacy, protection of personal data and against fraud. Furthermore, it covers additional aspects of interoperability, access to emergency services and compliance regarding the combination of radio equipment and software.

1.10.6   RoHS Directive (2015/863/EU)

It aims to protect human health and environment by restricting the use of certain hazardous substances (ten in total) in electrical and electronic equipment (EEE). The directive was published on 31st March 2015 and is legally binding from 22nd July 2016.

1.11   HARMONISED EUROPEAN *'EN'* STANDARD.

Harmonised European standard is defined as *a technical specification prepared by one of the European standards organizations under a mandate from the commission to establish a European requirement.* The *mandate* referred here is a contractual relationship, signed between the commission and the standardisation organisation, stipulating that the standardisation organisation will produce standards which provide a technical solution/interpretation of the essential requirements defined in a particular directive. The commission and the European standards organizations namely CENELEC, CEN and ETSI cooperate in order to produce a standard. CENELEC is the French acronym for *Committee' Europeane' de Normalization Electrotechnique* or *European Committee for Electro-technical Standardisation* and has its headquarters in Brussels. It prepares standards in the area of electro-technology. ETSI stands for *European Telecom Standards Institute,* is headquartered in Sophia-Antipolis and prepares telecommunication standards. CEN stands for *European Committee for Standardisation* and responsible for standards in all other fields.

1.11.1   About CENELEC

In 1959, five European IEC (International Electro-technical Commission) national committees namely Belgium, France, Germany, Italy and the Netherlands met and formed a body which became known as CENELCOM, the *European Committee for the Coordination of Electrical Standards in the Common Market Countries.* The aim was to harmonize national standards in areas where trade barriers existed. They mutually agreed to give priority to IEC standardization work whenever possible, to share information on new national work and cooperation in testing and certification. CENEL, the European *Committee for the Coordination of Electrical Standards* (the second direct ancestor of CENELEC) was formed in October 1960. CENEL studied IEC standards and their implementation within the countries involved. From 1st January 1973, CENELCOM and CENEL were disbanded and a new organization by the name CENELEC was established.

1.11.2 CENELEC Structure

CENELEC consists of various *technical committees or TCs* which are responsible for the preparation of standards within their scope. They are composed of national delegations designated by the CENELEC member countries. The CENELEC central secretariat located in Brussels, is in charge of the daily operations, coordination and promotion of all CENELEC activities.

1.11.3 CENELEC Membership

The members of CENELEC are the national electro-technical committees of all EU member states plus 3 EFTA members i.e. Iceland, Norway, and Switzerland. It also has members of affiliate countries that include countries like Poland, Turkey, Ukraine etc. In addition to this, CENELEC also counts 32 cooperating partners. Cooperation agreements signed with these industrial partners (European associations of manufacturers for a given sector) allow for direct cooperation and mutual contribution to each other's work. These agreements guarantee direct input from cooperating partners into the standardization work undertaken by CENELEC. They advise on and state standardization priorities, propose drafts as a contribution to the European standardization process and deliver expert advice on legislative consequences of new standards.

1.11.4 Other European Standards Organizations

CENELEC has two sister organizations at European level namely CEN (Committee' Europeanne' de Normalisation or European Committee for

Standardisation) and ETSI (European Telecommunications Standards Insititute). At international level, each of the European standards organizations has a counterpart. For CENELEC it is the IEC, that of CEN is ISO, and ETSI's counterpart is the International Telecommunications Union or ITU.

1.11.4.1 European committee for standardisation (CEN)

CEN is the biggest of the European Standardization bodies and shares with CENELEC the same internal regulations. Membership to both CEN and CENELEC is normally granted at the same time.

1.11.4.2 European Telecommunications Standards Institute (ETSI)

This other sister European organization makes standards for implementing the radio equipment directive (RED) in the EU for radio equipment which transmit and receive RF energy and includes fixed, mobile, radio, converged, broadcast and internet technologies.

1.11.5 The Making Of EN Standard

The development of a new EN standard is a fairly complex process and takes place in essentially the following four steps:

Drafting: Work on a standard begins with a draft. Drafts can come from any one of the four sources:

- **International Electro-technical Commission (IEC)** is one of the major sources of drafts. As one of its main policies is to develop standards with a global application, CENELEC works in close cooperation with the IEC. This is affected through the *Dresden agreement* of 1996 which was signed between CENELEC and IEC to accelerate the standards' preparation process in response to market demands, to expedite the publication and common adoption of international standards and also to facilitate the adoption of European standards internationally. CENELEC has used publications and draft documents issued by IEC as the most important source of reference documents for the preparation of its European standards. In fact, more than 80% of the drafts come from the IEC and about 66% of CENELEC standards are identical to those of the IEC.

- **CENELEC technical committees**.

  These are committees appointed by CENELEC itself for certain aspects of electro-technology.

- **CENELEC cooperating partners** like ECMA (European Computer Manufacturers Association), EECA (European Electronic Component Manufacturers) or ECSS (European Cooperation for Space Standardization).

- **National committees or "NCs"** (i.e. standardization committees of member states). To ensure that progress made at international level is not being overridden at national level, CENELEC has set a notification procedure called *Vilamoura procedure* which all members have to follow. In essence, this procedure obliges all CENELEC members to notify any national standardization project to all other members should there be an interest for a common standard at European level. If there is no reaction from member states in a maximum period of 3 months, then the country in question may continue developing this internal project.

Enquiry: When a suitable draft is available, it is submitted to the national committee (NC) of each member state for CENELEC enquiry, a procedure which lasts 6 months. The comments received are studied by the technical body working on the draft and if justified, incorporated into the document, where justified, before a final draft is sent out for vote.

Voting: The voting process usually takes 3 months. At this stage the members have weighted votes corresponding to the size of the country they represent. For instance, the larger countries like France, Germany, Italy and the UK have 10 votes each while the smaller ones have one or two weighted votes. The voting must yield a majority of national committees in favour of the document and at least 71% of the weighted votes cast should be positive.

Numbering: The shortest unambiguous reference to a European standard is its number. The number consists of the capital letters EN followed by a space and a number in Arabic numerals, without any space. The numbers 50000 to 59999 mostly cover CENELEC activities and 60000 to 69999 refer to the CENELEC implementation of IEC documents with or without changes. When IEC standard is adopted, the number is the same, only the letters *IEC* are replaced by the letters *EN*. For example, IEC 60950 becomes EN 60950.

When approved, the Commission publishes a *reference* of the standard in the Official Journal of the European Union (OJEU). Once published, the standard, now called European *harmonized* standard, provides *presumption of conformity* with the corresponding directive i.e. a product which complies with all the tests and evaluations given in the harmonised standard, is *presumed* to be in conformity with the corresponding essential requirements set by the directive.

Like the directives, a harmonized standard has to be *transposed* into a national standard i.e. EU member states should modify their law so as to include the standard in their law, after removing all conflicting national standards. For example, when Germany adopts a *EN* standard, it transposes the standard into a national standard designated as *DIN EN* where DIN stands for Deutsches institut fur normung (German standardisation institute). This ensures that when manufacturers in Germany are complying to a national standard i.e. *DIN EN* they are also complying with the *EN* standard and when their product goes to another European country, they are not required to test it again.

An important thing to note here is that the application of harmonized standards remains *voluntary* in the scope of the new approach directives which means that the manufacturer can use *other* methods (like technical justification) to demonstrate compliance with a directive. However, such methods do not grant presumption of conformity. Now, to keep the application of a standard voluntary especially after going through the rather complicated process of standard formulation is surprising to many. However, as explained earlier, the essence of CE marking is compliance with the essential requirements of the applicable directives and that the application of harmonised standards is just one of the ways (or shall we say the most acceptable way?) of doing this.

A list of standards which give presumption of conformity is normally published under a particular directive. The list can be freely downloaded from the EU website under the link *https://ec.europa.eu/growth/single-market/european-standards/harmonised-standards_en*. The standards are not available for free but need to be purchased. They can be somewhat costly and hence should be selected and bought carefully giving due consideration to the date of withdrawal (see section

1.13). Annexures I to IV give a list of standards under MDD, RED, LVD and EMC directives respectively. It is important to note that the list is not comprehensive i.e. only some standards that will be frequently referred to in this book are given in the list for reference purpose. Furthermore, since the list is routinely updated, readers are advised to download the latest complete list from the above link before selecting a standard.

1.11.6   Types EN Standards

EN standards can be grouped into three broad categories viz. basic standards, product standards and generic standards.

1.11.6.1 Basic standards

As the name suggests, these standards deal with basic phenomenon or requirement and methods to test equipment for such phenomenon / to meet that requirement. Basic standards are not dedicated to specific products or product families; they relate to general information, to a disturbing phenomenon or a safety requirement and to the measurement or testing techniques. For example, CENELEC produces the *EN 61000* series of basic standards which generally cover the so-called *safety aspects* of electromagnetic compatibility. In this series, one of the standards is EN 61000-4-2 which deals with the description, test equipment, test levels and test methods to test a particular equipment's immunity towards the phenomenon of *electrostatic discharge* (ESD). Another example is EN 61695-11-10 which is used to classify materials according to horizontal flammability ratings. Basic standards do not provide presumption of conformity, and hence do not find a mention in the list of standards against each directive.

1.11.6.2 Product specific (type-C) standards

Product specific standards provide requirements for a particular product and are prepared by product committees. Here the basic standards are taken as reference and limits/severity levels/performance criteria are specified for the type and phenomenon applicable for a particular product. For example:

- EN 62040-2 regarding EMC requirements for UPS under the EMC directive

- EN 62040-1 regarding safety requirements for UPS under the LVD.

1.11.6.3 Product family (type-B) standards

These standards provide requirements for a particular group or family of products delivering similar functions. Here also the basic standards are taken as reference and limits/severity levels/performance criteria are specified for the type and phenomenon applicable for a particular product family. Examples are:

- EN 60950 regarding safety of IT equipment family under the LVD

- EN 55032 regarding with EMC testing of multimedia equipment family under the EMC directive.

1.11.6.4 Generic standards

These are generally applicable to those products or product family for which no product standard exists. Here again basic standards are taken as reference and limits/severity levels/performance criterion are specified for the product or installation category or environment. e.g. EN 61000-6-1: EMC generic standard for immunity for residential, commercial and light-industrial environments.

1.12   THE CE MARKING PROCEDURE

The CE marking process is shown in Fig. 1.3. It starts with the identification of applicable directives. The procedure for identification is given in section 1.13. A product may give rise to different risks depending upon its application, its end use or the type of components it incorporates and hence more than one directive may be applicable. In case of doubt regarding the applicability of a particular directive it is prudent to download a copy of the directive then to refer the scope of the directive. It must be understood here that it is the responsibility of the manufacturer to ascertain whether or not his/her product falls under the scope of a particular directive. Once the directive(s) are identified, the manufacturer must follow one of the *routes* or *conformity assessment modules* prescribed by the directive for his category of equipment. There are eight conformity assessment modules viz. module-A to H.

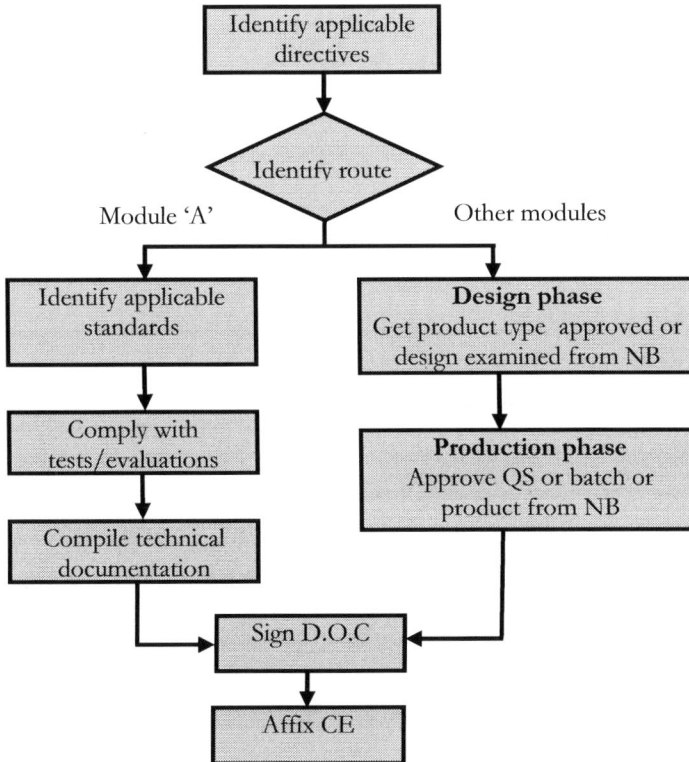

Fig. 1.3: CE Marking Procedure

1.12.1   Module A : D.O.C With Internal Production Control

The EMC, LVD, RoHS, MDD (for certain low risk products) and RED (if harmonised standards are fully applied) call for this module. Herein, the manufacturer has to identify the applicable harmonized EN standards, conformance to which gives him/her a presumption of conformity with the directives. The manufacturer has to then carry out tests on one representative sample as prescribed by the standard, either in-house (if equipment is available) or from a third-party test laboratory. It is not mandatory that the third party laboratory should be a *notified body (NB)* (which are laboratories based inside the EU which are designated by European authorities to carry some or all parts of conformity process) rather the manufacturer can choose any third party lab after confirming its credentials (like trained manpower, equipment, resources, experience or accreditation status).

After complying with all the tests and obtaining a test report, the manufacturer prepares and signs the *declaration of conformity* (D.O.C) (see section 1.14) wherein he/she *declares* that the product complies with the requirements of applicable directives and thereafter affixes the CE marking on the product. In support of the conformity claim, the manufacturer also compiles a technical file or technical documentation in a format prescribed by the directive (see section 1.15). Finally, the manufacturer has to ensure that his/her production process produces samples which are identical to sample that was tested (hence the name internal production control). For this, a description of the quality system (may be his/her internal quality system or quality system based on ISO 9000) and related procedures should be given in the technical documentation (see section 8.16, chapter 8).

As shown in Fig. 1.3, there are also other modules which are prescribed by RED and MDD (for certain high-risk products). They compulsorily involve European notified bodies (NBs) in one way or the other. The manufacturer may freely choose any NB of his choice. These other modules are discussed below.

1.12.2   Module-B : EC Type-Examination

If the applicable directive calls for this module, then in the *design* phase, the manufacturer lodges application with chosen notified body (NB), submits technical documentation and a sample of the product (called *type*) to the NB. The NB on its part examines and assesses documentation, carries out tests / inspection and on conformance issues an *EC type-examination certificate*. In addition, the manufacturer is under the obligation to inform the NB of any changes in product afterwards. In case of changes, the NB verifies whether these can affect conformity (it may repeat some tests if the need be) and issues supplements to the EC type-examination certificate. Since this is a *design phase* module, it cannot be applied in isolation and should be followed by any one of the *production phase* modules C to G as prescribed by the directive.

1.12.3   Module-B + C : Conformity To Type

If the applicable directive calls for this module, the manufacturer ensures that the manufacturing process assures compliance with the type as described in the

EC type-examination certificate, signs D.O.C, compiles technical documentation and affixes CE marking on the product. The NB takes no action in production phase.

1.12.4   Module-B + D : Production QA

The manufacturer runs quality system (QS) for production. The QS must ensure that all products conform to type as described in the type examination certificate and lodges application for assessment of his QS with NB. The manufacturer signs D.O.C, compiles technical documentation and affixes CE marking. The NB approves/controls/audits QS (may carry out tests) and generates report. In addition, the manufacturer is under the obligation to inform the NB of any changes in QS afterwards. The NB on its part, assesses whether these changes will affect the QS and conveys its observations to the manufacturer. The NB also carries out surveillance of the QS wherein it performs periodic, even surprise inspections, carries out tests, and generates test/inspection report.

1.12.5   Module-B + E : Product QA

The manufacturer runs quality system for final inspection and test. The rest of the process is the same as Module-B+D.

1.12.6   Module-B + F : Product Verification

The manufacturer takes all precautions so that manufacturing process produces products which confirm to type as described in type-examination certificate, signs D.O.C, compiles technical documentation and affixes CE marking. The NB examines/tests a random product from a batch, generates reports, affixes its identification number on each product and issues a *certificate of conformity*.

1.12.7   Module-B + G: Unit Verification

The procedure is the similar to as module B+F. The difference being that the NB examines each and every product, generates reports, affixes its identification number on each product and issues a *certificate of conformity*.

1.12.8 Module-H

This is the most stringent of all the modules. The manufacturer runs QS for design, manufacturing and QA. In the design phase, the manufacturer compiles a *design dossier* for his product and lodges application for assessment of his QS with NB and to examine the design dossier. The NB audits QS and generates report. It also examines the design dossier, carries out tests, generates *EC-design examination* certificate. The manufacturer signs D.O.C and affixes the CE marking and identification number of the NB on the product.

The manufacturer is also under the obligation to inform the NB of any changes in QS and product design afterwards. The NB on its part, assesses whether these changes will affect the QS and product conformity. After examining the changes in QS, the NB conveys its observations to the manufacturer. Design changes are also examined by the NB and upon approval the NB supplements the earlier design examination certificate. During the production phase, the NB can carry out periodic surveillance audits, even surprise inspections, may carry out tests and generate test/inspection report. Some directives (like the RED) also specify a variant of this module where the design examination is not required.

1.13 SELECTION OF DIRECTIVES AND STANDARDS

In this section we will discuss in detail about the delicate procedure of selecting applicable directives and standards. While selecting a directive, one can initially make a broad judgement considering the product type, the operating voltages, the operating environment, the end use, the type of components incorporated etc. The list can then be narrowed down after going through the scope and exceptions of each of the identified directives. While selecting a standard, one has to download the list of standards identified by each directive as discussed in section 1.11.5 and look for product specific standards (type-C standards) for the product in question and if available, apply that standard (i.e. perform tests/evaluation as per that standard). If product specific standard is not available then one has to look for product family standards (type –B standards) concerning a family of products delivering similar function. If neither is available, one has to apply the generic standards.

The correct selection and application of directives and standards is the sole responsibility of the manufacturer. Manufacturers outside EU have a little leeway since it is the importer (or person placing the product in the market) who is responsible for CE and hence many-a-times it is he or she who takes the final call regarding the selection. Hence for Indian manufacturers, it is prudent to consult the importer before finalizing directives and standards.

Further, for certain product categories (like those falling under the scope of MDD or RED), the manufacturer, as a rule, will have to perform the conformity assessment through a NB in which case, the NB will be responsible for selection of the correct standard. The following examples illustrate the identification of directives and standards for certain product categories.

– **Industrial use DC-DC converter (I/P: 24 V dc and O/P: 5 V dc, 12 V dc)**

This product involves electrical/electronic components, is liable to generate electromagnetic disturbance and is liable to be affected by it (see chapter 4). Hence EMC directive is applicable. The product is likely to contain certain hazardous substances like lead in solder hence the RoHS directive is also applicable. LVD is not applicable since the input and output voltages are outside the scope of the directive. Obviously MDD and RED are not applicable since it is neither a medical product nor does it contain any radio transmitter.

The route to conformity is module 'A' for both EMC and RoHS directives. NB involvement is not compulsory and it is up to the manufacturer to decide whether to avail the services of NB or not.

For identification of standards under EMC directive, according to the list in Annexure III, no product specific standard exists for DC-DC converter and no product family standard exists for similar product category. Hence the generic standards *EN 61000-6-3: EMC generic emission standard for industrial environment and EN 61000-6-2: EMC generic immunity standard for industrial environment* have to be selected.

For RoHS, if testing is being carried out by the vendor and the manufacturer is using technical documentation for demonstrating conformity, then the applicable standard is *EN 63000: Technical documentation for the assessment of electrical and electronic products with respect to the restriction of hazardous substances* has to be selected and due diligence with technical justification for meeting the essential requirements of the directive (see chapter 8) could be followed. If the manufacturer is also carrying out testing then the standard *EN 62321-1: Determination of certain substances in electrotechnical products* can also be used. The final documentation can be a combination of the above methods.

− **SMPS (I/P: 230V ac and O/P: 12 V dc, 5 V dc) for computer applications**

Here LVD is applicable since the input voltage falls within the range identified by LVD. In addition, EMC and RoHS directives are applicable for the same reasons as mentioned in previous example.

The route to conformity is module 'A' for all directive and services of NB are not compulsory.

For identifying applicable standards under LVD, one has to refer to the list of standards identified by the directive (see Annexure IV). Accordingly, there is no product specific standard for SMPS. Since the SMPS is for computer (I.T.) application, there is a product family standard namely *EN 60950: Safety of I.T. equipment* which can be selected and applied. Similarly, under EMC directive, after referring to list at Annexure III, it is found that no product specific standard exists for SMPS. However, there are product family standards for I.T. equipment namely *EN55032: EMC of multimedia equipment - Emission requirements* and *EN 55024: I.T. equipment -Immunity characteristics* which can be selected and applied. The *standard EN 55022: I.T. equipment - Radio disturbance (emission) characteristics - Limits and methods of measurement* is also available, but as of 5th March 2017, it stands withdrawn (see the column *date of withdrawal* in the downloaded list). The concept of *date of withdrawal* is explained in detail in the latter part of the section. The standard selection for RoHS is same as previous example.

− **High frequency surgical (Cautery) equipment**

This equipment uses high frequency radio waves for cutting & coagulating body tissue during medical surgery. Hence it falls under the scope of the MDD. In addition, RoHS directive is also applicable for the same reasons as mentioned in previous example. Since it is covered by MDD, it is it will not fall under the scope of LVD and EMC directive (see chapter 2 and 4). For MDD, the conformity assessment modules applicable are those for class IIb products as identified by the directive (see chapter 7). Hence for MDD, there is compulsory involvement of NB.

For RoHS, the route is module 'A'.

Under MDD, after consulting list of standards in Annexure I, it is found that there is a product specific standard available namely *EN 60601-2-2: High frequency surgical equipment* (which covers EMC and electrical safety) which can be selected. The standard selection for RoHS is same as discussed previously.

- **Automatic lathe machine involving motor drives, PLC, cutting heads and operating on 440 V ac, three phase supply.**

  The directives that are applicable are EMC directive, LVD and RoHS directives for reasons mentioned in the above examples. In addition, machinery directive is also applicable but it is outside the scope of this book.

  The route to conformity is module 'A' for all three directives.

  As far as standards are concerned, under EMC directive there is no product specific standard for lathes but there are product family standards for industrial machine tools namely *EN 50370-1: Electromagnetic compatibility (EMC) - Product family standard for machine tools - Part 1: Emission* and *EN 50370-2: Electromagnetic compatibility (EMC) - Product family standard for machine tools - Part 2: Immunity* which can be applied. For LVD there is neither a product specific standard for lathes nor a product family standard for machine tools. But there is a product family standard for machines namely *EN 60204: Safety of machinery - Electrical equipment of machines* which can be applied. The standard for RoHS is the same as with previous examples.

- **Handheld wireless trans-receiver (walkie-talkie) for analogue speech**

  The applicable directives are RED and RoHS. RED is applicable since the equipment incorporates RF transmitter. RoHS is applicable for the same reasons as mentioned in previous examples. LVD and EMC directive are not applicable since the RED itself contains EMC and safety requirements.

  The route to conformity is Module B+C or Module H (without design examination)

  Under RED, the following standards are applicable (see chapter 7)

  Article 3.1a: Health and safety -*EN 62368-1: Safety of electrical and electronic equipment within the field of audio, video, information and communications* and *EN 62311 : Assessment of electronic equipment related to human exposure to electromagnetic fields.*

  Article 3.1b: EMC - *EN 301489-5: EMC standard for radio equipment and services. Part 5: Specific conditions for Private land Mobile Radio (PMR)*

  Article 3.2: Spectrum related matters -*ETSI EN 300 296-2: EMC and Radio spectrum Matters (ERM), Land Mobile Service.*

  The standard for RoHS is the same as with previous examples.

One very important aspect to consider while referring to the list of standards is the column *date of withdrawal from OJ (end of presumption of conformity)*. As discussed earlier, harmonised standards give a presumption of conformity with the essential requirements of the directives. Many a times, it so happens that a standard is published at a particular date with reference the technology prevailing on that date. As the technology moves on, that particular standard will not be able to address requirements vis-à-vis new technology. Consider for example the EMC standard *EN 55022* for I.T. equipment which was published in 2010. At that time the concept of multimedia equipment was in its infancy. However, with the progress of technology many of the I.T equipment also integrated other media like facsimile, wireless communications, audio visual devices making it a multimedia equipment. Around the year 2012 a standard *EN 55032: EMC of multimedia equipment - Emission requirements* was published while in the year 2017 a standard *EN 55035: EMC of multimedia equipment - Immunity requirements* came into being. Hence it was decided that as of 5th March 2017, the standard EN 55022 would stand withdrawn from OJEU and all the I.T equipment will have to conform to the standard *EN 55032*.

Hence while selecting a standard from the list, the column *date of withdrawal from O.J.* should be given special attention. If the date of placement of the product in the EU market is later than the date of withdrawal, an alternative related standard has to be selected. Also, all product models placed after the date of withdrawal should conform to the requirements of the new alternate standard whose date of withdrawal is later than the date of placement of the model.

1.14   CONTENTS OF DECLARATION OF CONFORMITY

All directives provide a format for D.O.C which is more or less the same with minor modifications. The D.O.C must contain, at bare minimum, the name and address of company, name and model number of the product, model variants (if the declaration is to cover all model variants), photograph of the product (if available), list of directives applied, numbers, title and year of standards applied, a statement *that the product conforms with the essential requirements of the directive(s)* and that *the declaration is issued under the responsibility of the manufacturer.*

Additional information (if any, which the manufacturer may like to include), the signature of person authorised to sign on behalf of the manufacturer (along with name and designation), the place and date of issue. The D.O.C should be on company letterhead with contact numbers. It must be kept at the disposal of surveillance authorities (i.e. must be produced if asked to do so) for at least 10 years (5 years for MDD) from the date of placement of the last product model in the EU. It must be drawn in one of the official languages of the community and must be enclosed in the technical documentation and must also be translated into the language of the European country in which the equipment is placed. The formats of D.O.C as recommended by directives namely EMC, LVD, MDD, RoHS and RED are discussed in section 8.1 chapter 8 concerning technical documentation.

1.15   TECHNICAL FILE/DOCUMENTATION.

As per the directives, a manufacturer should compile technical documentation in support of CE marking. This documentation may be part of the quality system documentation where the directive provides for a conformity assessment procedure based on a quality system (modules D, E, H and their variants) otherwise it can be a separate document. The technical documentation must be kept within the EU and at the disposal of European surveillance authorities for at least 10 years (5 years for some directives like MDD) from the date of last placement product model in the EU market. This is the responsibility of the manufacturer or the authorised representative established within the community market or in their absence, the importer or the person placing the product on the market. The technical documentation may be demanded anytime by surveillance authorities and as such should unambiguously describe the equipment and the conformity process. Each directive specifies the contents of the documentation in accordance with the products concerned. The documentation can also be in soft copy form.

The technical documentation is distinct from equipment manual or instructions. The technical documentation is in support of and documents the CE marking process. It is not addressed to the user but to the customs/surveillance authorities. The technical file need not accompany each and every product. On the contrary, the manual should accompany each and every product as it is addressed to the user / operator. The contents of the technical documentation are discussed in detail in chapter 8.

1.16   AFFIXING THE CE MARKING.

The article 30 of regulation 765/2008 (regarding *setting out the requirements for accreditation and market surveillance*) gives clear guidelines regarding the size and placing of the CE marking which is followed by all directives. The CE marking should be affixed before placing the product in the community market. The marking should be legible and durable and it should meet the durability test prescribed by safety standards (see section 3.36.12.6). The regulation provides a graphic as to how the marking should look like. The graphic depicts two semicircles as shown in the Fig. 1.4.

Fig. 1.4: CE graphic.

The length of the centre stroke in the semicircle representing the "E" must be 80% of the radius of the semicircle representing the "C". The CE marking must be at least 5mm high to ensure that it is legible (although some directives like the RED allows reduced dimensions). If the CE marking is reduced or enlarged the proportions given in the graphic must be respected.

| CE marking with NB identification number | CE marking placed directly on the product | CE marking placed on name plate |

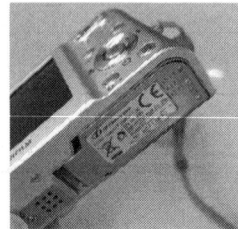

Fig. 1.5: CE Marking Placement

As a rule, the CE marking should be affixed directly on the product or on its nameplate. However, where this is not possible or not warranted on account of the nature of the product, it must be affixed to the packaging, if any, and to the accompanying documents. Some directives (like the MDD) require services of NB to be availed compulsorily for certain class of product and they also call for placing the identification number of NB along with the CE marking

1.17   PRODUCTS IMPORTED FROM OUTSIDE EU.

A manufacturer established in a third country outside EU (e.g. India) is responsible, in the same way as a manufacturer established in a EU member country, for CE marking. If the manufacturer is based outside the EU, then according to the directive on product liability (85/374/EEC), the liability (i.e. the responsibility to pay for damages) is placed on the *producer*. The producer can be authorised representative appointed by the manufacturer or in his absence, the importer. A distributor /assembler/employer can also be considered an importer. Manufacturers, importers, assemblers and authorised representatives are termed as *economic operators*.

1.17.1   Manufacturer.

A manufacturer is any person (signing the D.O.C) responsible for designing, assembling, packaging or labelling a product with a view to placing it on the community market under his own name. The definition contains two cumulative conditions -the person has to manufacture (or have a product manufactured) and -to market the product under his own name or trademark. So, if the product is marketed under another person's name or trademark, that person will be considered as the manufacturer. At the design stage, the manufacturer is responsible for designing and manufacturing the product in accordance with essential requirements laid down by all applicable directives, for carrying out conformity assessment in accordance with the procedure(s) laid down by the applicable directive(s), for compiling technical documentation, drawing the D.O.C. and for affixing the CE marking on the product. The manufacturer has to ensure that the product bears model, type, batch number, serial number and other elements allowing its clear identification. Further, the product name plate should also bear manufacturer name, trademark and postal address. In case the manufacturer is based outside the community market, the product should also bear the name, address and contact number of the importer/ authorised representative/distributor/assembler. In case the product size is small and it is not possible to include the manufacturer and importer details on the nameplate, then the details should be on the packaging or in the accompanying documentation (i.e. the manual). The manufacturer should ensure that all products placed in the community market must be accompanied by a manual containing instructions and safety information. The product labelling, instructions and manual should be in a language easily understood by the end user or the surveillance authorities. What language could be easily understood in a particular member state is to be decided by that member state. It is therefore advisable to the manufacturers or importers that the manual should also be in official language of that member state where the product is going to be placed. Since CE marking is based on testing of one sample, in the *production phase*, the follow up action is required as below:

-   The manufacturer should put in place certain quality system checks to ensure that all samples subsequently coming out of the production line are identical to the sample that was tested. Such checks may include inspection, sample testing etc. and maintenance of relevant records.
-   In case the manufacturer has any reason to believe that the products already placed in the market are not in conformity, he/she should bring them into conformity, or if the need arises, recall or withdraw the product from the market.
-   The manufacturer should also take in to account the changes in the harmonised standards to which conformity of a product has been declared. If the manufacturer has reasons to believe that the product may not be in conformity with any new requirement of a standard, then he should take appropriate steps to bring the product into conformity and ensure that all samples placed after the date of the new standard are in conformity. This might require changes in product design, re-testing, updating of D.O.C etc.

1.17.2 Authorised Representative.

The authorised representative is any person, based inside the community market (i.e. a citizen of one of the countries of the EEA), who is explicitly designated by the manufacturer through a written contract and acts instead of the manufacturer, to take up some of manufacturer's obligations. Only the *administrative* tasks can be delegated to the authorised representative. Obligations regarding conformity of the apparatus with essential requirements and compiling of technical documentation remain with the manufacturer. The authorised representative has to *ensure* that these procedures have been carried out by the manufacturer.

The authorised representative can perform such tasks as signing D.O.C., affixing the CE marking and keeping technical documentation at the disposal of surveillance authorities if permitted by the directive in question. For many directives like LVD, EMC and RED the authorised representative is not compulsory, although having one has obvious advantages especially for non-European manufacturers who are new to the CE compliance process. Some directives like medical devices directive (only for class I devices, see chapter 7) and the in-vitro devices directive make it compulsory for the manufacturer to appoint an authorised representative. There are several individuals or companies who can become the manufacturer's representative for a certain fee.

1.17.3 Importer.

The importer is a *person* based inside the community market (i.e. a citizen of one of the countries of the EEA) who places a product from a non-European country on the community market. In the absence of authorised representative, the importer assumes the liability for CE marking. The importer should not be seen simply as a re-seller of products, rather he has a key role to play in guaranteeing the compliance of imported products. The importer must therefore ensure that:

— The manufacturer has met all his obligations regarding product conformity, that the appropriate conformity assessment procedure has been carried out by the manufacturer.

— The manufacturer has compiled technical documentation and the apparatus bears the CE marking and is accompanied by required documents like instructions and safety information (e.g. manual).

— The product (if not possible, in the accompanying documentation) bears the importer's name, registered trade mark and postal address.

The importer is also responsible for continued compliance, product labelling, safety instructions etc. and should keep the declaration of conformity and technical documentation till 10 years (5 years for MDD) after the last sample of the product has been placed in the market and also should be able to provide the same to the European surveillance authorities, when asked for.

1.17.4 Distributor.

The distributor is a person in the supply chain based inside the community market, who takes subsequent commercial actions, after the product is placed in the community market. If the distributor directly places a product from a non-European country in the community market then the obligations of the distributor are the same as that of the importer.

1.17.5 Assembler.

The assembler is a person, based inside the community market, who assembles different components (which cannot function individually) already placed on the market to realize the final function. Assembler is required in lifts and pressure vessels directive. If the assembler assembles a product by directly buying components from non-European entity, the obligations of the assembler will be the same as that of the importer.

1.17.6 Employer.

An employer is considered to be any person who has an employment relationship with a worker (who any person employed by an employer), and has responsibility for the undertaking or establishment. The employer must take all measures necessary to ensure that the work equipment (for example machinery and apparatus) made available to the workers is suitable for the work carried out, and may be used by workers without impairment to their safety or health. Further, the employer has an obligation to provide information and training for workers as regards the use of work equipment for instance, make correct use of apparatus and personal protective equipment.

1.17.7 End User

The term *end user* is not included in any of the directives and hence is not subject to obligations of the economic operators (importers/distributors/ assemblers) as discussed above. End user includes consumers and professional users.

1.18   MARKET SURVEILLANCE

Market surveillance is required to ensure that products placed on the community market meet essential requirements of applicable directives and do not endanger health and safety. Member states are under obligation appoint appropriate surveillance authorities (usually public authorities) as per article 10 of EC treaty. They are entrusted with powers, resources and knowledge necessary for proper performance of their tasks of monitoring products placed on the market. In case of products presenting a risk (or other form of non-compliance) the surveillance authorities can take appropriate action to remove the risk and enforce conformity. Under regulation No 765/2008, these national market surveillance authorities have clear obligations to proactively control products placed on the market, to organise themselves and ensure coordination between themselves at the national level and to cooperate at the EU level. Economic operators (importers/distributors/assemblers) also have the clear obligation to cooperate with the national market surveillance authorities and to take corrective action where necessary.

The authorities carry out surveillance in the following ways:

– By visiting commercial (including malls), industrial and storage premises regularly.
– By visiting work places and premises where product is put into service.
– By organising random and spot checks
– By pulling samples for verification of compliance.

Apart from these, information about dangerous products can also be obtained from the citizens themselves. If a citizen is confronted with a dangerous product, he/she can make an online complaint to their respective national authorities responsible for consumer safety. Member states ensure that the public is aware of the existence, responsibilities and identity of national market surveillance authorities and of how those authorities may be contacted. Upon receipt of the complaint, the surveillance authority compiles and processes information for further action. Another source of complaints is the customs and border control authorities that seize products from EU entry points and state borders.

Every potentially dangerous product so identified is studied and then classified according to the risk it may represent. When there are sufficient reasons to believe that a product presents a risk, it may also be subjected to risk assessment and testing in order to evaluate this risk. All European member states have test laboratories which test products that are seized from their territory. Any non-compliance observed is classified into two categories -non-substantial (formal) or substantial.

Non-substantial non-compliance includes absence of CE marking, wrongly designed or wrongly placed CE marking, incomplete or absence of D.O.C, incomplete or absence of technical documentation etc. In such cases, the manufacturer or his authorised representative (or in their absence the importer/distributor/assembler) is asked to bring product into compliance within a certain time frame.

In case of substantial non-compliance (like if the product is unsafe and does not conform to essential requirements) the measures correspond to the risk the product poses. It can range from a simple warning to withdrawal of the product from the market or product recall or even legal action against the manufacturer or person responsible for placing the product in the community market (i.e. authorised representative/importer/distributor/assembler/employer). Thereafter appropriate prohibition notices are issued and product can be banned in all member states.

1.19   RAPID INFORMATION EXCHANGE (RAPEX).

The RAPEX or rapid information exchange programme is a system of linked servers (called the GRAS-RAPEX) put in place to share information about unsafe products. It covers all non-food consumer products (except pharmaceutical and medical products). The aim is to bring together all national surveillance authorities under one single network, since isolated national measures are insufficient to deal with the entire EEA region. The principle of the RAPEX system is fairly simple and straight forward. As soon as a product is deemed dangerous by one of the member states, the national authority there takes appropriate action to eliminate the risk like withdrawal from the market, recall from consumers or issue warnings. The information is conveyed on-line to the commission which, after analysis and translation, is transmitted rapidly to the other member nations. The national authorities in those nations then take similar measures in their region. In this way, the entire community market is protected and dangerous products are prevented from reaching the consumer. Information on banned products is available on-line on the Commission's website and is updated every week.

1.20   CONTROL OF PRODUCTS BY CUSTOMS

The customs authorities are ideally placed at entry points (like ports, airports and land borders) to stop non-compliant and unsafe products from entering the community market. To check conformity with the directives in the case of products imported from third countries, the customs authorities are required to be closely associated with the market surveillance authorities and information exchange systems (like RAPEX). Customs authorities have the following responsibilities under regulation 765/2008:

- To suspend the release of products when there is a suspicion that the products presents a serious risk to health, safety, environment or other public interest and/or do not fulfil documentation and marking requirements and/or the CE marking has been affixed in a false or misleading manner.
- Thereafter, the customs have to immediately notify the competent national market surveillance authority. The authorities have *three* working days to perform a preliminary investigation of the products to decide if they can be released (if they do not present a serious risk to the health and safety) or they must be detained for further checks to ascertain their safety and conformity.

If the market surveillance authority ascertains that the product presents a serious risk or it does not comply with directives, it must prohibit their placing on the EU market. The market surveillance authority (as per regulation 756/2008) then requests the customs authority to mark the commercial invoice accompanying the product (and any other relevant accompanying document) with the words *dangerous product — release for free circulation not authorised* for product presenting serious risk OR mark *product not in conformity — release for free circulation not authorised* for products that do not comply with relevant directives. Thereafter, the market surveillance authority must use the RAPEX system to inform the surveillance authorities of all other states to stop entry of the product in their respective state or to weed out the product if already in their market.

If the market surveillance authority ascertains that the product in question does not present a serious and immediate risk and is in conformity with relevant directives, the product is released for free circulation. If, within three working days of the suspension of release for free circulation, the market surveillance authority has not notified customs of any action taken by them, the product has to be released for free circulation.

CONCLUSION

CE marking is by far the simplest of all certification processes. For certain low risk products, the EU has made the process as easy as possible. Another important feature of the marking is that it is not owned by the EU government and hence neither do they charge any fee or royalty for using the marking nor does the manufacture has to seek permission for using the marking. Also, as the reader must have noticed by now, CE marking is not given by any EU authority -not even the notified body. It is put on the product by the manufacturer himself and the manufacturer himself declares conformity. This comes as a surprise to most and one is compelled to think as to how the market is protected from spurious and falsely marked product. Well, the market surveillance and the RAPEX are in place to weed out products that are falsely CE marked. In addition, the EU also has mutual recognition agreements (MRAs) with countries which are major exporters to the EU (particularly China) which also keeps a tab on products.

References

Guide to implementation of directives based on new approach and global approach ("The Blue Guide") published by the European Commission.
Directive 2014/30/EC relating to electromagnetic compatibility.
Directive 2014/35/EC relating to electrical equipment designed for use within certain voltage limits (i.e. the low voltage directive).
Directive 2014/53/EC regarding placing radio equipment on the community market.

Directive 2015/863/EC regarding protection restricting the use of certain hazardous substances in electrical and electronic equipment (EEE).

Directive 2014/53/EC regarding placing of medical devices on the community market.

Regulation 765/2008 relating to setting out the requirements for accreditation and market surveillance relating to the marketing of products.

● ● ●

2

THE LOW VOLTAGE DIRECTIVE

2.1   INTRODUCTION

The directive on *electrical equipment designed for use within certain voltage limits* popularly known as the *low voltage directive* (or LVD in short) is perhaps the oldest directive and has been in force since 1973. It would not be an exaggeration if someone refers to it as the mother of all directives –which it is. The directive has been revised many times, the latest revision being the 2016 version of the directive codified as 2014/35/EC published on 26th February 2014 and replaces the old directive 2006/95/EC. The new directive is a recast of the older version which means that all major provisions of the older version remain as they are. The new directive came into force on 20th April 2016. As discussed in section 2.3, the LVD ensures health and safety and requires products to be manufactured such that they will not pose electrical hazard (primarily) so as to endanger safety of persons, domestic animals and property. Non-electrical hazards like mechanical, radiation, chemical etc. are also applicable under certain circumstances. In this chapter we are going to discuss about the contents of the LVD, which is important in understanding the concepts explained in the chapters that follow.

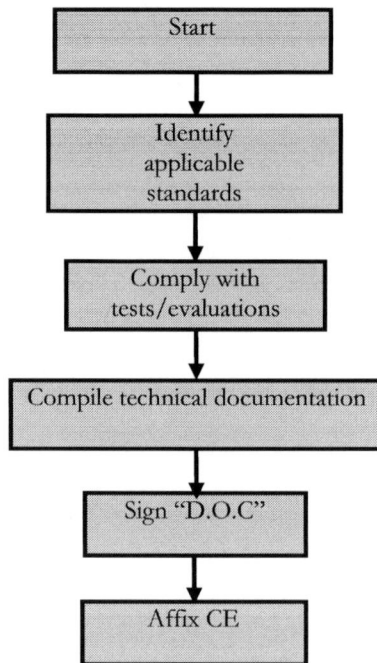

Fig. 2.1: Compliance Module 'A'

For all products falling under its scope, the directive prescribes module 'A' (self-declaration with internal production control) as the route to conformity (see Fig. 2.1). The procedure is given in annexure III of the directive, where in the manufacturer evaluates product safety as per applicable EN standard and upon compliance compiles a technical documentation, signs the declaration of conformity (D.O.C) and affixes the CE marking on the product. The manufacturer then ensures continued compliance i.e. ensures that all samples of the same model subsequently coming out of the production line are identical to the sample that was tested. This can be done by having an internal quality system or a quality system based on ISO 9000 where in appropriate procedures to that affect have been defined.

2.2   CONTENTS OF THE LOW VOLTAGE DIRECTIVE

As with any other directive, the LVD being a legal document starts with the so-called *recitals* typically beginning with the word *whereas*. It is then followed by *articles* which give information regarding a particular aspect of electrical safety. The directive ends with various annexures giving details of procedures to be followed for documentation, CE marking, risk analysis etc. Let us take a look at the various articles one by one.

2.3   ARTICLE 1 : SCOPE

This article deals with the subject matter and scope of the directive and calls for providing high level of protection to persons, animals and property and is applicable to all electrical equipment operating between 50-1000 V ac and 75-1500 V dc. Products that are outside the scope are given in Annexure II. Accordingly, the directive is not applicable to electrical equipment covered by other specific directives like those used in an explosive atmosphere (covered by ATEX directive), electrical equipment for radiology and medical purposes (covered by medical devices directive), electrical parts for goods and passenger lifts (covered by the lifts directive), electricity meters plugs and socket outlets for domestic use, electric fence controllers, radio transmitters and amateur radio equipment (covered by RED), specialised electrical equipment for use on ships, aircraft or railways, which complies with the safety provisions drawn up by international bodies in which the member states participate.

2.4   ARTICLE 2 : DEFINITIONS

This article provides definitions of important terms used in the directive. Of importance to Indian manufacturers is the definition of importer who, as we have seen, is any natural or legal person established within the EU (i.e. who is a citizen of one of the European member states) who places electrical equipment from a third country on the European market. In simpler terms, manufacturers outside the EU can make their product available in the EU market through the importer who being a citizen of one of the member states of the EU is liable to pay for damages, in case the product is found to be unsafe.

2.5   ARTICLE 3 : MAKING AVAILABLE IN THE MARKET AND SAFETY OBJECTIVES.

As we have seen, making available in the market means supply of any equipment for distribution, consumption or use in the EU i.e. any transfer of a product between economic operators. The article identifies *essential requirement* of the directive and states that any equipment made available in the EU, when properly installed and maintained, should be constructed using good engineering practices and should not endanger health and safety of persons, domestic animals and property.

The hazards against which that the directive seeks protection are listed in annexure I of the directive. The hazards arising from the equipment (when used in applications for which it was made) include harm or injury caused by direct or indirect contact with the equipment (shock hazard), temperature, arcs, radiation and non-electrical dangers (like chemical contamination). The directive also states that the insulation should be suitable for the purpose for which the equipment is being used. The directive also seeks to reduce the hazards due to external influences on the equipment like conditions of overload. The equipment should also not pose mechanical hazards.

2.6   ARTICLE 4 : FREE MOVEMENT

This article ensures one of the *four freedoms* i.e. freedom of movement of goods within the European Union. It states that if equipment meets the requirements of this directive, the European member states cannot create obstacles impeding the availability of the product in their market. This means that a compliant product can freely move within the EU and freely enter any member state.

2.7   ARTICLE 5 : SUPPLY OF ELECTRICITY

This article prohibits electricity distribution companies from imposing requirement stricter than those mentioned in the directive.

2.8   ARTICLE 6 : OBLIGATION OF MANUFACTURERS

This article elaborates the responsibilities and obligations of electrical equipment manufacturers. These have already been discussed in section 1.17.1.

2.9   ARTICLE 7 : AUTHORISED REPRESENTATIVES

This article allows the manufacturers to appoint authorised representatives who, through a written contract, can carry out some obligations of the manufacturer.

2.10   ARTICLE 8 : IMPORTERS

If the manufacturer based outside the community market has not appointed an authorised representative, all the liability of the CE marking lies on the importer (or distributor or assembler). The duties of the importer have been discussed in the previous chapter.

2.11   ARTICLE 9 : DISTRIBUTORS

This clause gives the obligations of the distributor which are the same as that of the importer.

2.12   ARTICLE 10 : OBLIGATIONS OF MANUFACTURERS APPLYING TO DISTRIBUTORS & IMPORTERS

This article states that, if the distributors or importers place a product in the market in his own name, then that importer or distributor will be considered as a manufacturer.

2.13   ARTICLE 11 : ECONOMIC OPERATORS

As per the definitions given at the beginning of the directive, manufacturers, importers, authorised representatives and distributors are considered as economic operators. Each economic operator is liable to provide to the market surveillance authorities, information regarding who has supplied them the equipment and to whom a particular economic operator has supplied equipment to. For example, if an importer buys equipment from a non-European manufacturer and gives it to a distributor, the importer is liable to provide information of the manufacturer and the distributor to the market surveillance authorities.

2.14   ARTICLE 12, 13 AND 14 : PRESUMPTION OF CONFORMITY

The article 12 states that any equipment complying with harmonised EN standards is presumed to comply with the essential requirements of this directive. The new directive has a provision that if no harmonised EN standard exists for a particular product of a product family, then IEC standards or national standard of the European country (in which the product has been manufactured) can also give presumption of conformity. Some of the most frequently referred safety standards are:

EN 61010-1: Safety Requirements for Electrical Equipment for Measurement, Control, and Laboratory Use.

This standard specifies general safety requirements for electrical equipment for professional, industrial process and educational use. This includes any peripheral electrical equipment or accessories or computing devices that may be used in conjunction with measurement and test equipment, control devices or laboratory use devices. For instance, electrical measurement and test equipment may include any measurement and recording devices such as computers, oscilloscopes, main frame testers, etc. Non-measuring electrical equipment such as signal generators, power supplies, transducers, transmitters, etc., are also within the scope of this standard. Electrical control equipment can be any equipment that controls an output as a function of some input control variable such as programmable logic controllers (PLCs) etc. Electrical laboratory equipment is any equipment which measures, indicates, monitors or analyzes substances, or is used to prepare materials. This standard applies to equipment designed to be safe at least under the following conditions: Indoor use, altitude up to 2000 m, temperature 5°C to 40°C, maximum relative humidity 80% for temperature up to 31°C, decreasing linearly to 50% relative humidity at 40°C. The mains supply voltage fluctuations should not exceed ±10% of nominal value (other supply voltage fluctuations as stated by manufacturer).

EN 60065: Safety Requirements for audio, video and similar apparatus for household and similar use

This standard specifies safety requirements for household equipment intended for domestic and indoor use. Examples of this type of equipment are radios, TVs, professional sound/video systems, amplifiers, video cameras, electronic musical instruments etc. The standard applies to equipment for use for a 2000m maximum altitude. This standard does not apply to equipment designed for a rated supply voltage that exceeds 433 Vrms (between phases for a 3-phase supply) and 250 Vrms in all other cases. Rated supply voltage is the supply mains voltage for which the manufacturer has designed the apparatus.

EN 60601-1:General Safety Requirements for Medical Electrical Equipment

The general requirements for the safety of medical electrical equipment under the MDD are addressed by this standard and it further serves as a basis for product standards. The product standards take priority in case of any conflicting requirements of the general standard and a product standard. The equipment in this standard is capable of operating in an ambient temperature range of +10°C to +40°C, relative humidity range of 30% to 75%, and atmospheric pressure range of 500 hPa to 1060 hPa.

EN 60950: Safety of Information Technology Equipment, Including Electrical Business Equipment

EN 60950 is a well-known international standard which specifies the requirements for equipment connected directly to a tele-communication network. This type of equipment includes work stations, personal computers, printers, facsimile machines, cash registers, mail processing machines, document shredders, plotters, postage machines, modems and many more associated with telecommunication. EN 60950 is not limited to, but it is often associated with office equipment. The rated voltage of EN 60950 equipment is not to exceed 600 V. Many countries have adopted EN 60950 as their own standard for office and related equipment. There are various regional standards based on or identical to the above safety standards. These are given in the following table:

Table 1.1: Application Vs. Global / Regional Safety Standards

Application	International	European	U.S.A	Canada	Germany
Organisation	IEC	EN	UL	CSA	DIN/VDE
Industrial	60204	60204	508	14-M91	--
Information Technology	60950	60950	60950	60950	60950
Medical	60601	60601	2601-1	601	750
Audio Video	60065	60065	8730-1		860
Measurement, Control, lab use	61010-1	61010-1	1262	1010	0410, 0411

EN 60664-1: Insulation Coordination for Equipment within Low-Voltage Systems

This standard addresses the insulation coordination for equipment within low voltage systems and is therefore a very crucial standard. EN 60664-1 is a basic safety standard and is a guide for technical committees responsible for different equipment in order to rationalize their requirements so that insulation coordination is achieved. Insulation coordination is the mutual correlation of insulation characteristics of electrical equipment taking into account the voltage environments to which the equipment will be subjected to in their normal use. It classifies environment and identifies *pollution degrees* (see section 3.10.8.4), presence of transient peak voltages (called transient over voltages) and also classifies insulation material as per their tracking index (see section 3.10.9.2). The standard also includes methods of electrical testing with respect to insulation coordination.

2.15   ARTICLE 15: DECLARATION OF CONFORMITY

After complying with the tests and evaluations (as mentioned in the applicable harmonised standards) the manufacturer signs as declaration of conformity. The requirements and contents of the D.O.C are given in Annexure IV of the directive have been explained in chapter 8.

2.16   ARTICLE 16,17: AFFIXING THE CE MARKING

This article tells how CE marking should be affixed on the product. The requirements have been dealt with in section 1.16.

2.17   ARTICLE 18, 19 AND 22: MARKET SURVEILLANCE AND NON-COMPLIANCE

As seen in the earlier chapter, European authorities carry out market surveillance to weed out non-compliant products. This article says that, if the surveillance authorities establish that a particular product placed in the community market is non-complaint and presents a risk, then the concerned economic operator (importer, authorised representative, distributor or the employer) who has placed the product in the community market, is required to bring the product into compliance, recall or withdraw the product from the market depending upon the severity of the risk. If the economic operator does not take corrective action within a stipulated period, the surveillance authorities can take appropriate action to prohibit sales, withdraw already placed products or even take legal action against the economic operator.

2.18   ARTICLE 20: SAFEGUARD PROCEDURE

As seen in the earlier chapter, European authorities carry out market surveillance to weed out non-compliant products. This article gives a procedure as to what measures are to be taken in case of non-compliant products. The types of non-compliances are identified in article 22 (see section 2.21). When the market surveillance authorities come to know or have reasons to believe that that the electrical equipment presents a risk to the health and safety then they have the authority to carry out a full safety evaluation of the product. If after the evaluation, it is proved that the product does not comply with the requirements of the directive, the market surveillance authorities instruct the relevant economic operator (importer/ manufacturer/ distributor) to take all appropriate corrective actions to bring the electrical equipment into compliance with those requirements within a certain time period.

In case the surveillance authorities consider that the non-compliance is not restricted to their country, they have the obligation to inform other member states and to the Commission regarding the results of the evaluation and the measures that the economic operator has been instructed to take.   If the economic operator fails to take appropriate corrective actions to bring the electrical equipment into compliance, the market surveillance authorities have to take appropriate measures to restrict the equipment from being placed in the market or to withdraw the electrical equipment from the market or to recall it within a reasonable period. The surveillance authorities are obliged to convey the measures taken to other member states and to the commission. The other member states, on their part, are required to take similar measures to restrict/withdraw/ recall the product.

2.19 ARTICLE 20: SAFEGUARD PROCEDURE

When the Commission is informed about a product non-conformity, it decides whether the measures taken against the erring economic operator is justifiable or not. If the measures are justifiable, then all member states should take appropriate measures to restrict/withdraw/recall the product (as the case may be) and inform the Commission accordingly.

2.20 ARTICLE 21: WHEN COMPLIANT EQUIPMENT PRESENTS A RISK.

If, in case if the surveillance authorities find that a product presents a risk in a given scenario, but (after relevant testing) find that the product is compliant to relevant standards, then the economic operator (importer, authorised representatives or distributor) is required to take additional measures to remove a risk being present in that special scenario.

2.21   ARTICLE 22: FORMAL NON-COMPLIANCE.

This article talks about *formal* or *non-substantial* non-compliances (other than the *substantial* non-compliances dealt with in article 19).   These can include but not restricted to:

- CE marking has not been correctly affixed, proportions not respected   or not affixed at all.
- The D.O.C has not been drawn up or incorrectly drawn up.
- Technical documentation is either incomplete or not available.
- Manufacturer (or importer/distributor if the manufacturer is non-European) has not indicated on the electrical equipment (if that is not possible, on its packaging or in its documentation) their name, registered trade name/mark and the postal address at which they can be contacted (the contact details being in a language easily understood by end-users and market surveillance authorities).
- The product is not accompanied by manual and/or safety instructions or labeling etc.

If such non-compliances are reported, the manufacturer or the importer is obliged to correct them. If this is not done then the member state concerned is authorized to take all appropriate measures to restrict or prohibit or recall or withdrawn the product from the market.

2.22   ARTICLE 23: COMMITTEE PROCEDURE.

This article gives an internal procedure (and is not of immediate consequence to the economic operator) that the European commission should be assisted by a committee on electrical equipment which consists of experts in field of electrical safety. This committee provides guidance on the legislation and future development and amendments.

2.23 ARTICLE 24: PENALTIES.

This article empowers the member states to lay down rules for pursuing legal cases against economic operators of non-complaint product and imposing penalties thereafter. It also empowers the member states to impose criminal penalties on serious non-compliance.

2.24 ARTICLE 23,25,26,27, 28: TRANSPOSITION.

As we have seen earlier, directives are not directly applicable to member states, rather once a new directive is published; each member states are obliged to modify their law so that the directive becomes part of their law. The articles 23 and 25 to 28, instruct member states to transpose the provisions of the new directive into their law. The new LVD was published on 26th February 2014 and the commission had given a transition period up to 19th April 2016 to enable the member states to include the provisions of this directive into their law and make those mandatory as of 20th April 2016. The member states were also instructed to repeal the earlier directive with a condition that they should not impede the making available of a product placed in the market before 20th April 2016.

2.25 TECHNICAL DOCUMENTATION.

The LVD, through its annexure III, specifies the minimum contents of technical documentation. The documentation should make it possible to assess the electrical equipment's conformity to the relevant requirements, and should include an adequate analysis and assessment of the risk(s) (see section 3.38.2). The technical documentation should specify the applicable requirements and cover, as far as relevant for the assessment, the design, manufacture and operation of the electrical equipment. It should be kept at the disposal of surveillance authorities (i.e. should be produced when demanded) by the manufacturer or the importer for ten years after the placement of the last sample of the product type. If the manufacturer is non-European, the documentation should be with the importer. The contents of the technical documentation are discussed in detail in the chapter 8.

2.25.1 Risk Assessment And Rationale For Conformity

This section of the technical documentation should provide description risk analysis and assessment carried out for the equipment for demonstrating compliance to the LVD. A detailed procedure for risk analysis is discussed in section 3.38.2. The documentation should also include a list of the standards (harmonised or otherwise) applied in full or in part and an explanation (rationale) as to how the hazards have been eliminated.   Where standards have not been applied, descriptions of the solutions adopted to satisfy the essential requirements. The section may also include results of design calculations made, examinations carried out, etc. This will prove to surveillance authorities the due-diligence on part of the manufacturer and his sincere efforts towards designing a safe product.

CONCLUSION.

The low voltage directive is by far the oldest and the simplest of all directives in the sense that the simplest module i.e. Module 'A' is applicable which does not compulsory involve a Notified Body. Now, when a manufacturer declares compliance with this directive through the D.O.C, it means that he has read, thoroughly understood and accepts the legal requirements of the directive. This chapter has sought to explain the requirements of the directive in the simplest way possible. If understanding legal requirements was tough, a bigger challenge awaits the manufacturer since technical compliance to harmonised standard, which requires product evaluation and risk assessment, is a different matter altogether. The chapters that follow throw light on various design measures the manufacturer should take so that his product complies with the technical requirements given in the safety standards.

Reference

The Low Voltage Directive 2014/35/EC of the European Council and Commission.

. . .

3

DESIGN FOR SAFETY

3.1   INTRODUCTION.

The low voltage directive is the primary directive for safety compliance of electrical equipment. Some people assume that the term low voltage means safe voltage, but as we have seen the LVD applies to products that operate at typical line (mains) voltages (e.g., 230/400 volts) that present shock and fire hazards. Safe voltages are those less than 42.4 V ac (peak voltage) or 60 V dc, depending on the standard or term applied and are referred to as extra-low voltage or safety extra-low voltage (ELV/SELV). The LVD applies to all safety aspects of electrical equipment, including protection from mechanical and other hazards.

In this chapter we will take a look at the basics of electrical safety of equipment. Proper application of the guidelines in this chapter and chapters that follow, along with the standards, will ensure that the product complies with European standards. Let us get familiarised with some important terms that we would be frequently referring to in this section.

Safety means *freedom from unacceptable risk*. Equipment is sometimes referred to as product and includes machinery also. The term *appliance* is generally applied to household and similar products and includes commercial appliances. *Operator* is an authorized person designated to operate a machine as intended, except service personnel. Safety requirements assume that operators do not mean to create a hazardous situation and are not aware of electrical and other hazards. One also must assume that the operator does not normally possess tools reserved for service and maintenance purposes. Sometimes the terms *user* and *consumer* are also used instead of operator and hence are equivalent terms. Apart from this, equipment must protect janitors and casual visitors too. An *operator access area* is any area that, under normal operating conditions, allows access without the aid of a tool, for example, by a person's hand or fingers alone. Opening a hinged door by hand, without a tool, makes the area behind the door an operator access area and all hazards should be adequately guarded or the door interlocked to remove hazards before access.

Service personnel are those that have training and experience and are considered be aware of potential hazards while performing a task and can take measures to minimize the danger to themselves and others. Service personnel have access to maintenance areas and have to be reasonably careful in dealing with the obvious hazards. A *tool* is reserved for service personnel and defined as *any object that can be used to operate a screw, latch, or similar fixing means*. The *service access area* is an area, other than the operator access area, that service personnel can access even with the equipment in power ON condition.

As a designer one should first give careful consideration to the selection of components and construction requirements, which minimizes the risk that the product will fail its first test, thus, causing costly product redesign. One should not - rely on test results to be the stimulus for the safe design of the product rather safety should be considered at the design stage, before addressing other design elements. A product's conformity to the European safety standards relies on the use of proper component and construction principles. Testing should be performed only after a sound design is in place.

3.2   THE SINGLE FAULT CONCEPT.

The European safety standards stress an important principle called the *single fault* concept that is at the core of safety philosophy. This principle states that even under a single fault (i.e. a wire coming loose/neutral open/component failure/ short/open circuit/insulation failure etc.) at least one level of protection (LOP) or insulation must be maintained (after the first fault) to ensure adequate protection of the user. This concept is sometimes referred to as *double improbability* which relies on the fact that the failure of one LOP is high but the failure of the second LOP (after the first) is extremely low and that there is always a second means of protection or insulation should the first one fail. This means that two levels of protection should be provided such as by insulation, grounding, shielding or safety interlocks.

This will be clear by taking the example of insulation which is one of the methods to isolate circuits carrying hazardous voltages from user accessible parts (like switches or enclosures). If only one layer of insulation is used, it provides protection under single fault conditions. To take care of second failure as a consequence of a first failure, safety standards demand a redundant system with at least two levels of protection. Thus, in case of accessible components/parts, they must be insulated from hazardous voltages by a double-level system. One method is single insulation plus protective earthing of a conductive enclosure. Here if the insulation between hazardous voltage and enclosure fails, the earthing provides a second level of protection. Another method is to use two layers of insulation (double insulation) so that in the event of failure of one layer of insulation, there is still another layer for protection.

With the single-fault concept we can determine the protection or insulation needed to satisfy the electrical safety standards. In addition, it is through single-fault analysis that the number and type of abnormal tests are determined. During safety evaluation, a detailed safety fault analysis is performed to determine which faults can occur to establish the proper insulation type(s) and identify the abnormal testing required. Safety circuits for machines sometimes require special components such as relays, contactors, interlocks and E-stops. Common terms associated with these machine components are control reliable, fault tolerant and fail-safe, which means that they fail to a safe condition after a single fault. Most of the safety standards call for the simulation of single faults (discussed in detail in section 3.20 and 3.36.13) such as electrical shorts/open, component failures, mechanical faults, locked rotor test, blocked ventilation vents and short circuit/overload of transformers/power supplies as applicable. The equipment's temperatures and electrical outputs are monitored during these tests and the equipment should not become hazardous.

3.3   PRINCIPLES OF SAFETY.

To meet the relevant safety standards and manufacture safe products, it is essential that designers understand the principles of safety. The safety concepts and requirements given in chapters that follow are not an alternative to the detailed requirements of the standards, but are intended to provide designers with an appreciation of the principles on which these requirements are based. The product's design should protect users/operators and service personnel from any possible mishap by using insulation/shields for hazardous voltages, segregation of SELV from hazardous voltages, warning labels and interlocks. A design should protect service personnel from unexpected hazards, such as accidental or inadvertent touching of live electrical components and other hazardous parts during servicing. A safety standard is intended to reduce the likelihood of death, injury and property damage due to the following hazards:

- Electric shock hazard
- Flammability hazard
- Mechanical hazards
- Other hazards (like chemical, radiation, implosion etc.).

These hazards and design considerations to address and mitigate them is discussed in subsequent chapters.

3.4   ELECTRIC SHOCK HAZARD.

It is a common knowledge that 230V ac wall-outlet voltages can cause severe shock or death. But rather than the voltage, it is the current (which is forced by the voltage) that causes electric shock when it flows through the human body. Having said that, people who know this generally underestimate how little current it takes to electrocute a person. Take the example of a 15 W night light which, at a mains voltage of 230 V ac, can draw a current of about 65 mA. This, a common man may think, is pretty low to cause any harm –but he does not know how wrong he is, till he is told that current above 10 mA may be enough to cause severe trauma and even death! And that is true. We shall see how.

3.5   SEVERITY OF SHOCK HAZARD

The factors that determine the severity of an electrical hazard and its effect on the human body are voltage, current, resistance, frequency, duration and pathway. The voltage forces a current to flow which can damage the heart or brain or cause involuntary muscle contractions. Current determines the extent of the damage and can cause heating of external and internal human-body tissues and organs. Human-body resistance varies depending on how dry or moist the body is and on the current's path through the body. Current passing though the arm generates more thermal damage than through the abdomen because the arm has a smaller cross-sectional area than the abdomen. Frequency also influences the danger; AC causes more ventricular fibrillation than DC, but both can lead to injuries. Duration affects the severity of heating of human tissues and organs; the longer the contact, the greater the damage.

Now, voltages of 100 to 250 Vac in wall outlets are the most common and can be lethal. This voltage range can cause significant current flow through the body. Electrical current follows the path of least resistance to ground. The human body is approximately 70% water and makes an excellent conductor. Human-body resistance varies from as low as 10 kΩ to 100 kΩ depending upon grounding, age, size, and gender. Seemingly inconspicuous things like perspiration on the body also affect its resistance. The resistance even varies from body part to body part. Arteries, nerves, and muscles have low resistance, whereas bone, fat and tendons have relatively high resistance. The human brain, heart and nervous system are the most sensitive. Hence the resistance may vary depending upon which path the current takes through the body. A greater chance of mortality exists for a hand-to-hand current pathway through the heart, and lesser for a hand-to-foot path. An old adage tells us to place one hand in our pocket when working near hazardous electricity so that current does not pass through your chest. A better recommendation is not to touch voltages on wall outlets!

Let us see the actual effect of current on the human body. Currents up to 0.5 mAac (2 mAdc) region can be perceived by people and may cause tingling sensation. Higher currents up to 3.5 mAac are still tolerable but can cause a startle reaction or quick movement that can result in a person falling or striking a secondary object, resulting in injury. As the current increases from 3.5 mAac up to 10 mAac (25 mAdc) the startle reaction gives way to contraction of muscles. The muscular contractions and accompanying sensations of heat increase as the current increases. Sensations of pain develop, and voluntary control of the muscles that lie in the current pathway becomes increasingly difficult. As current approaches 10 mAac, the victim cannot let go of the conductive surface being grasped. At this point, the individual is said to *freeze* to the circuit. This is frequently referred to as the *let-go* threshold and the person cannot let go of the current carrying object. Currents higher than 10 mAac (typically near about 40 mAac) can cause fibrillation of the heart. Ventricular fibrillation is defined as *very rapid uncoordinated contractions of the ventricles of the heart resulting in loss of synchronization between heartbeat and pulse beat.* Once ventricular fibrillation occurs, it will continue and death will ensue within a few minutes. Use of a special device called a de-fibrillator is required to save the victim.

It must be noted that safety standards make certain presumption to approximate the effects of electricity on the human body. Firstly, the human body is always presumed to be connected to earth and that the user is not using personnel protection equipment whatsoever. Secondly, the area of electrical connections to the human body is presumed to be 8000mm^2 (unless otherwise stated) which is the maximum area of a typical human hand. Thirdly, 10 mm^2 is considered to be a small area of contact (provided the area can only be touched and not gripped, example is a finger approaching a connector pin) while above 10mm^2 is considered to be a large area of contact.

3.6   TYPES OF CIRCUITS

Protection against high currents can be obtained by limiting the currents. Such circuits in which the current is limited are called as limited current circuits (LCC). In such circuits, the current does not exceed 0.7 mA peak/0.5 mArms or 2 mAdc, for frequencies less than or equal to 1 kHz, during normal operation or single fault conditions. For frequencies above 1 kHz, the limit is determined by multiplying 0.7 mA with the value of frequency (in kHz). However, the current is not allowed to exceed 70 mA under any circumstances at all frequencies above 100 kHz. It should be noted here that these limits can still cause burns if one touches a sharp edge or corner, as current density may be high due to concentration of charges.

Since it is the circuit voltage that drives this current, lets us now talk a little about circuit voltages. Safety standards categorize voltages to be either hazardous or safe according to the maximum voltage possible at all points within the circuit, during both normal operating conditions and under single fault conditions. Within this criterion, there are three classifications of circuits. One is the limited current circuit (LCC) which we have discussed above and the others are SELV circuits, ELV circuits, hazardous voltage circuits and hazardous energy level circuits. Let is now take a look at these.

3.6.1   Safety Extra-Low Voltage (SELV) Circuits

Safety Extra-Low Voltage (SELV) is a voltage that remains   below 42.4 Vpeak or 60 Vdc under normal condition and 71 Vpeak or 120 Vdc under single fault condition for maximum 200 ms (i.e. the voltages under a single fault can exceed 42.4 Vpeak up to 71 Vpeak for a short duration but return to normal limits within 200 ms). SELV voltage is normally used for circuit or components that may be touched by the user (i.e. I/O connectors, wiring for peripherals, some exposed PCB traces etc.). Accessibility to SELV circuit is permitted (i.e. the circuit can be touched). SELV can be derived from a hazardous voltage source but has to be isolated from the hazardous voltage by double level of protection (double insulation or single insulation plus earth).

3.6.2   Extra-Low Voltage (ELV) Circuits

Extra-low Voltage (ELV) is a voltage that remains below 40 Vpeak or 60 V dc under normal conditions but can achieve hazardous value under single fault conditions. Consequently, circuits bearing such voltages are termed as ELV circuits. ELV can be derived from a hazardous voltage source but may not be isolated from the hazardous voltage i.e. separated only by single level of protection like single insulation.

3.6.3   Hazardous Voltage Circuits

Voltages above 42.4 V peak or 60 Vdc are termed as hazardous voltages and circuits or parts carrying such voltages are called hazardous voltage circuits and are considered to be *live*.

3.6.4   Telecommunication Network Voltage (TNV) Circuits.

As the name suggests such voltages are observed in telecommunication networks. TNV circuits are defined in EN 60950 and are further classified as TNV1, TNV2 and TNV3.

TNV1 circuits are those where the voltages can exceed SELV limits but are constrained by either accessibility or duration. The normal operating voltage can be up to 42.4 Vpeak ac or 71 Vdc while the voltage under single fault condition can be up to 71 Vpeak ac or 120 Vdc. The difference between TNV1 voltage and SELV voltage is that in TNV1, there is no time limit of 200 ms, the accessible contact area is limited to that of a connector pin (i.e. the entire circuit is not accessible but only connector pins can be touched) and there could be a possibility of transients from the telecommunication network and cable distribution systems up to 1500 V but of short duration. Example of such a circuit is a 4 wire ISDN internal line from EPABX that operates on 48 Vdc.

TNV 2 circuits are those whose voltage can be up to 71 V peak ac or 120 V dc under both normal and single fault condition (voltage exceeds TNV1 even under normal condition) but there is no possibility of transients. Example of TNV2 circuit is a 2-wire telecom line running inside a building.

TNV 3 circuits are those whose voltage can be up to 71 V peak ac or 120 V dc under both normal and single fault condition (voltage exceeds TNV1 even under normal condition) and there could be a possibility of transients from the telecommunication network and cable distribution systems up to 1500V but of short duration. Example of TNV3 circuit is a 2-wire telecom line from the main exchange to the building.

Table 3.1. TNV circuits

Circuit	Voltages		Transients
	Normal	**Single Fault**	
TNV 1	42.4 Vac peak or 71 Vdc	71Vpeak ac or 120 Vdc	Yes
TNV 2	71 Vpeak ac or 120 Vdc	71Vpeak ac or 120 Vdc	No
TNV 3	71 Vpeak ac or 120 Vdc	71Vpeak ac or 120 Vdc	Yes

3.6.5   Hazardous Energy Level Circuits

These are circuits where the stored energy levels are greater than or equal to 20 joules or an available power level of 240VA or more at a potential of 2 V or more.

3.7   EXAMPLES OF CIRCUIT TYPES.

Let us consider some examples by which we can understand the different type of circuits. The Fig. 3.1 shows an ISDN Network Termination Adapter (NTBA). The basic function of the circuit is transformation of a 2-wire digital interface (Uko) at 100 Vdc (typical) into a digital 4 wire interface (So) at 40 Vdc (typical). Another function is the feeding of power to the 4 wire So interface.

During normal operation, the power is fed at 40 Vdc via an internal 230V power supply (drawing AC power from the mains and rectifying it). In case the mains fail, power is drawn via the Uko-interface at 100V DC and converting it to 40 Vdc using a DC-DC converter. The 2-wire line is a telecommunication line with length in excess of 5 km. The voltage can range from 48 Vdc to 120 Vdc with typical value being 100 Vdc. At the source, one wire could be connected to earth. The 4–wire line is an in-house telecommunication line. The voltage can range from 36 Vdc to 42 Vdc with typical value being 40 Vdc. The line has floating voltage and is user accessible (i.e. can be touched).

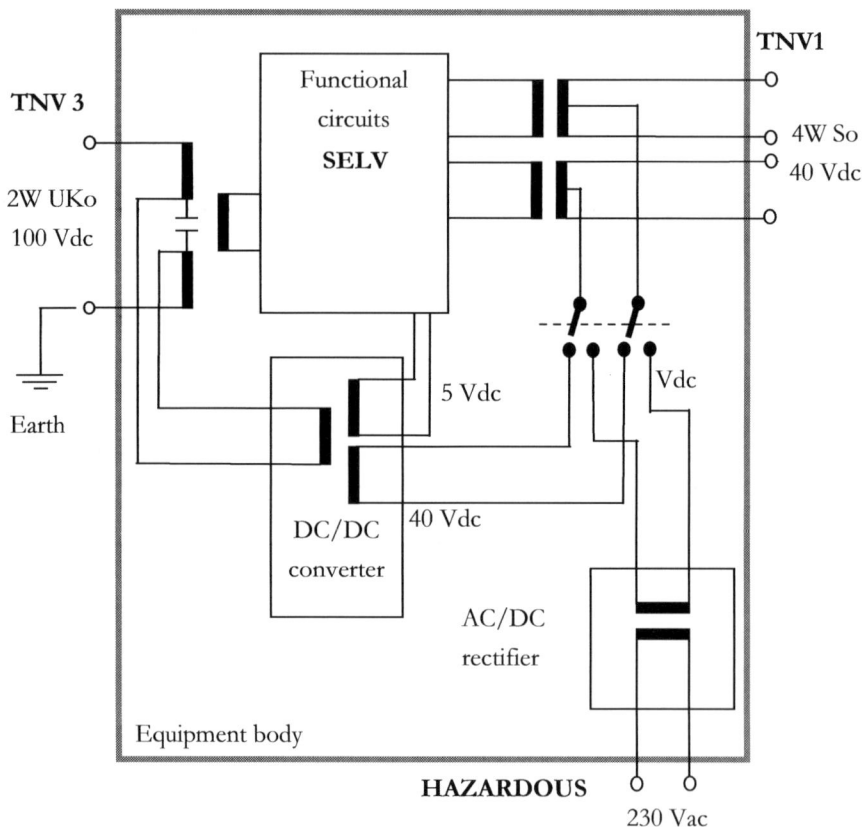

Fig 3.1: Circuit Classification For I.T. Product

The power for the 4-wire interface is drawn from the in-house 230 Vac mains via a rectifier. This is a class II supply in overvoltage category II (see section 3.10.10). The 5 Vdc required for the operation of the functional circuit is drawn from 100 Vdc via a DC-DC converter. We will now classify the circuit types for the above product. The 2-wire line (UKo) is a telecommunication line with length in excess of 5 km. The voltage under normal condition can range from 48 Vdc to 120V dc. Considering the length of the line, it can have transients in excess of 1500 V. It is clear therefore that this is a TNV3 circuit. The 4-wire line (So) has a voltage that can range from 36 Vdc to 42 Vdc. But since this is an in-house telecommunication line, it can accumulate transients and hence is a TNV1 circuit. The 230 Vac is a hazardous circuit, while the functional circuit runs on 5 Vdc and so is a SELV circuit.

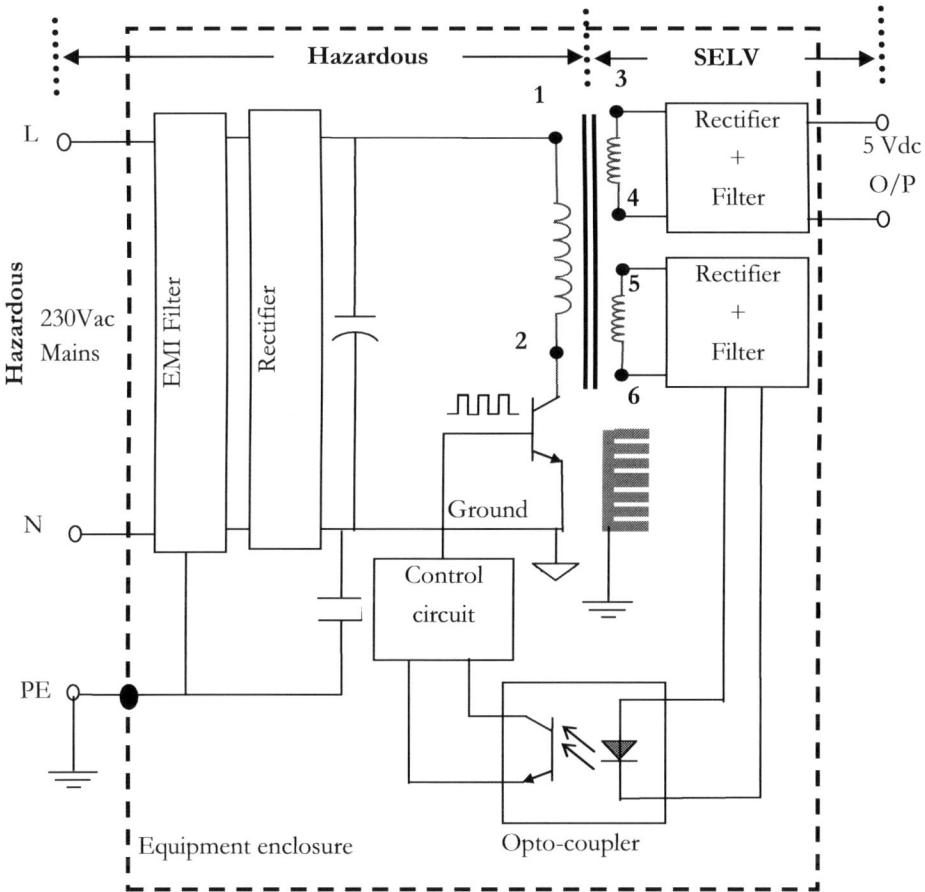

Fig 3.2: Circuit Classification For SMPS

The second example is a switched mode power supply (SMPS). As shown in the Fig. 3.2, it is fed with a 230V, 50 Hz mains supply which is filtered by the EMI filter and given to a rectifier which converts the AC into DC. The DC is then switched by a switching transistor to convert it into a high frequency square wave.

The voltage is down converted by the transformer and rectified to provide a 5 Vdc output.

A part of the output is fed back via the opto-coupler (which isolates the output from the input) to the control circuit to generate the PWM pulses for the switching transistor. Let us now classify the various voltages. The voltage at the input mains is obviously a hazardous voltage. The voltages at points 1 and 2 are also hazardous. The output is at 5V dc on pins accessible to the user and so is SELV. The opto-coupler is bridging SELV on one side and hazardous voltage on the other.

3.8   SEGREGATION OF CIRCUITS

It is worthwhile to note that only SELV and LCC circuits allow the operator unrestricted access to bare circuit components. TNV-1 circuit allows restricted contact to the operator. On the other hand, TNV-2, TNV-3, ELV, hazardous voltage and hazardous energy circuits are considered as dangerous and unsafe and must be protected against operator contact.

Electric shock can occur when there is a breakdown of insulation (i.e. failure of insulation under electric stress when the discharge completely bridges the insulation) between SELV circuits/ accessible parts and parts that carry hazardous voltage. To avoid shock, hazardous voltage circuits and SELV circuits (user touchable) must be properly segregated by using following means:

- Restricting access.
- Isolating (by insulation and/or circuit separation).
- Separating by earthed screens (grounding).
- Using protection components/devices.

Where the above means are not practicable or as an added measure, suitable warning signs should be used to alert the operators and service personnel about hazards. Let us consider the measures one by one.

3.9   RESTRICTING ACCESS

Safety standards such as EN 60950 call for protection of the operator in operator access area against access (or contact) to bare parts carrying hazardous voltage, ELV circuit, solid insulation (functional and basic) and wiring for ELV & hazardous voltage. Unrestricted access is permitted to limited current and SELV circuits.

Hazardous voltage circuits have to be covered so that access is not possible under normal as well as single fault conditions as shown in Fig. 3.3. The following diagrams illustrate accessibility requirements

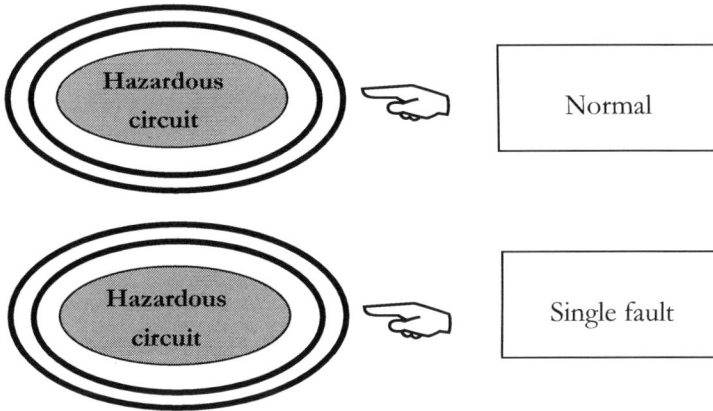

Fig 3.3: Accessibility Of Hazardous Circuits

As shown in Fig. 3.4, TNV-2 and TNV-3 circuits have to be covered so that access is not possible in normal operating condition.

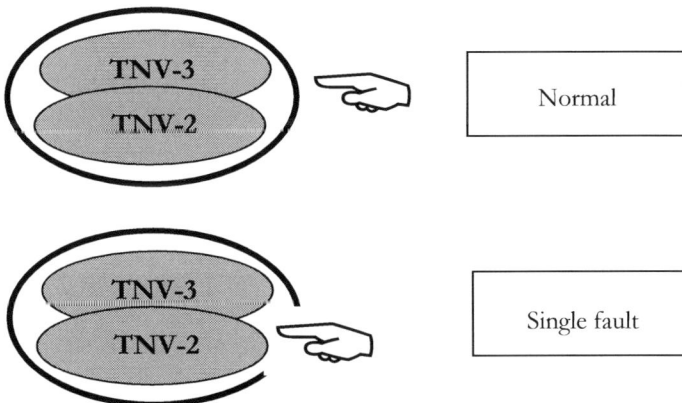

Fig 3.4: Accessibility Of TNV-2 And TNV-3 Circuits

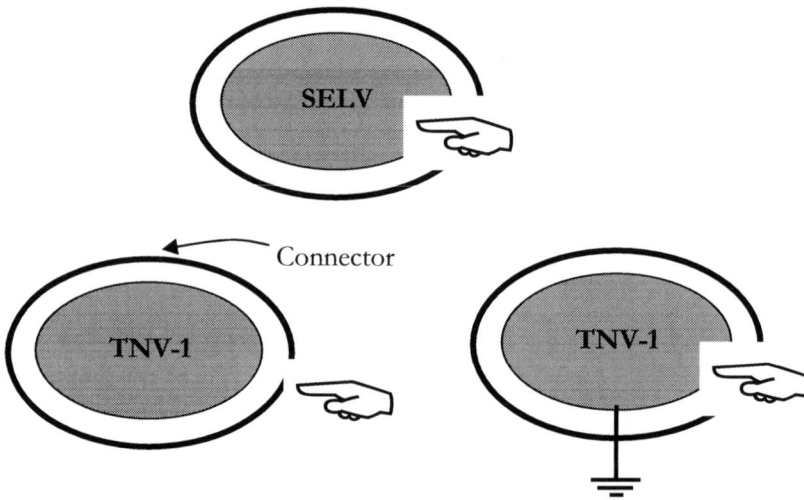

Fig 3.5: Accessibility Of SELV And TNV-1 Circuits

Accessibility to SELV circuits (see Fig. 3.5) is permitted at all times while accessibility to TNV-1 circuits is permitted at all times only if the circuits are earthed, if not then only connector pins should be accessible.

3.9.1   Tests For Accessibility

An accessible part is determined by means of the straight and jointed test fingers defined as test probes 11 and B of EN 61032, and the small finger probes defined as test probes 18 and 19 of EN 61032. Following is the list of some test probes used in this standard.

- Un-jointed (rigid) test probe 11 (adult test finger)
- Jointed test probe B (adult test finger)
- Small finger test probe 18 (36 months up to 14 years; child test finger)
- Small finger test probe 19 (up to 36 months; child test finger)

Most safety standards specify that the jointed test finger is applied, without appreciable force and in any possible position, to the surface and openings of the equipment. Where access behind a door, panel, removable cover, etc. is possible without the use of a tool, or access is directed, with or without the use of a tool, the test finger is applied to surfaces and openings in those areas.

The straight un-jointed test finger is applied with a force of 30 N in any possible position to the surface and openings of the equipment. If the un-jointed test finger enters the equipment, the jointed test finger is pushed through the aperture to determine if any parts behind the aperture are accessible. The test is repeated with the appropriate small finger test probes 18 and 19 of EN 61032, applied with a force of 10 N, if the equipment is likely to be accessible to children.

3.10   SEGREGATION BY ISOLATION

Isolation means that there is no electrical connection between two or more circuits or between circuits and accessible parts. Its purpose is to limit transient over-voltages and to electrically segregate circuits that, if connected directly, could allow the flow of harmful voltage, current, energy, or charge. Isolation prevents unacceptable current from flowing as a result of a potential ac or dc difference between the circuits. Safety isolation also minimizes the possibility of transient voltage arc-over or arc-through insulation to user-touchable circuits and enclosures. Galvanic isolation is the most popular method. Transformers and opto-isolators are examples of galvanic isolation. Transformers use separate windings to magnetically convert power from inputs to outputs. Opto-isolators convert input signals into light and then convert it back for the output signals.

3.10.1 Isolation By Insulation.

Isolation can be achieved by introducing insulation between accessible circuits and circuits carrying hazardous voltages. Insulation separates live parts of a product from the user and also prevents live parts from coming into contact with each other. For protection against direct and indirect contact with live parts, insulation is safer than grounding or fusing. The protective measures that we are going to discuss are given in the respective product standards and apply to all products and machines. The types of insulation applied between circuits of different potentials and metal parts depend on the equipment class, possible faults, and the parts in question. The voltage and circuit accessibility are also important for proper selection of the insulation type. There are following basic types of insulation:

3.10.1.1 Operational insulation.

Also referred to as functional insulation, it is that insulation which is necessary only for circuit operation. It is assumed to provide no safety protection. It is usually used to separate SELV circuits from earth. Examples are lacquer, solvent-based enamel, ordinary paper, cotton, oxide film etc. It also includes displaceable insulation such as beads and sealing compounds (other than self-hardening resins).

3.10.1.2 Basic insulation.

It provides basic protection against electric shock with a single level (see Fig. 3.6). However, this category does not have a minimum thickness specification for solid insulation and is assumed to be subject to pinholes. Basic insulation by itself does not ensure safety. Safety is ensured by providing a second level of protection such as supplementary insulation or protective earthing. Basic insulation is usually used to separate ELV circuits from hazardous voltage circuits. For an air gap to serve as basic insulation, it should be at least 2 mm. It should withstand hi-pot test (see section 3.36.4) of up to 1500 Vrms depending on the standard being referred.

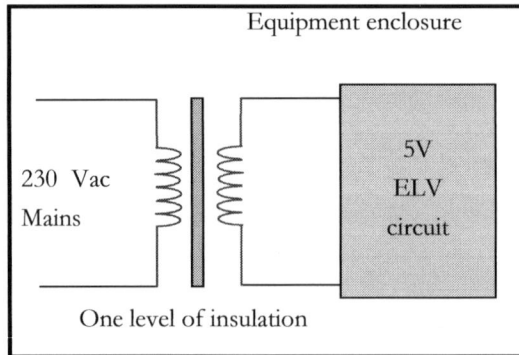

Fig 3.6: Basic Insulation

3.10.1.3 Supplementary insulation

This type is normally used in addition to basic insulation to provide a second level of protection in the event that the basic level fails. A single layer of insulating material must have a minimum thickness of 0.4 mm (EN 60950, EN 60065) to 1mm (EN 60335) to be considered supplementary insulation. It should withstand hi-pot test of up to 1500 Vrms (see section 3.36.4).

3.10.1.4 Double insulation

This is a two-level system consisting of basic insulation plus supplementary insulation. Basic insulation plus ground is equivalent to double insulation. It is used to separate SELV circuits from hazardous voltage circuits.

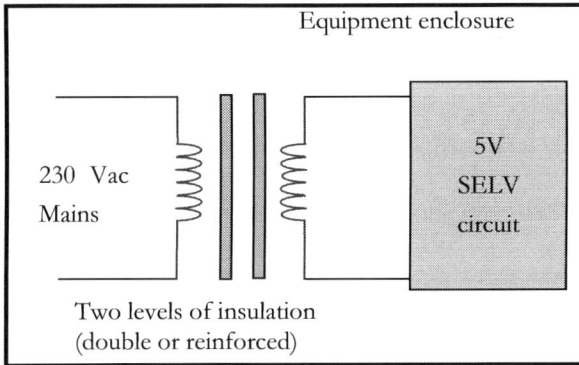

Fig 3.7: Double Insulation

3.10.1.5 Reinforced insulation

This is a single insulation system which provides a degree of insulation equivalent to double insulation. It may also comprise several layers to meet thickness and/or test requirements. A single layer of insulating material must have a minimum thickness of 0.4 mm (EN 60065) to 2mm (EN 60335) to be considered reinforced insulation. It should withstand hi-pot test (see section 3.36.4) of up to 3000 Vrms. For air gap to serve as reinforced insulation, it should be at least 10mm. As we have seen, a single level of protection is acceptable if the circuit is not accessible. In case of accessible components/parts, they must be insulated from hazardous voltages by a double-level system. One method is double or reinforced insulation while other method is single insulation plus protective earthing as shown in the Fig. 3.8.

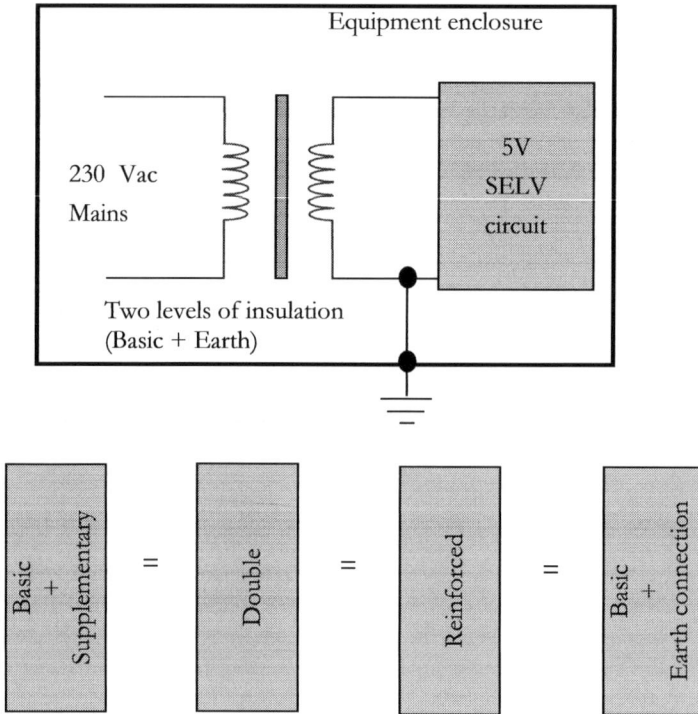

Fig 3.8: Earthing As a Level Of Protection

3.10.2 Levels Of Protection (LOP)

Safety standards dictate certain levels of protection (LOP) or means of operator protection (MOOP) between various types of circuits that we have discussed in section 3.6. The method of protection is one insulation plus another layer of insulation or earth connection or one protection component/device. The Fig. 3.9 shows a general arrangement of insulation requirement in a typical circuit. Internal floating ELV needs to be protected from primary hazardous voltage by one level of protection (i.e. basic insulation). Accessible terminals need to be protected from internal floating ELV by one level of protection. However, if the ELV circuit is grounded then no protection is required for accessible parts since grounding provides one level of protection. Now, accessible terminals need to be protected from internal primary carrying hazardous voltages by two levels of protection. Similarly, internal secondary circuits carrying SELV voltages should be protected from primary hazardous voltage by two levels of protection. External grounded metal must be protected from primary hazardous voltage by one level of protection (since the ground already provides one level of protection). However, if the metal is ungrounded then must be protected from primary hazardous voltage by two levels of protection. For TNV circuits, in addition to type of insulation, hi-pot test (see section 3.36.4) and impulse test (see section 3.36.9) is also required.

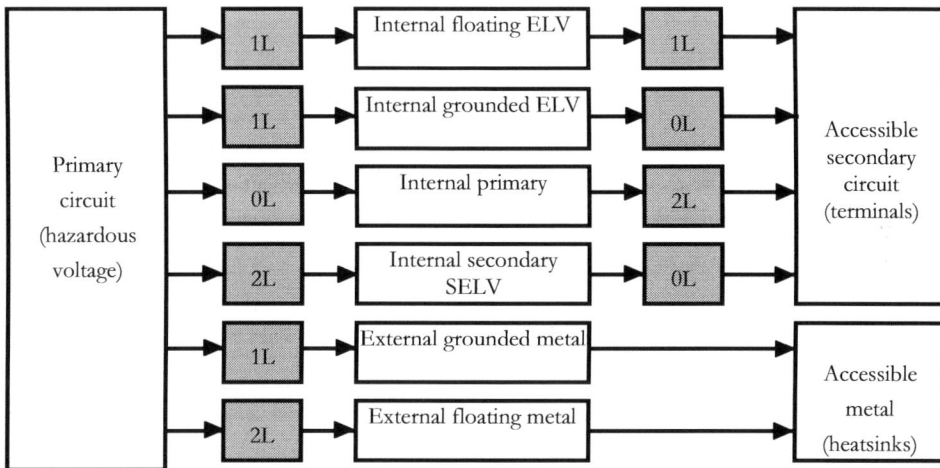

Fig 3.9: Levels Of Protection

3.10.3 Insulation Coordination

There are normally certain rules to be followed as regards to the type of insulation depending on the voltage and type of circuits. When these rules are followed, one can say that the insulation between different circuits has been coordinated. Standards, notably the EN 60664-1 (Insulation coordination for equipment within low-voltage systems), address insulation coordination for equipment within low voltage systems. Insulation coordination is the mutual correlation of insulation characteristics of electrical equipment taking into account the voltage environments to which the equipment will be subjected to, in their normal use. In other words, the insulation coordination indicates the selection of the type of electrical insulation for particular voltage characteristics of the equipment with regard to its application and takes into account the working voltage, rated voltage and rated insulation voltage of the equipment. It also specifies the requirements for distances through air (clearance) and along the surface of a dielectric (creepage), between various circuit types (including distances inside an insulator) as a means for providing safety and provides methods of electrical testing with respect to insulation coordination.

The insulation coordination uses a preferred series of rated impulse voltages for equipment energized directly from the low-voltage mains. Consideration is also giving to recurring peak voltages and to the extent partial discharge (see section 3.10.8.3) can occur in solid insulation. The conditions in the immediate vicinity of the equipment are taken into account and quantified by what is called as *pollution degree* (see section 3.10.8.4). Equipment is dimensioned based on overvoltage category according to the expected use of the equipment. This standard also uses a number called the comparative tracking index (CTI) (see section 3.10.9.2) to classify the insulation materials.

3.10.4 Insulation Requirements Of Circuits

The insulation requirements between different un-carthed circuits (as per EN 60950) are given in the table 3.2. Accordingly following can be summarised:

Functional insulation (F)

Required between circuits of the same type (like SELV to SELV).

Basic insulation (B)

- Between TNV1 and TNV3.
- Between SELV and TNV2.

Basic insulation + 1.5kV impulse/1kV hi-pot test without creepage (B*)

- Between SELV and TNV1.
- Between TNV1 and TNV2.

Basic insulation + 1.5kV impulse/1kV hi-pot test with creepage (B#)

- Between SELV and TNV3.
- Between TNV2 and TNV3.

Double or reinforced insulation (D or R)

- Between SELV and limited current circuits and between SELV and hazardous voltage circuits.

- Between hazardous voltage circuits and limited current circuits.
- Between hazardous voltage circuits and limited current circuits on one hand and SELV and TNV circuits (1,2,3) on the other.

Table 3.2: Insulation Between Different Circuits

Circuit	SELV	TNV1	TNV2	TNV3	LCC	Hazardous
SELV	F					
TNV1	B*	F				
TNV2	B	B#	F			
TNV3	B #	B	B*	F		
LCC	D or R	D or R	D or R	D or R	F	
Hazardous	D or R	D or R	D or R	D or R	D or R	F

Note 1: *If any of SELV or TNV1 circuit is earthed, basic insulation is not required with the exception that with earthed SELV and TNV3 creepage/clearance requirements apply.*
Note 2: *See section 3.36.4 for hi-pot test & section 3.36.9 for impulse test (10/700 μs).*

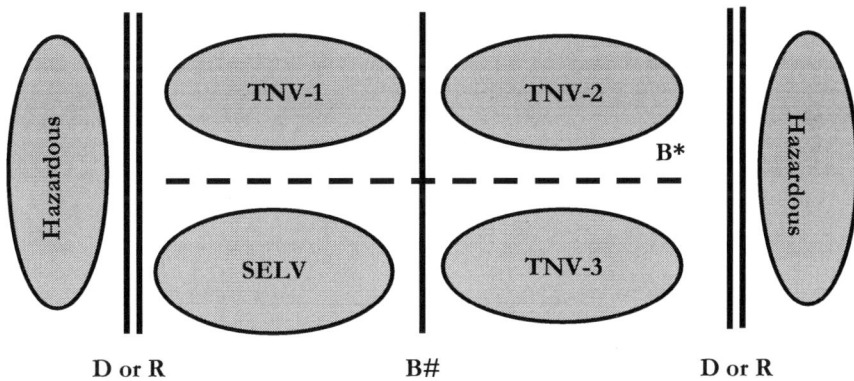

Fig 3.10: Insulation Between Circuits (as per EN 60950)

Let us consider insulation requirement depending on circuit types for the product ISDN Network Termination Adapter (NTBA), which we have considered earlier. As shown in the Fig 3.11, the 2-wire line (UKo) is a TNV3 circuit while the functional circuit (which runs on 5 Vdc) is a SELV circuit. Hence between these two circuits basic insulation with creepage is required. The 4-wire line (So) is a TNV1 circuit. Hence between this and functional circuit (which is SELV), basic insulation without creepage is required. The 230 Vac obviously is a hazardous circuit, and hence double or reinforced insulation is required between this circuit and the 4-wire line (So) (which is a TNV1 circuit). As far as the DC-DC converter is concerned, it is between a TNV3 circuit on one hand and a SELV and TNV 1 circuit on the other. It appears therefore that basic insulation will suffice. However, careful examination will show that in case of insulation breakdown, almost 140V DC (100 Vdc + 40 Vdc) may appear across the SELV functional circuits, exposing the operator to hazardous voltage. Hence double or reinforced insulation is required for the DC-DC converter.

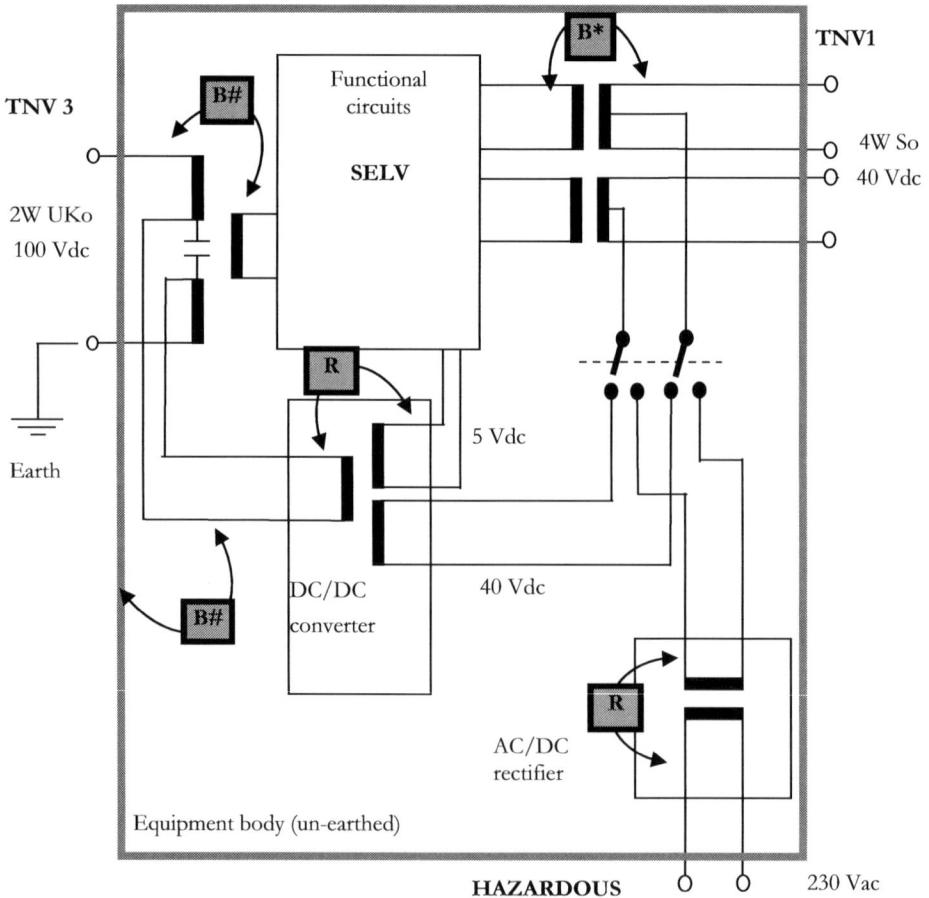

Fig 3.11: Insulation Requirements for I.T. Product

As far as our second example of SMPS is concerned, considering that the metallic enclosure is earthed, basic insulation is required between the live and the enclosure (see Fig. 3.12). Again, the transformer primary carries hazardous voltage and the secondary is at SELV. Hence the insulation between primary and secondary has to be double or reinforced. The opto-coupler bridges SELV and hazardous voltage and hence it should also have reinforced insulation. Functional insulation is required between output contacts (which are at SELV) & body and between ground & earth.

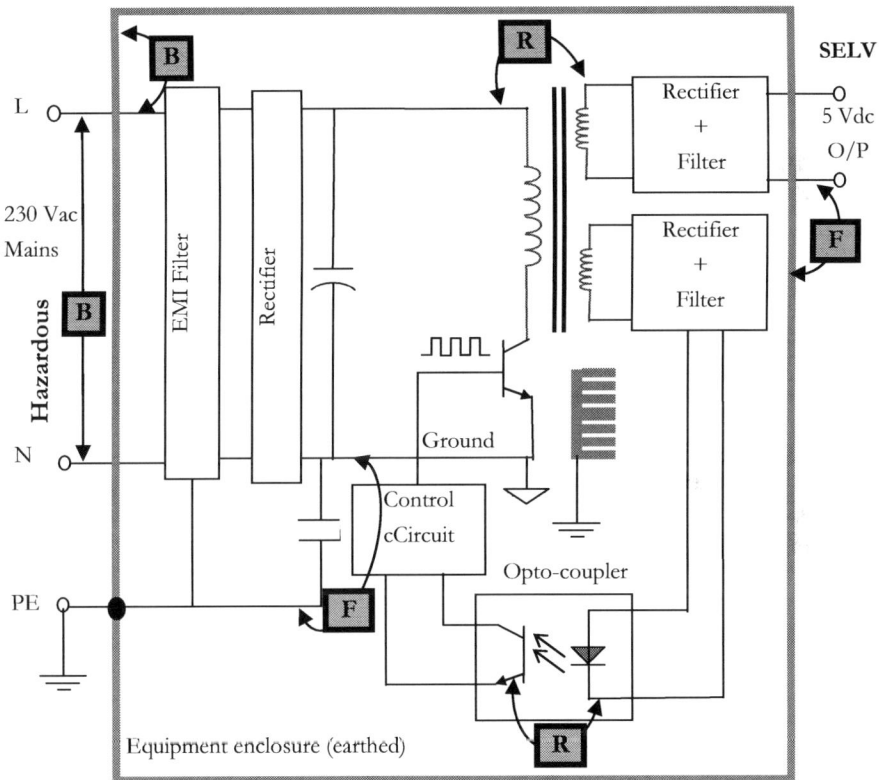

Fig 3.12: Insulation Requirements for SMPS

3.10.5   Tests For Insulation

Apart from the insulation requirements which we have seen in previous section, safety standards also specify that the electric strength of the insulating materials used within the equipment should be *adequate* (EN 60950). Hence the insulation should be subjected to dielectric strength (also called HV test or hi-pot test) at a certain severity level depending upon type of insulation. As per EN 60950, before carrying out the test, the equipment is operated under normal load until the temperature has stabilized.

Temperature limiters and thermostats are permitted to operate as long as they do not interfere with the normal operation of the equipment. However thermal cut-outs are not permitted to operate. The hi-pot test is carried out immediately after thermal stabilisation while the equipment is still in heated condition according to specified test voltage. The table 3.3 gives the recommended test voltages as specified by EN 60950. For basic insulation (equipment with an earth connection), the test voltage is applied primarily between mains supply leads connected together and the earthed body and across all other circuits separated by basic insulation.

For reinforced and double insulation (equipment without earth connection), the test voltage is applied between primary circuit and SELV circuit and the chassis. The test voltages in this case are higher than those for basic insulation simply because there is no earth used and the safety is provided by the insulation only. For test procedure refer section 3.36.4. Many-a-times, for components such as coated PCB, safety standards prescribe humidity preconditioning at certain temperature and humidity values before carrying out the HV test. This is discussed in section 3.36.4.5.

Table 3.3: Test Levels For Insulation (I.T. Products)

Insulation	Points Of Application		
	Primary Circuit To Body Primary Circuit To Secondary Circuit & Between Parts In Primary Circuits		Secondary Circuit To Body Between Independent Secondary Circuits
	Working Voltage U, Peak Or dc		Working Voltage U
	$U \leq 210V$	$210V < U \leq 420V$	$U \leq 42.4$ VPk Or 60 V dc.
	Test Voltage AC r.m.s		
Functional	1000	1500	500
Basic, Supplementary	1000	1500	No test
Reinforced	2000	3000	No test

Let us now see how these are applicable to the equipment i.e. NTBA which we had considered earlier (see Fig 3.13).

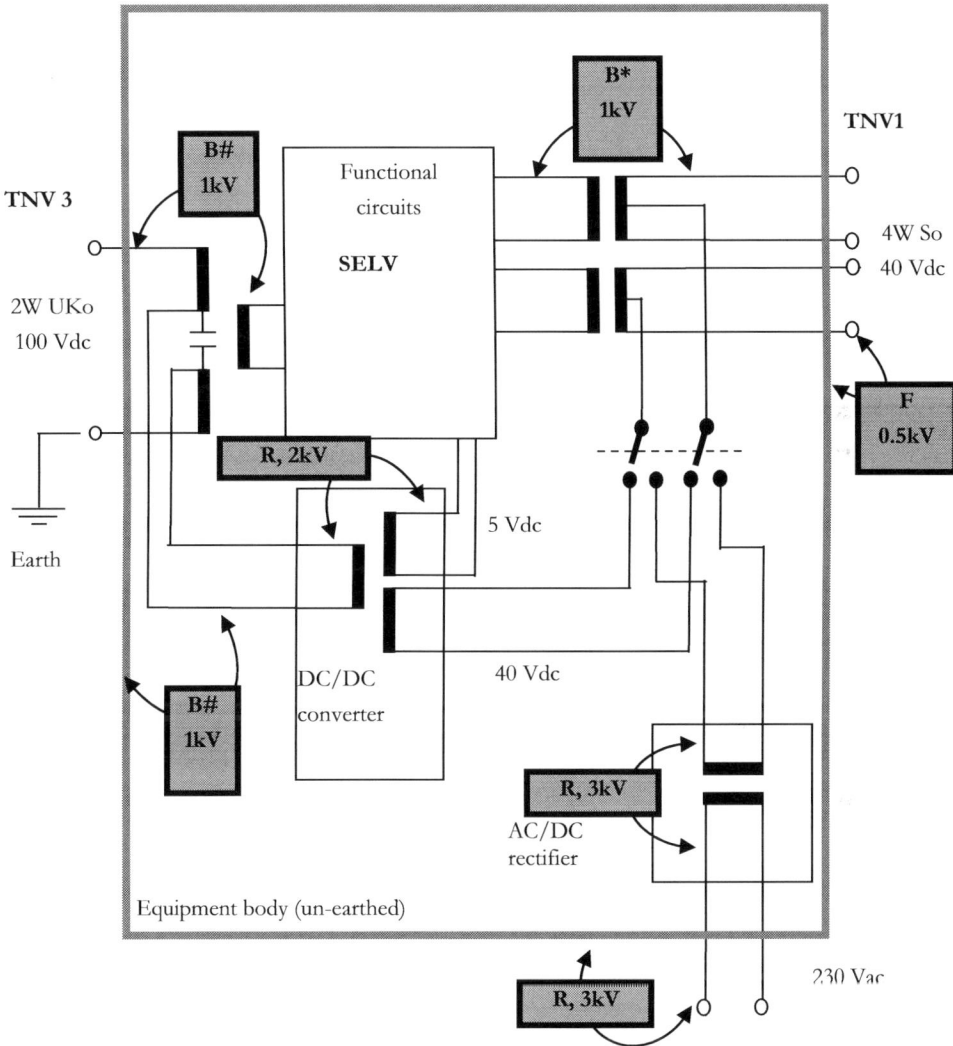

Fig. 3.13: HV Test Levels For I.T. Product

3.10.5.1 Insulation test levels primary to secondary circuits

The insulation between UKo (which is a TNV3 circuit) and the functional circuits (which are at SELV) is basic. Hence it should be subjected to a dielectric test of 1 kV. Same is the requirement between the primary of the DC-DC converter to enclosure which has basic insulation. There is reinforced insulation between input and output of the DC-DC converter, hence it has to be subjected to a hi-pot test of 2 kV (see table 4.3). Again, between the primary and secondary of the AC-DC converter (transformer) reinforced insulation is required and hence it has to be subjected to a hi-pot test of 3 kV. The same test is applicable between the 230 Vac input and the enclosure.

3.10.5.2 Insulation test levels secondary circuits

The insulation between functional circuits (SELV) and two wire So O/P (TNV1) is basic and hence as per the table 3.3, 1 kV test is required. While the insulation between So O/P (TNV1) and enclosure body is functional and hence a 0.5 kV test is required.

As far as our second example of SMPS is concerned, the HV test levels will depend upon the end application. Let's us consider SMPS used in information technology applications.

3.10.5.3 Information technology SMPS

If the SMPS is used in information technology product, then the applicable standard is EN 60950 and the HV test levels will be dictated by the table 3.3. For EN 61010, the test voltages are the same for 230 Vac mains operated equipment. The insulation between the mains and equipment body is basic, so a HV test of 1.5 kV is required (done between live + neutral shorted and earthed body). While performing the test, components between line and earth like MoVs or Y-capacitors (which are a part of the EMI filter) are removed if they are certified. Insulation between ground and earth and that between O/P DC and earth (equipment body) is functional and hence a 0.5 kV test is sufficient. Insulation between primary and secondary of the transformer is reinforced and hence a 3 kV test is required. Also, a 3 kV test is required between the 230 Vac I/P mains and the DC O/P. The opto-coupler is bridging SELV and hazardous circuit and hence a 3 kV test is required between it's input and output.

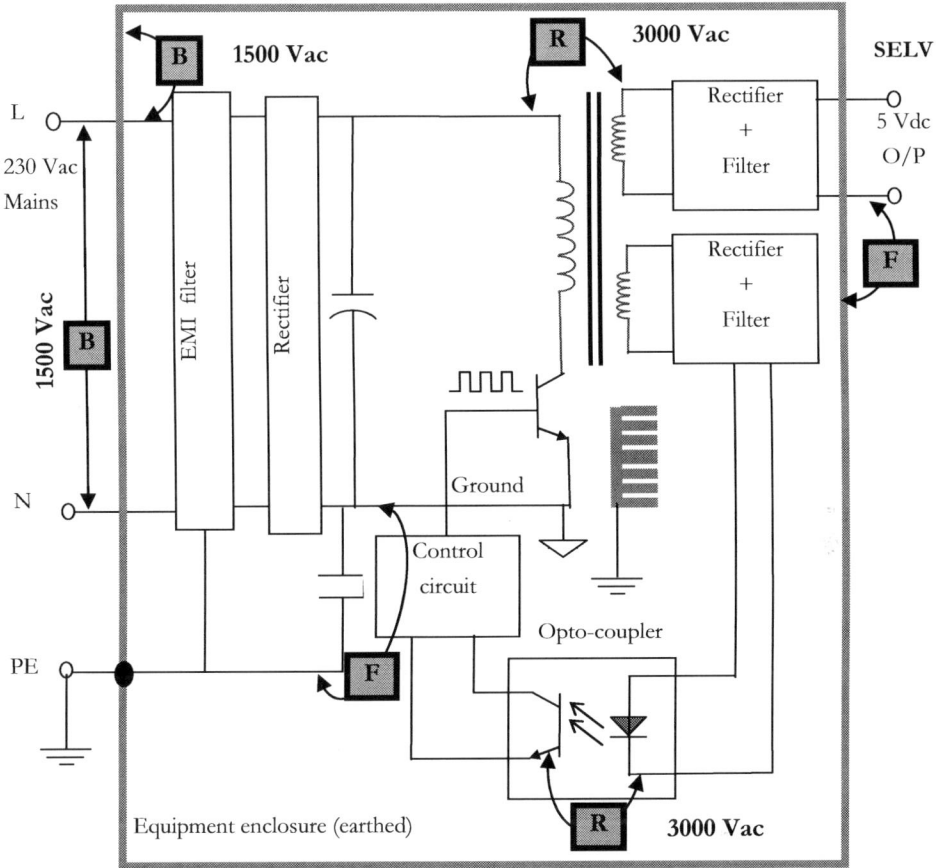

Fig. 3.14: HV Test Levels For I.T. SMPS

3.10.6 Properties Of Insulating Material

Insulating material is characterized by electric strength, thermal strength, mechanical strength, dimensions and other properties. The choice and application of insulating material should take into account the needs for electrical, thermal and mechanical strength, frequency of the working voltage and working environment (temperature, pressure, humidity and pollution). For solid insulation (see 3.10.8.3) only non-hygroscopic, flame resistant materials should be used. In case of wiring insulation, some material compounds may contain plasticizers (for imparting flexibility) but with a side effect of increased flammability. Semiconductor devices and other components that are moulded in solid insulating material typically are independently qualified and inspected in the manufacturing process.

Compliance is checked by inspection and, where necessary, by evaluation of the data for the material. If the data does not confirm that the material is non-hygroscopic, the hygroscopic nature of the material is determined by subjecting the component or subassembly employing the insulation in question, to the humidity treatment. The insulation is then subjected to the relevant electric strength test while still in the humidity cabinet, or in the room in which the samples were brought to the prescribed temperature.

If an air gap serves as reinforced insulation between parts at hazardous voltage and un-earthed conductive accessible parts of floor standing equipment (or that of non-vertical surfaces of desk top equipment) then (as per EN 60950) the minimum clearance distance is 10mm while for air gap used as basic insulation (between hazardous voltage and an earthed conductive accessible part), it should be 2 mm (3mm for contacts of disconnect device).

3.10.7 Circuit Separation By Distance

Separation of the internal wiring, such as between primary and secondary wires of adequate thickness or layers of insulation, is required even within the equipment. Whenever possible it is best to physically separate all primary and secondary wires so that there is no possibility of their coming into contact with each other. When this is not possible, as when primary and secondary wires touch, care must be taken to ensure that the proper insulation level (reinforced) is maintained. As a general rule, when primary and secondary wiring touch, only two layers of insulation is required (one on each wire) if one layer has a minimum thickness of 0.4 mm and the other wires' insulation passes a hi-pot test (1,500 Vac). To obtain reinforced insulation with only two layers (one on each wire) it is best to use 600 V-rated insulation (0.7 mm thickness) for the primary wires since 300 V wire insulation has typically less than the required 0.4 mm thickness. Whenever the minimum 0.4 mm thickness requirement is not met, reinforced insulation may be accomplished with the use of three layers of insulation, passing the appropriate hi-pot test (1,500 Vac for two layers). Either the primary or the secondary wiring may be sleeved with an additional layer, thereby achieving the total of three layers. The primary wires are usually sleeved with solid tubing, such as heat shrink or clear PVC. Porous insulation and spiral wrap do not count as insulation since they typically fail the hi-pot tests. Adequate separation must also be maintained between wiring and exposed circuits such as exposed terminals on components or PCB traces. Secondary wires contacting primary circuits on power supplies or primary wires contacting exposed secondaries are areas that must also meet the reinforced insulation requirements. Where sleeving is used as supplementary insulation on internal wiring, it must be positively retained in position and is considered as such if it can be removed only by breaking or cutting or if it is clamped at both ends.

3.10.8 Creepage, Clearances & Distance Through Insulation

Well then, we now come to two highbrow terms associated with electrical safety compliance as per low voltage directive –creepage and clearance. And believe me, they are the most misunderstood too. These are much talked about terms which are frequently and nonchalantly used by experts who want to impress novices with their technical acumen!

We have seen that circuits can be isolated by using insulation. However, electrical breakdown can occur under certain conditions between two conductive parts either through air or along a surface, creating a conductive path which can lead to hazard. Insulating materials, such as printed-wiring-board and component surfaces, can become partially conductive when exposed to deposits or deterioration from humidity, contamination, chemicals etc. which can lead to a phenomenon called *tracking* (i.e. creation of a conductive path) over the material's surface. Similarly, arc-over or flash-over can take place between two circuits separated by an air gap because of ionization of the intervening air. This especially can happen due to sudden high voltage transients. Factors affecting air ionization include humidity, pollution, temperature and altitude. At sea level the rule of thumb is that 1mm of air gap can flashover at 3 kV.

Hence, circuits carrying hazardous voltage and SELV circuits/operator accessible parts can be separated by maintaining certain distance between them. The separation is given in terms of creepage and clearance which indicate spacing between components that are required to withstand a given voltage called as *working voltage*. Let's us take a look at these terms in detail.

3.10.8.1 Creepage

As shown in Fig. 3.15, creepage is the shortest distance along the surface of an insulation material between two conducting parts (such as PCB tracks or component leads). It is associated with tracking. For a component, this shortest distance may be either over the package or under the package or side of the package. To visualize creepage, one can imagine an ant *creeping* (i.e. crawling) on a surface. The ant will move along the surface always hugging all the undulations. Creepage is a slow phenomenon determined by dc or rms voltage rather than peak events and is heavily influenced by the surface conditioning of the insulating material. Inadequate creepage distance may take days or even months before the insulation fails.

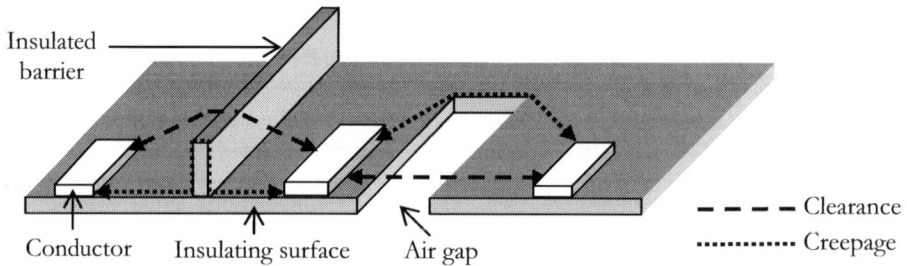

Insulated barrier

Conductor   Insulating surface   Air gap

– – – – Clearance
··········· Creepage

Fig. 3.15: Creepage And Clearance

3.10.8.2 Clearance

Clearance (see Fig. 3.15) is defined as the shortest distance through air between two conducting parts. A very good analogy is a flying insect which will go from one point to the other directly through the air. For component leads, clearance distance can be visualised as the *line-of-sight* distance. Clearance is heavily influenced by the quality of the air. Normally 1 mm of air gap will breakdown at 3 kV at sea level. At such a high voltage, electrons are stripped of their atoms and become free, causing air to become conductive. The presence of humidity, moisture, conductive particles can cause the breakdown to happen at lower voltages as there are already some free electrons to start with. Another important factor is the altitude. Now this comes as a surprise as one may think what does altitude have to do with insulation breakdown? Well as the altitude increases, the air density decreases and the dielectric properties of air change. The air is no longer a good insulator (until it reaches vacuum) which causes it to breakdown at lower voltages. This is called the Paschen effect and the Paschen's curve describes electric discharge voltage as a function of atmospheric pressure. As shown in the Fig. 3.16, there are four curves. Let us consider an air gap of 10 cm. As per curve I it will break down at about 30,000 volts at sea level. At 50,000 feet (approx. altitude at which commercial jets fly) the same 10 cm air gap will breakdown at about 6,000 volts (Curve III). So as the operational altitude increases, the clearance distance should also be increased to avoid insulation breakdown.

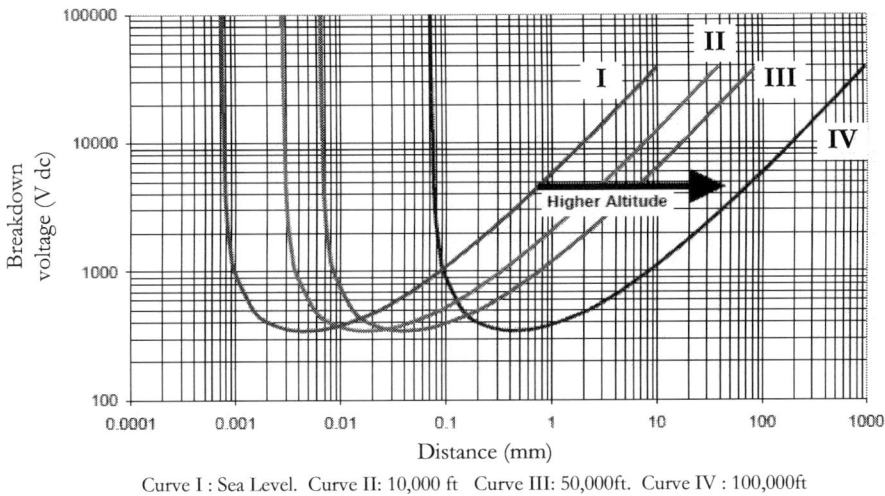

Curve I : Sea Level. Curve II: 10,000 ft Curve III: 50,000ft. Curve IV : 100,000ft

Fig 3.16: Paschen's Curves

3.10.8.3 Partial discharge and distance through insulation

Electrical breakdown that occurs across clearances is influenced by the property of air between two conductive surfaces while that across creepage is influenced by the property of the dielectric substrate (although the flashover occurs through air). Now, electrical breakdown can also occur through void-free *solid insulation* of components and between traces on inner layers of multilayer PCBs. The term solid insulation is used to describe many different types of construction, including monolithic blocks of insulating material or insulation subsystems composed of multiple insulating materials, organized in layers or otherwise. The electric strength of a solid insulation of certain thickness is considerably greater than that of air of the same thickness. Hence while addressing solid-insulation requirements, safety standards may specify insulating distances or *distance through insulation* which are typically smaller than the distances through air. Spacing examples include opto-isolator input-to-output-pin spacing, connector pin-to-shell spacing, printed-wiring-board trace-to-trace spacing, wiring-to-enclosure spacing, transformer primary to secondary spacing etc.

Design of solid insulation requires attention to what is called as *partial discharge* phenomenon. Partial discharge is electric breakdown that partially bridges the insulation and not fully. To understand partial discharge, let us consider a case when a solid insulation system is constructed from layers of solid materials. In such a scenario, there are likely to be gaps or voids between layers. These voids will introduce discontinuities in the electric field so that a disproportionately large part of the electric field is located in the void, causing ionization within the void, resulting in partial discharge only in the void. In such condition the insulation does not fail upon a single discharge but does so after repetitive cycles of partial discharge over a long period of time running into months or even years. These partial discharges will influence the adjacent solid insulation and may reduce its service life. Thus, partial discharge, is symptomatic of incipient damage, a cumulative phenomenon which will eventually lead to catastrophic damage over time. In particular, a solid insulation subjected to high frequencies is more sensitive to partial discharges.

3.10.8.4 Pollution degree

As seen earlier, all safety standards call for different insulation types which have to be maintained for protection against shock. However, ingress of dust (due to pollution) and humidity can cause an arc or flashover between two circuits separated by insulation. The flashover depends on what is called as *pollution degree* as well as the working voltage (which is the highest measured RMS voltage between any two points). Pollution degree is a number characterizing the expected pollution of the micro-environment (i.e. the environment in the immediate vicinity of the insulation). Let us take a look at these one by one.

Pollution degree 1: This is the most benign pollution environment where there is either no pollution or even if present it is permanently dry and non-conductive. It has no real impact. This type of environment can be an air-conditioned clean environment like a laboratory clean room or an office. It is also associated with components and assemblies which are sealed to exclude dust and moisture.

In order to be declared as pollution degree 1, a sample must be subjected to the following tests:

The sample is subjected ten times to the following sequence of thermal cycling:

- 68 hours at 85 ± 2 °C;
- 1 hour at 25 °C ± 2 °C;
- 2 hours at 0 °C ± 2 °C;
- not less than 1 hour at 25 °C ± 2 °C.

The sample is permitted to cool to room temperature and is then subjected to the humidity conditioning (for tropical areas: 120 hours at relative humidity of 93±3% at temperature between 40±2 °C non condensing). This is immediately followed by the electric strength test. Compliance is checked by inspection of the cross-sectional area, and there should be no visible voids, gaps or cracks in the insulating material.

Pollution degree 2: Generally, non air-conditioned office or home environment is regarded as pollution degree "2" and is characterized by non-conductive particles which occasionally and temporarily become conductive due to moisture. Household appliances such as electric cookers, laundry washing machines, dishwashers, stereos, amplifiers, radios, and electric musical instruments fall under this category. It can also include IT products that are used in household environment.

Pollution degree 3: In this environment, the equipment is subject to conductive pollution or dry non-conductive pollution which becomes conductive (permanently) due to moisture condensation. Industrial shop floor is a typical example of pollution degree 3.

Pollution degree 4: In this environment, the equipment is exposed persistently to conductive pollution, moisture, rain or snow. Equipment installed outdoors without shelter is an example of such environment.

The above pollution degrees have been specified by EN 60664-1. Depending upon the pollution degree and working voltages, the standard specifies creepage / clearances in tabular form. These are then used by the product standards as a basis for specifying creepage and clearances for their particular operating environment. Hence, the product standard may or may not consider all pollution degrees. For example, the standard EN 60065 (Safety requirements for audio video and related apparatus for household and similar general use) identifies only pollution degree "2" as this is the environment for household appliances and so does the standard EN 60601-1 (general safety requirements for medical electrical equipment). The standard EN61010-1(safety requirements for electrical equipment for measurement, control and laboratory use) identifies pollution degree "1" and "2". On the other hand, the standard EN 60950 (safety of information technology equipment including business equipment) identifies pollution degree "1", "2" & "3".

3.10.9 Creepage Distance

Let's take a look at the calculation/selection of creepage distances. We will consider I.T. equipment, medical electrical equipment, household appliances and measurement & control use products. Creepage depends on the pollution degree, RMS working voltage and material group. We have seen pollution degree in the previous section. Let us see the other terms.

3.10.9.1 RMS working voltage

This is the most important parameter which indicates what the maximum continuous operating voltage that a device or insulation can be exposed to without any danger of the insulation being catastrophically destroyed. Working voltage is *the highest voltage to which the insulation or the component under consideration can be subjected to when the equipment is under normal condition.* Normal condition means the operational condition as specified by the manufacturer with all protection intact. The working voltage between a primary circuit and a secondary circuit, or between the primary and ground, is taken as the upper limit of the rated voltage range for the supply. In case of TNV circuit connected to a telecommunication network, the normal operating voltages should be taken into account. If these are not known, they should be assumed to be the following values:

– 60 Vdc for TNV-1 circuits;

– 120 Vdc for TNV-2 circuits and TNV-3 circuits.

3.10.9.2 Material group

Material group is defined by value of comparative tracking index (CTI). Tracking is an electrical breakdown on the surface of an insulating material when a large voltage difference gradually creates a conductive path across the surface of the material by forming a carbonized track. UL 746 defines CTI as the voltage which causes tracking after 50 drops of 0.1% ammonium chloride solution have fallen on 3mm thickness of a material (the test being done under test conditions defined by the standard). Higher the CTI better the material. Material Group I have CTI greater than or equal to 600 (Tracking occurs at 600V). Material Group II have CTI between 400 and 600 while Material Group IIIa have CTI between 100 and 400 and Material Group IIIb have CTI between 100 and 175. The requirements are summarised as below:

Material Group I   : $600 \leq$ CTI

Material Group II   : $400 \leq$ CTI < 600

Material Group IIIa : $175 \leq$ CTI < 400

Material Group IIIb : $100 \leq$ CTI < 175

The table 3.4 gives creepage distances (in mm) as specified by EN 60950.

Table 3.4: Creepage Distances (in mm) as per EN 60950

Working voltage up to and including	Pollution degree 2				Pollution degree 3			
	Material gr II		Material gr IIIa		Material gr II		Material gr IIIa	
Vr.m.s. or d.c.	O/B/S	R	O/B/S	R	O/B/S	R	O/B/S	R
50	0.9	1.8	1.2	2.4	1.7	3.4	1.9	3.8
100	1.0	2.0	1.4	1.8	2.0	4.0	2.2	4.4
125	1.1	2.2	1.5	3.0	2.1	4.2	2.4	4.8
150	1.1	2.2	1.6	3.2	2.2	4.4	2.5	5.0
200	1.4	2.8	2.0	4.0	2.8	5.6	3.2	6.4
250	1.8	3.6	2.5	5.0	3.6	7.2	4.0	8.0
320	2.2	4.4	3.2	6.4	4.5	9.0	5.0	10.0
400	2.8	5.6	4.0	8.0	5.6	11.2	6.3	12.6
500	3.6	7.2	5.0	10.0	7.1	14.2	8.0	16.0
600	4.5	9.0	6.3	12.6	9.6	19.2	10	20
1000	7.1	14.2	10	20	14.0	28.0	16	32

O : Operational Insulation, B: Basic Insulation, R : Reinforced Insulation

Note that in the table, reinforced insulation has creepage distances that are twice the value of the voltage specified for basic insulation. Let us now consider creepage requirements for our I.T. equipment ISDN network termination adapter (NTBA), which we have considered earlier (see Fig. 3.17). Since we have a 230Vac mains connected circuit, the PCB should be of material group IIIa (or better). Say the material group of PCB is IIIa (standard FR4 laminate). The equipment works in a typical office environment which may not be air-conditioned, the environment is of pollution degree "2". The insulation between TNV3 (working voltage up to 120V) and SELV is basic. Referring to the table 4.6 the minimum creepage distance between these two circuits should be 1.5mm.

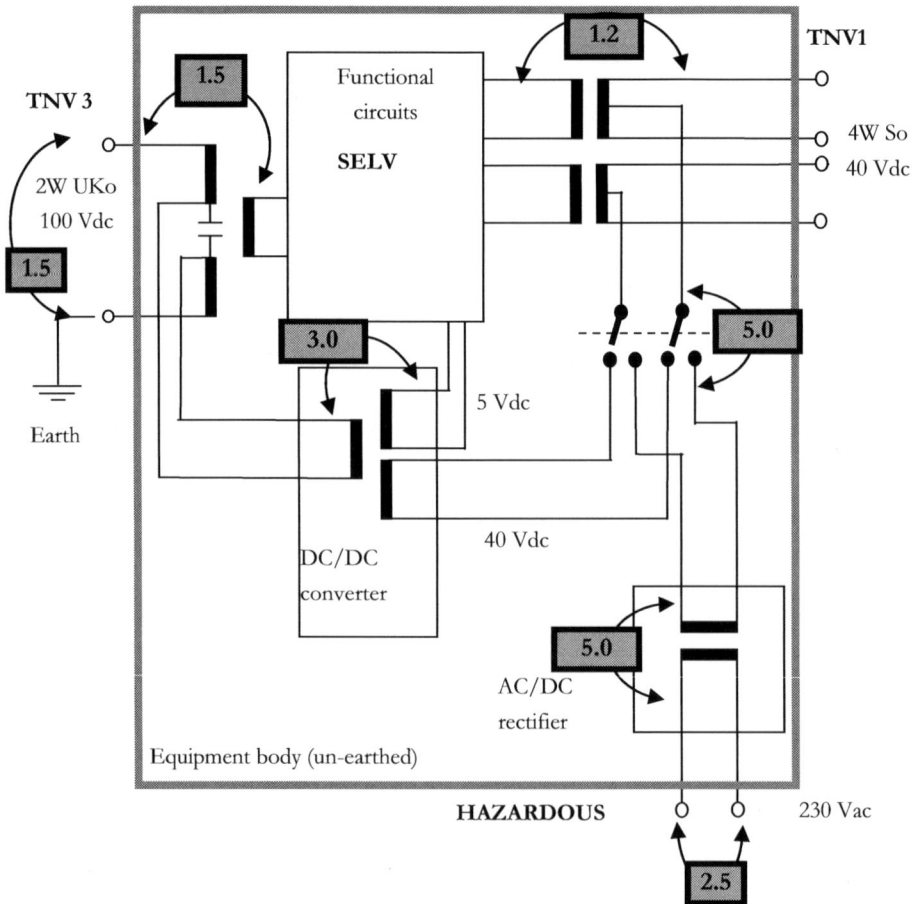

Fig. 3.17: Creepage Distance in mm for I.T Product

This is the distance that has to be maintained, for example, between edges of primary and secondary pads of the TNV3 circuit transformer.

The creepage distance between SELV and TNV1 (40 Vdc) circuit is 1.2 mm. Considering reinforced insulation between primary and secondary of the DC-DC converter and a working voltage of up to 120 V, the creepage distance here will be minimum 3 mm. The distance has to be maintained between edges of input and output pads. The working voltage between primary and secondary of the AC-DC transformer/rectifier is 230 Vac and hence the creepage distance (between edges of primary and secondary PCB pads) here will be minimum 5 mm considering reinforced insulation.

Between edges of the input and output pads of the relay, the creepage distance should be 5.0mm since in the event of insulation failure of the AC-DC transformer, hazardous voltage may appear at the relay pins and hence reinforced insulation may be required. While that between the pads of the 230 Vac and those of 2 W TNV3 should be 2.5 mm and 1.5 mm respectively considering basic insulation. It must be noted that the creepage distance requirements will be reduced if a PCB material is of group II (which is better than that of group IIIa due to higher CTI).

The distance through insulation (insulation thickness or internal creepage) for transformers and semiconductor devices is discussed in 3.10.11. Accordingly, for I/P and O/P transformers, there is no minimum thickness requirement considering basic insulation. For DC-DC converter and the AC-DC rectifier transformer, the insulation thickness should be minimum 0.4 mm considering reinforced insulation. Other alternatives are discussed in 3.10.11.

Now consider the example of SMPS used in I.T applications. Again, the PCB should be of material group IIIa (or better) and the pollution degree is "2". The creepage distance between live and earth pads of the PCB should be 2.5 mm minimum considering basic insulation. For the transformer, considering a working voltage of 230 V and reinforced insulation, the creepage distance along the PCB is 5 mm. It is quite possible that the voltage at the *hot-end* of the transformer may be more than 230 V, hence a precise creepage distance value can be arrived at by measuring the actual voltage and then dimensioning the PCB accordingly. The creepage distance along the PCB for the opto-coupler is 5 mm considering working voltage of 230 V and reinforced insulation.

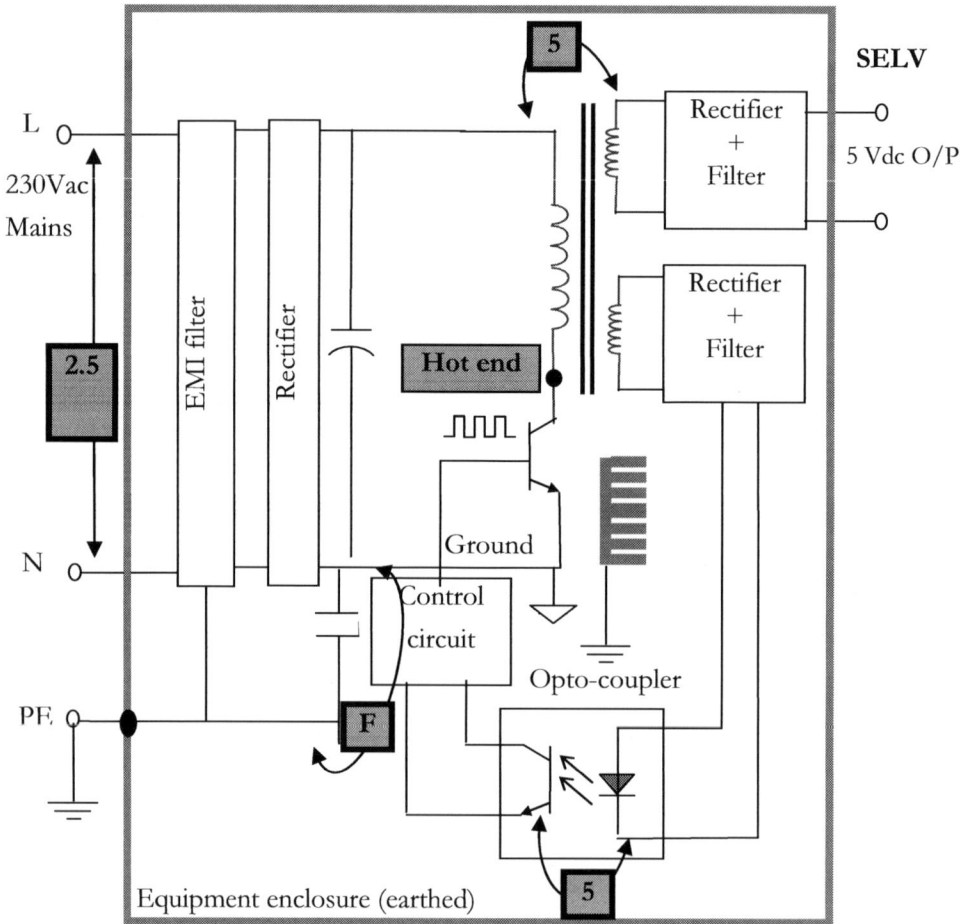

Fig. 3.18: Creepage Distance in mm for I.T. SMPS

Having seen creepage distance specified for I.T products, we will now see the requirements for other product categories. The table 3.5 shows creepage requirements for medical electrical equipment as given by EN 60601-1, while those for audio video product (as per EN 60065) are given in table 3.6. For these two equipment categories, only pollution degree "2" and installation category II is applicable. The example of distances for measurement control laboratory use equipment (EN 61010) is given in table 3.7.

Table 3.5: Creepage & Clearance as per EN60601-1

Reference Voltage dc	Reference Voltage ac	Insulation	Creepage mm	Clearance mm	Test Voltage
15	12	Basic	1.7	0.8	500
		Reinforced	3.4	1.6	500
34	30	Basic	2	1	500
		Reinforced	4	2	500
75	60	Basic	2.3	1.2	1000
		Reinforced	4.6	2.4	3000
150	125	Basic	3	1.6	1000
		Reinforced	6	3.2	3000
300	250	Basic	4	2.5	1600
		Reinforced	8	5	4200
450	380	Basic	6	3.5	1900
		Reinforced	12	7	4800
600	500	Basic	8	4.5	2200
		Reinforced	16	9	5400

Notes:

1. Reference voltage is that voltage to which the relevant insulation is subjected in normal use OR rated supply voltage whichever is greater.

2. The dielectric strength of the electrical insulation should be able to meet the test voltages (V AC rms for 1 minute) indicated in the table.

Table 3.6: Creepage Clearance distances as per EN 60065

Mains Voltage ac	Insulation	Creepage mm	Clearance mm	Test Voltage
34	Basic	0.6	0.6	707
	Reinforced	1.2	1.2	
70	Basic	1	1	1410
	Reinforced	2	2	
354	Basic	3	3	1000
	Reinforced	6	6	
500	Basic	4	4	2000
	Reinforced	6	6	

Note: *The dielectric strength of the electrical insulation should be able to meet the test voltages indicated in the table (V peak for 1 minute).*

Table 3.7: Creepage Clearance distances as per EN 61010

Voltage	Values of Creepage (mm)								
(~ ~ ') Vᵣ ₘ ₛ	PCB		Other Insulating material						
or d.c. (V)	Pollution degree 1	Pollution degree 2	Pollution degree 1	Pollution degree 2			Pollution degree 3		
	All Mat. Groups	Mat. gr I, II, IIIa	All Mat. Groups	Mat. gr I	Mat. gr II	Mat. gr III	Mat. gr I	Mat. gr II	Mat. gr III
≤ 150	0.5	0.5	0.5	0.8	1.1	1.6	2.0	2.2	2.5
>150≤ 300	1.5	1.5	1.5	1.5	2.1	3.0	3.8	4.1	4.7

Notes:

1. Value of clearance is 0.5 mm for ≤ 150 V and 1.5mm for >150≤ 300 V

2. The creepage values are for basic and supplementary insulation. For reinforced insulation, the values are twice those for basic insulation.

3.10.10 Clearance distances

We will now take a look at the calculation or selection of clearance distances. Clearance depends upon the peak working voltage, pollution degree and mains transient voltage (also referred to as transient over-voltage). Peak working voltage is *the peak value of a working voltage, including any dc component and any repetitive peak impulses generated in the equipment* (includes peak-to-peak ripple that exceeds 10 % of the average value). Mains transient voltage is the *highest peak voltage expected at power input to the equipment arising from external transients on mains*. Since clearance is distance through the air, and that air is a dielectric, there can be a dielectric breakdown in case of a sudden transient voltage (like a pulse) that may enter an equipment from outside OR in case of peak voltage that may be generated inside the equipment.

Clearances should be maintained such that the mains transient voltage or the peak working voltage does not cause a flashover which can result in a hazard. Now, for equipment to be supplied from an AC mains supply, the value of the mains transient voltage depends on the AC mains supply voltage and *overvoltage category*.

There are basically four overvoltage categories (see Fig 3.19) depending upon the installation environment of the equipment, which are as follows:

Overvoltage Category I

Such equipment are connected to circuits where measures are taken to limit transients (like filters, transient suppressors etc.) to an appropriately low level. These measures ensure that the transients are sufficiently limited so that their peak value does not exceed the relevant rated impulse voltage for equipment (as mentioned in the manual). Examples of such equipment are information technology equipment used in a laboratory and supplied via an external filter or supplied by mains from a motor- driven generator. It also includes peripherals which are not directly supplied from mains (like USB external hard drive) since the mains may already be filtered by the PC.

Overvoltage Category II

This category includes pluggable or permanently connected equipment that are supplied from a building wiring where no transient suppression measures are taken. As we will see, most of the equipment fall in this category. Examples of such equipment are household appliances, portable tools, home electronics and information technology equipment used in the building.

Overvoltage Category III

These are the equipment that are an integral part of a building wiring and for cases where the reliability and the availability of the equipment is subject to special requirements e.g. socket-outlets, fuse panels, switch panels, power monitoring equipment and industrial equipment with permanent connection to the fixed installation.

Overvoltage Category IV

This category includes equipment that will be connected to the point where the AC mains supply enters the building or is for use at the origin of the installation. Electricity meters, primary overcurrent protection equipment and communications information technology equipment for remote electricity metering are some examples of such equipment.

Similar to creepage distances, safety standards prescribe clearance distances in tabular form. Let us consider the product NTBA to illustrate the calculation of clearance distance. The product comes under over voltage category II and pollution degree 2. EN 60950 gives different clearance requirements depending on whether the circuit is primary or secondary. The tables 3.9 to 3.12 give clearance distances as specified by EN 60950 for primary circuits and between primary and secondary circuits for equipment operating up to 2000 m. Only part of the table up to 420 V is reproduced here. The table 3.8 gives values of mains transient over voltage

Table 3.8: Mains Transient Overvoltage

Voltages	Maximum Transient Overvoltage (V)			
L-N (V)	Overvoltage Category (I)	Overvoltage Category (II)	Overvoltage Category (III)	Overvoltage Category (IV)
≤ 50	330	500	800	1500
≤ 100	500	800	1500	2500
≤ 150	800	1500	2500	4000
≤ 300	1500	2500	4000	6000
≤ 600	2500	4000	6000	8000

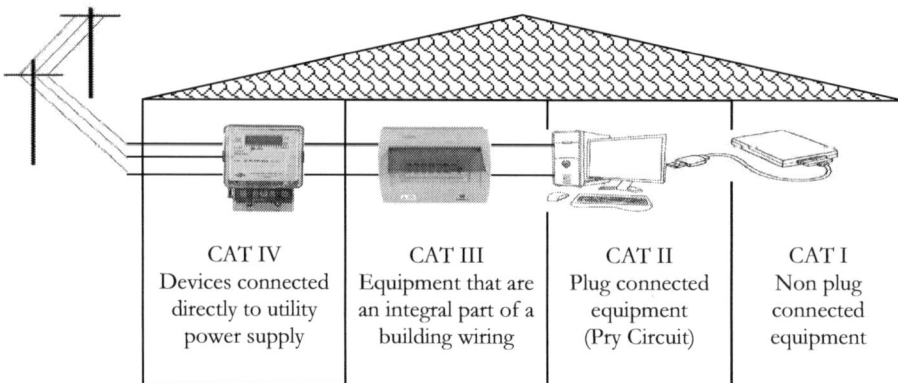

CAT IV	CAT III	CAT II	CAT I
Devices connected directly to utility power supply	Equipment that are an integral part of a building wiring	Plug connected equipment (Pry Circuit)	Non plug connected equipment

Fig 3.19: Illustration Of Over-Voltage Categories

Table 3.9: Clearance distances (in mm) for I.T equipment

Clearances (in mm) between primary and between primary and secondary circuits						
Peak working voltage up to and incl.	Mains transient voltage 1500V					
	Pollution degree					
	1,2			3		
	F	B	R	F	B	R
71	0.4	1	2	0.8	1.3	2.6
210	0.5	1	2	1.8	1.3	2.5
420	F: 1.5	B: 2	R: 4			

Table 3.10: Clearance distances (in mm) for I.T equipment

Clearances (in mm) between primary and between primary and secondary circuits						
Peak working voltage up to and incl.	Mains transient voltage 2500V					
	Pollution degree					
	1,2			3		
	F	B	R	F	B	R
71	1	2	4	1.3	2	4
210	1.4	2	4	1.5	2	4
420	F: 1.5	B: 2	R : 4			

Table 3.11: Clearance distances (in mm) for I.T equipment

Clearances (in mm) in secondary circuits						
Peak working voltage up to and incl.	Highest transient over-voltage above 800V up to and including 1500V					
	Pollution degree					
	1,2			3		
	F	B	R	F	B	R
71	0.5	1	2	0.8	1.3	2.6
210	0.5	1	2	1.8	1.3	2.5
280	0.5	1	2	1.8	1.3	2.5
420	F: 0.8 B: 1.4 R : 2.8					

Table 3.12: Clearance distances (in mm) for I.T equipment

Clearances (in mm) in secondary circuits						
Peak working voltage up to and incl.	Highest transient over voltage above 1500V up to and including 2000V					
	Pollution degree					
	1,2			3		
	F	B	R	F	B	R
71	1.5	2	4	0.8	1.3	2.6
210	1.5	2	4	1.8	1.3	2.5
280	1.5	2	4	1.8	1.3	2.5
420	F: 1.5 B: 2 R : 4					

The standard EN 60950 considers a circuit connected to DC mains supply as a secondary circuit. Further for TNV-1 and TNV-3 circuits, the mains transient over voltage is assumed to be 1500V and that for TNV-2 circuit as 800V. Let us consider the example of NTBA for clearance requirements as shown in Fig 3.20. Accordingly, the input 2-W for our product the NTBA is a TNV-3 circuit, transient over voltage is 1500 V and the insulation is basic. Hence according to table 3.9, the clearance distance is 1mm.

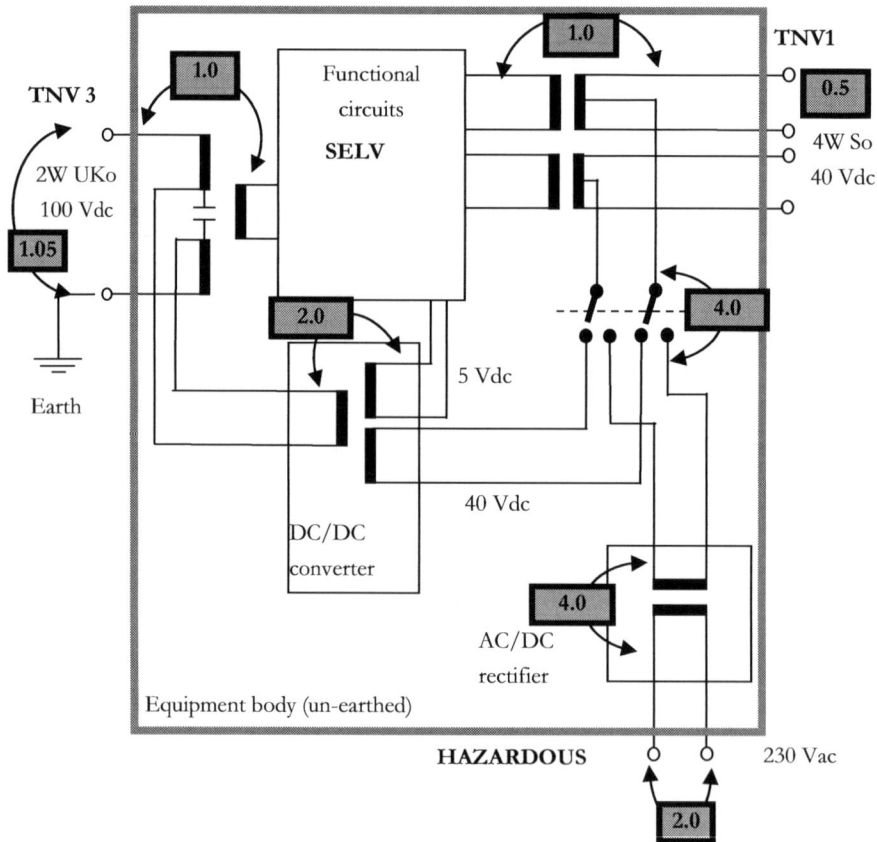

Fig 3.20: Clearance Distance For I.T Product (in mm)

For the 230 Vac input primary side (basic insulation and over voltage category II (2500V), according to table 3.10 the clearance distance is 2.0 mm considering basic insulation between live and neutral (voltage is 230 Vac rms (393 Vpeak) and hence the 420 V row is applicable) while the clearance distance between primary and secondary of the transformer (reinforced insulation) is 4mm. The clearance between primary and secondary of DC-DC converter (table 3.9) is 2mm assuming reinforced insulation. The clearance between input pins, output pins and between input and output pins of the relay is 4 mm assuming reinforced insulation (see table 3.10). The clearance between output pins of the 4-W So output is 0.5 mm assuming functional insulation and peak working voltage of less than 71 V (table 3.11). The clearance distance between the primary and secondary of the 4W output transformer is 1mm (table 3.11).

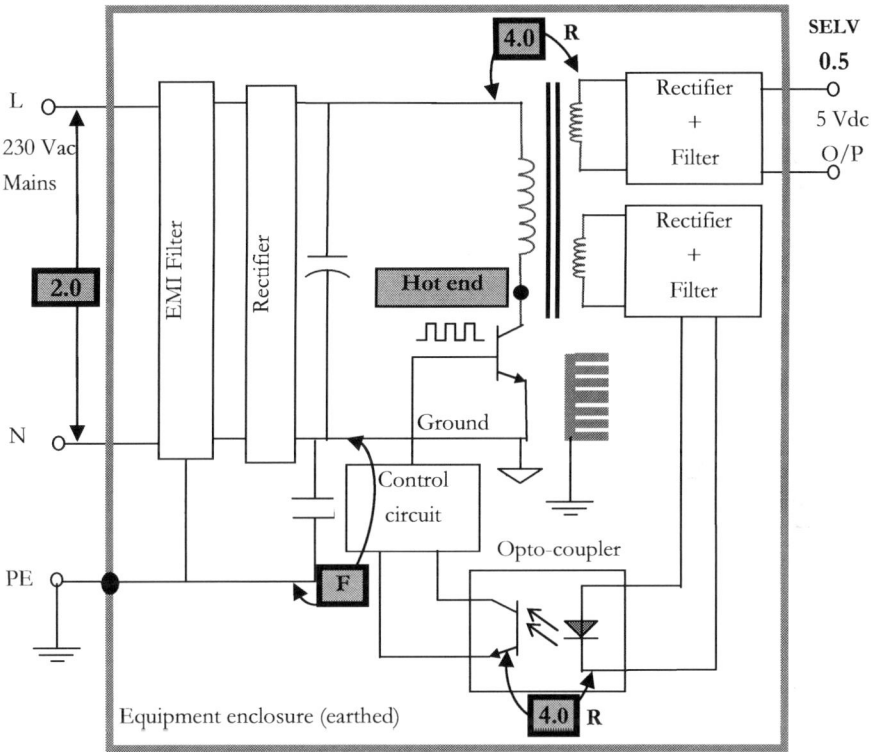

Fig 4.21: Clearance Distance in mm for I.T. SMPS

For the other example of SMPS, assuming an overvoltage category of II and pollution degree 2 as is the case with majority of cases of SMPS for I.T. applications, the clearances at the 230 Vac rms input (393 Vpeak) is 2 mm (table 3.10 row corresponding to 420V and basic insulation). Between primary and secondary pins of the power transformer, the clearance distance is 4 mm assuming reinforced insulation and a peak working voltage of 420 V (Table 3.10). A more precise distance can be arrived at by measuring the peak working voltage at the *hot end* of the transformer. The clearance between the input and output pins of the opto-coupler is 4 mm assuming reinforced insulation (Table 3.10).The standard EN 60950 specifies minimum clearances if air gaps are used as insulation. Accordingly, if an air gap serves as reinforced insulation between parts at hazardous voltage and un-earthed conductive accessible parts of floor standing equipment (or that of non-vertical surfaces of desk top equipment) then the minimum clearance distance is 10 mm while for air gap used as basic insulation between hazardous voltage and an earthed conductive accessible part should be 2 mm (3 mm for contacts of disconnect device). The requirement however does not apply to the air gap between the contacts of thermostats, thermal cut-outs, overload protection devices, switches of micro-gap construction, and similar components where the air gap varies with the contacts.

For connectors (not complying to EN 60083, EN 60309, EN 60320, EN 60906-1 or EN 60906-2) the clearances between the bonding surface of a connector and conductive parts within the connector (that are connected to a hazardous voltage) should comply with the requirements for reinforced insulation. However, basic insulation requirements are to be met for connectors that are fixed to the equipment and located inside of the outer enclosure of the equipment and are only accessible after removal of a user-replaceable subassembly (that is in place during normal operation).

3.10.10.1 Measurement of creepage and clearance.

The procedure for measurement of creepage and clearances is given below. Before measurement, the sample needs certain preparation. Many standards require internal components to be pushed with a test finger by a force of 10N before making measurements. This is to simulate the realities of production where components may not always be mounted in the perfect position. Similarly, enclosures need to be moderately pushed by a force of 30N before making measurements. Many tools are available for measurement of creepage and clearance depending upon how small or large the distance is to be measured. A calibrated metal ruler can provide reliable results for large measurements. For verifying compliance for smaller distances, slip gauges or spacing keys can be used. By using spacing keys that correlates required distance, the user can quickly determine whether a particular key can fit between the required traces, components or circuits. They ensure a higher level of quality for making the measurement and help to verify and make sure that all points along the width of the creepage and clearance are checked, reducing the risk of missing a non-compliant creepage and clearance distance that may be hard to reach or measure. When in doubt, more precise tools like digital verniers can be used. For measurements on a PCB, small monocle type optical comparators with calibrated scale can also be used. Nowadays 3-D product design softwares are used for product design which have the added feature of incorporating creepage and clearance distances in the design.

One more important aspect is taking into consideration small gaps, grooves and ridges on components or PCB that one may encounter during measurement. As a general rule, for pollution degree 2, gaps less than 1 mm should be ignored during measurement due to potential build-up of dust within an equipment. As shown in the Fig 3.22a, the gap is equal to or more than 1 mm and hence should be considered during creepage measurement (shown dotted) i.e. creepage will be more than clearance while as shown in Fig 3.22b, the gap is less than 1 mm and hence should be ignored during creepage measurement (i.e. creepage will be equal to clearance). Another example may be a "V" shaped groove (Fig 3.22c) or slot where the creepage is not measured to the bottom of the "V" but stops and cuts across the "V" at the 1 mm width point.

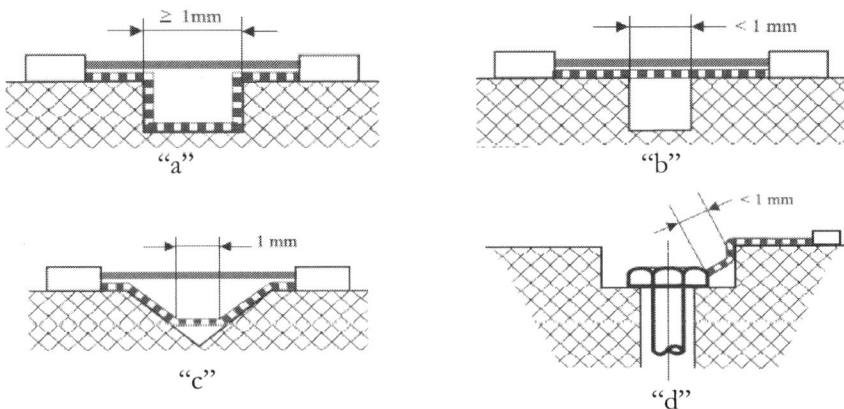

Fig 3.22: Creepage Measurement Over Grooves And Slots.

Although we have considered here the "1mm" rule, different standards may provide different requirements and it is always prudent to check requirement of a particular standard. Safety standards also provide guidance in the form of annexures that contain diagrams that help the manufacturer to understand this requirement.

3.10.11 Distance Through Insulation (Internal Clearances).

As discussed earlier, in addition to creepage and clearances, safety standards also dictate requirements for insulation thickness or distance through insulation for solid insulation. The dimensions of the solid insulation should be such that it does not break down in presence of peak voltages that may be generated within the equipment or overvoltage and transients that may enter the equipment. Normally safety standards prescribe that solid insulation (except PCBs) should either meet minimum thickness requirement (when used as supplementary or reinforced insulation) OR should comply with prescribed HV tests. The following examples are considered in EN 60950.

3.10.11.1 Solid insulation in a single layer

As per EN 60950, there is no thickness requirement for solid insulation if peak working voltage is below 71V. If the working voltage is more than 71V, there is no minimum thickness requirement for basic and functional insulation whereas supplementary and reinforced insulation should have a minimum thickness of 0.4 mm provided by a single layer.

3.10.11.2 Solid insulation in multiple layers

If the insulation is in the form of multiple layers or films there is no dimensional or constructional requirement if used as functional insulation or basic insulation. However, if used as supplementary or reinforced insulation, then two or more layers should be used and the insulation should be located well within the enclosure where it is not subject to handling during operation or servicing. If the layers are separable, there should be a minimum of two layers, each of which should pass the dielectric strength test for supplementary insulation (1.5 kV for 230 V mains operated equipment) or reinforced insulation (3 kV for 230 V mains operated equipment). If three layers are used then all combinations of two layers together should pass the dielectric strength test. As far as the material is concerned, the layers can be of different materials or of different thicknesses, or both. The table 3.13 summarizes the layer requirements for EN 60950 and EN 60601-1.

Table 3.13: Insulation thickness

	EN 60950		EN 60601-1	
	B,S	R	B.S	R
Single material	N/A	0.4 mm	N/A	1mm
Two layers	N/A	Each pass 3kV	N/A	0.3 mm each
Three layers	N/A	Two layer pass 3kV	N/A	Two layers pass 4 kV

If the layers are non-separable, the dielectric test voltage is double the voltage prescribed for separable layers if two layers are used (3 kV for supplementary and 6 kV for reinforced insulation). If more than two layers are used, it is 1.5 times the voltage prescribed for separable layers.

The Fig. 3.23 exemplifies the insulation requirements in a typical concentrically wound transformer used in SMPS.

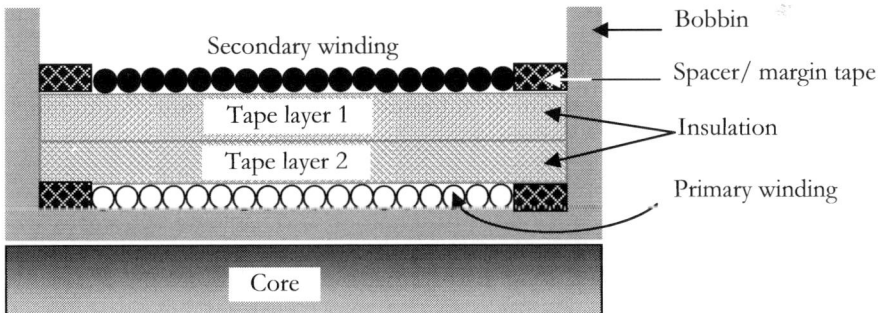

Fig 3.23. Insulation Of A Transformer

As per safety standards, windings of a transformer are considered a part of the circuit to which they are connected. During safety evaluation of the transformer, same requirements apply to a transformer as applicable for the circuit and therefore have to comply with the dielectric tests and creepage/clearance requirements applicable to the circuit. Hence careful thought must be given to the construction of the transformer. There are two tape layers between the primary and the secondary and each need to pass 3 kV considering reinforced insulation. If the insulation is provided by a single layer, it should be at least 0.4 mm thick. The margin tape or the spacer maintains the required creepage (5 to 8 mm) between the windings and the bobbin end flanges.

Other considerations include winding end turns retained by positive means and using double fixing to avoid reduction in creepage due to displacement of windings during installation of connection to external wires.

Allowable constructions are where the windings are isolated from each other by:

- Placing them on separate limbs of the core OR
- Windings on one-piece spool with a partition wall OR
- Concentric windings separated by appropriate insulation wound on a spool OR
- By insulation applied in thin sheet form on the core.

3.10.11.3 Insulating compound as solid insulation.

In cases where a component uses potting or encapsulation using insulating compound and that this compound completely fills the component casing, then a minimum distance of 0.4 mm should be maintained and a single sample should pass the test for pollution degree 1 (see section 3.10).

3.10.11.4 Semiconductor devices

If insulating material completely fills the casing of the device, there is no minimum requirement of distance through insulation for supplementary as well as reinforced insulation. However, three samples of the component should pass the following tests:

Sample 1: Thermal conditioning 68 hours at 85 ± 2 °C followed by a dielectric test at 1.6 times that specified in section 3.10.5 (for reinforced insulation 4.8 kV)

Sample 2 and 3: Humidity conditioning (48 h at 40 ± 2 °C and RH 93+3 %) followed by a dielectric test at 1.6 times that specified in section 3.10.5 (for reinforced insulation 4.8 kV).

3.10.11.5 Requirement for Opto-couplers.

To minimize the likelihood of exposing the end user to injury from hazardous voltages, isolation is required in electronic circuits. Electrical isolation is typically achieved by one of three methods: magnetic, capacitive or electro-optical. Magnetic isolation (using an isolation transformer) is probably the longest-established method of electrical isolation, providing high levels of isolation at high frequencies in a robust package. Among the downsides of this method of isolation are a large device footprint when compared with other methods and suitability only for AC signal coupling. Due to these characteristics, magnetic coupling is for the most part limited to high-power AC applications. The second common method of electrical isolation is capacitive coupling. The advantages of capacitive coupling are high switching speeds and a relatively small package footprint, but to eliminate the need for a floating power supply on the secondary side, a large capacitance is required to transfer energy from the primary to the secondary side. Thus, the electrical isolation value of this technique is greatly diminished by the need for efficient energy coupling. Optical isolation has advantages over the former methods without the drawbacks. Mainly, optical isolation offers high electrical isolation values, an effective *Lakshaman rekha* that the *demons* of hazardous voltages are incapable of crossing. Hence it comes as no surprise that opto-couplers have been traditionally used to provide electrical isolation from hazardous voltage, the most common application being in the feedback loop of a typical SMPS.

Considering such applications, safety standards put opto-couplers in critical component category and they have to meet the requirements of reinforced insulation. All opto-couplers should meet the requirements of EN 60747-5-5. EN 60950 requires an opto-coupler to meet the dielectric strength test of 3 kV. The same external creepage and clearance requirements apply to opto-couplers as for the circuits for which they are used. As per EN 60950, if insulating material completely fills the casing of the opto-coupler, there is no minimum requirement of distance through insulation for supplementary as well as reinforced insulation. However, if the insulation does not completely fill the component, the LED and the photo diode should be separated by a dielectric of thickness 0.4mm minimum considering reinforced insulation. The Fig. 3.24 and Fig. 3.25 show cross section of two types of opto-couplers. The silicone dome type is rated up to 6kV while the planar types are rated up to 4kV.

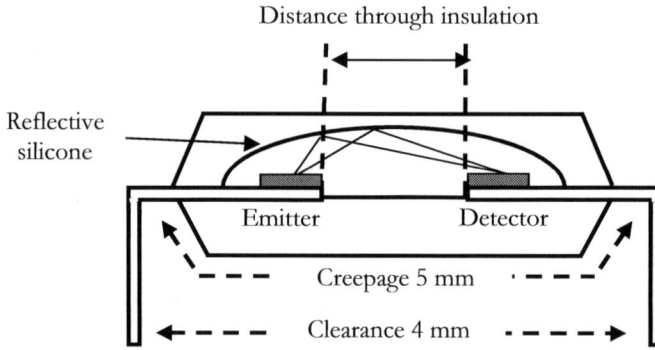

Fig 3.24: Creepage And Clearance For Silicone Dome Opto-coupler

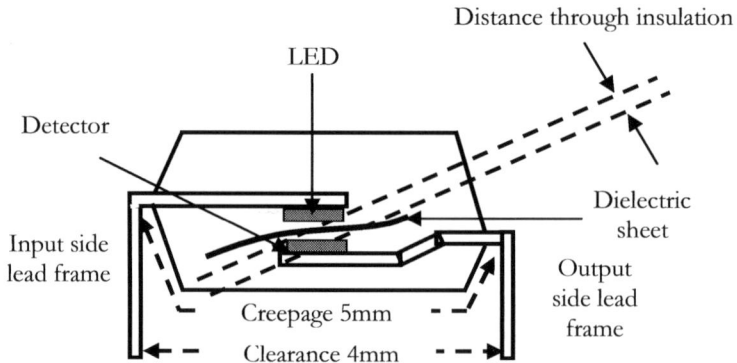

Fig. 3.25: Creepage And Clearance For Silicone Planar Opto-coupler

3.10.11.6 Insulation requirements for PCBs

Safety standards also prescribe insulation requirements for PCBs. The standard EN 60950 prescribes requirements for tracks on outer surfaces of a PCB, tracks on same inner surface and tracks located on different inner surfaces.

Accordingly, tracks on outer surfaces of a PCB should comply with creepage and clearance requirements as given in section 3.10.8. Insulation between conductors on same inner surface should comply with creepage and clearance requirements for pollution degree 2. If not than one sample of the board should comply with HV tests as given in section 3.10.5. While insulation between tracks located on different inner surfaces should be minimum 0.4 mm thick.

3.11   PROTECTION BY EARTHING.

As we have seen, earthing is one of the means of offering protection against shock hazard in addition to insulation and separation. Earthing can be considered as a case of grounding. Grounding can be defined *as a connection, whether intentional or accidental, between an electrical circuit or equipment and the earth or to some conducting body that serves in place of the earth.* Earthing is thus a special case of grounding where the ground is essentially the planet Earth. Earthing can be for safety purposes or for functional purposes. The former case is for providing safety against electric shock and is called *protective earthing* (PE) and while the latter case is for purposes other than safety (circuit function) and is called as *functional earthing.*

Protection against electric shock is illustrated by the circuit in Fig 3.26. It shows the simplified circuit of a power supply whose input is 230V hazardous voltage while the output is operator accessible and at SELV. As seen in previous section, the SELV should be separated from hazardous voltage by two levels of protection. In this example one level of protection is provided by basic insulation (which separates the primary and secondary of the transformer) and   other level is provided by connecting one end of the SELV to protective earth (PE). In case of a single fault in the form of a short circuit between the primary and secondary, the hazardous voltage can appear on the operator accessible SELV leading to a shock hazard.

However, since the SELV is earthed at one end, the fault current flows through the secondary winding into this earth. Now two things happen, one is that even if the operator touches SELV output, it will be at earth potential negating the danger of an electric shock while the other thing is that this fault current will operate the protective device (a fuse or a circuit breaker) isolating the SELV from hazardous voltage.

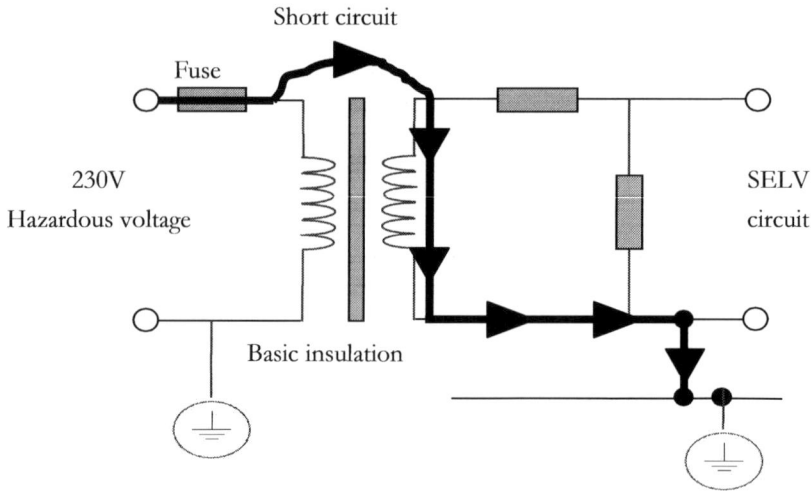

Fig. 3.26: Protection Against Electric Shock By Earthing

3.11.1 Protective Earthing

Protective earthing is used for providing safety against electric shock. Connection to protective earth is required for the following parts of equipment:

- Accessible conductive part that can assume hazardous voltage in case of single fault.
- Accessible parts separated from hazardous voltage by single insulation.
- For pluggable equipment having connections to TNV circuits.
- Circuits, transformer screens and components (such as surge suppressors) that cannot assume hazardous voltage but are required to be earthed in order to reduce transients that might affect insulation.
- Circuits having high leakage current or touch current.

3.11.2 Functional Earthing

Functional earthing, if used, should be separated from hazardous voltage by two levels of protection which could be double (or reinforced) insulation or basic insulation and screen (or conductor) connected to protective earth. Functional earth can be connected to protective earth.

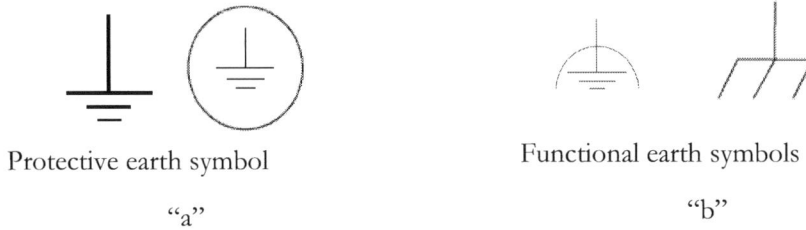

Protective earth symbol

"a"

Functional earth symbols

"b"

Fig. 3.27: Earthing Symbols

3.11.3 Earthing Symbols/Marking.

Protective earth should be denoted with the symbol as given in Fig. 3.27a while functional earth with the symbol as per Fig 3.27 b.

3.11.4 Earthing Conductor

Earthing or connection to earth is normally provided by wires or conductors. Safety standards like the EN 60950 identify two types of earthing conductors viz. *protective earthing* conductors and *protective bonding* conductor. Protective earthing conductor is a conductor in the building installation wiring or a conductor in the power supply chord which connects the protective earthing (PE) terminal of an equipment to the earthing point of building. Whereas protective bonding conductor is a conductor inside an equipment (or a combination of conductive parts including mounting plate etc.) which connects a part of the equipment (module or sub-assembly) to the protective earthing (PE) terminal of an equipment. The Fig. 4.28 illustrates the difference between protective earthing conductors and protective bonding conductor.

3.11.5 Colour Of Insulation

Reliability and proper identification are key to the protective earth (PE) in equipment. Safety standards prescribe a unique colour scheme for identification of the earth wire. Accordingly, insulation of protective bonding conductor and protective earthing conductor should be green with a yellow stripe.

The green/yellow combination is strictly reserved for the aforementioned grounds and should not be used for low voltage (< 50 V), EMC/RFI grounds or functional earthing. However, in case of an insulated earthing braid (used to extend earthing to doors) transparent insulation is allowed by some safety standards. Also, if the protective bonding conductor is a part of ribbon cables or printed wiring or if it is in the form of a bus bar, any colour is permitted provided that there is no mis-interpretation.

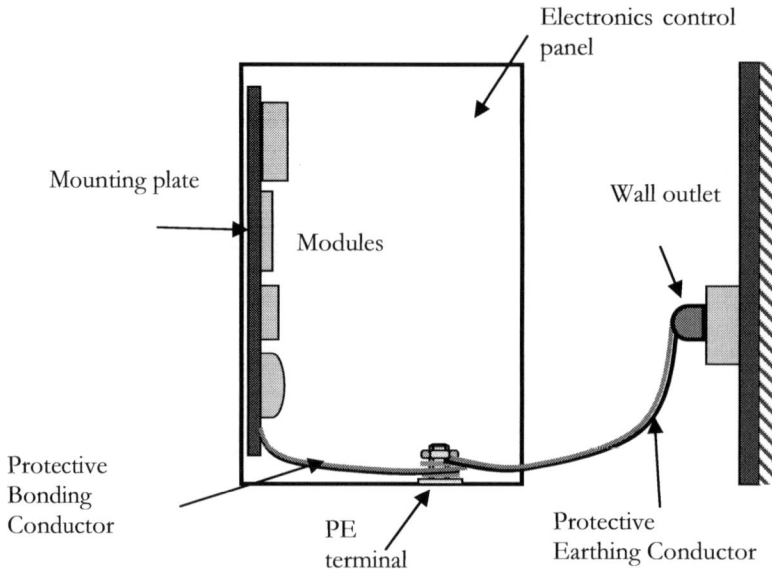

Fig. 3.28: Earth Wire Connection

3.11.6 Size Of The Earth Wire

The size of the earthing wire depends upon the protective current rating of the equipment i.e. the rating of over current protective device in place for protecting a circuit. For pluggable equipment type 'A' (non-industrial plug) the size of the protective earthing wire depends on the rating of overcurrent protective device as given in table 3.14 subject to a minimum of 16A. For pluggable equipment type 'B' (industrial plug) and permanently connected equipment, the size of the protective earthing wire depends on the rating of overcurrent protective device as given in table 3.14. The compliance for the same is normally checked by inspection.

Table 3.14: Size Of The Earth Wire

Protective Current Rating	Minimum conductor sizes
Of The Circuit Under Consideration Up To And Including (A)	Cross-sectional area mm^2
20	Size not specified
25	1,5
32	2,5
40	4,0
63	6,0
80	10
100	16
125	25
160	35
190	50
230	70
260	95
300	120
340	150
400	185
460	240

3.11.7 Earthing Terminals

All accessible metal parts, such as panels and doors, providing guarding from hazardous voltages, must be properly bonded to the main protective earthing terminal to ensure a low-resistance joint. For equipment with detachable power chord, the earthing terminal of the appliance inlet is considered as the protective earthing terminal. Bonding is defined as *a permanent joining of metallic parts to form an electrically conductive path that not only ensures electrical continuity but also has the capacity to conduct safely any current imposed on the joint.* Thus, bonding is that means which serves to give a low-resistance connection to earth.

To ensure a low earthing resistance, masking of paint on mating surfaces and ground straps between metal doors and the chassis helps to ensure an adequate safety ground. Grounding through hinges of metal doors is not considered adequate. Ground straps or wires are necessary for all hinged metal doors that cover hazardous voltages to ensure an adequate safety ground. An isolated ground point or terminal is preferred and often required in some standards, along with the protective earth symbol (Fig 3.27a) for connection of the PE wire only (green-yellow insulation). Only one PE symbol is allowed in the product, unless there is more than one power cord. The chassis ground symbol may be used for other safety or chassis grounds. In addition, all ground terminations, such as ring lugs, must be reliably fixed and not allowed to turn.

The Fig. 3.29 shows typical bonding of earth wire to the PE terminal which is in the form of a welded stud. A screw with a captive nut (instead of the weld) can also be used provided that the thickness of the sheet metal wall (on which the screw is threaded) is more than twice the pitch of the screw thread. It should be noted that in practical case all the nuts and washers would be tightened, they have been shown loose for better understanding. Here the PE terminal –essentially a bolt screwed to position is held in place by a captive nut or a stud welded to the enclosure cabinet, the latter being a better option. For ensuring safety, the stud should penetrate the enclosure and come out from the other side where it is welded. While fixing the earth wire, any insulating surface finish (paint or powder coating) is removed from the contact area or masked during the powder coating process.

The contact area should be slightly larger than the surface area of the lug. A spiky or shake proof washer is first put into place to ensure a good life time bond. Over this is placed the lug of the protective bonding conductor. A shake proof washer is then placed over the lug after which the nut is tightened. The lug of protective earthing conductor is directly put on the nut followed by a shake proof washer followed by another nut. The order of stacking of the two lugs is normally not specified by the standards. If the bond is likely to be exposed to moisture or corrosive environment, then a coating of paint or grease is applied over the bond. To avoid corrosion that may take place between dissimilar metals (due to electrochemical potential) the bolt, screw, nut, lug, washer should be of the same metal.

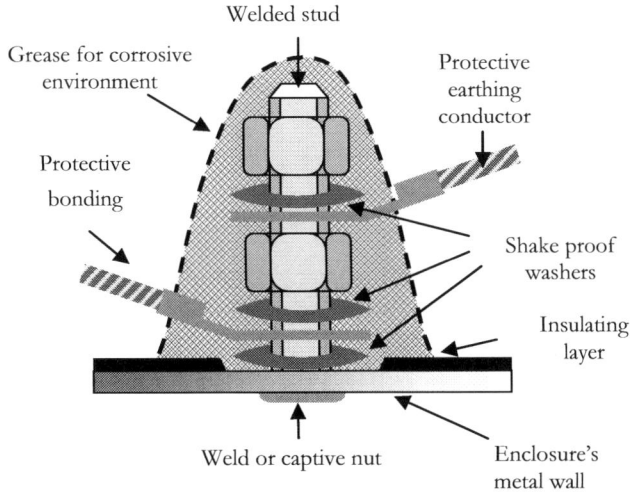

Fig. 3.29: Bonding Of Earth Wire

As per EN 60950, if they are of dissimilar metals, then they should be such that the electrochemical potential they develop should be less than 0.6 V. The choice of metal is given in Annex J of the standard.

3.11.8 General Considerations

Following are some general considerations as regards to earthing in an equipment:

- Fixing of the PE conductor should be such that it cannot be disconnected during the servicing operation to replace any parts. This means that the grounding stud or screw should not be used to hold any other replaceable part in place.

- In a system of interconnected equipment, the protective earthing connection should be ensured for all equipment requiring such a connection.

- Protective earthing conductors and protective bonding conductors should not contain switches or overcurrent protective devices.

- Protective earthing connections should be such that disconnection of a protective earth at one point in a unit or a system does not break the protective earthing connection to other parts or units in a system.

- Protective earthing connections should make earlier and break later than the supply connections. Examples could be a connector of a part that can be removed by an operator or a plug on a power supply cord.

- Protective earthing should not rely on cable distribution system like conduits, cable trays etc.

It is worthwhile to perform a ground bond test between all user-touchable metal parts and protective earth terminal. The ground bond test is mandatory if the protective bonding conductors do not meet the size requirements. It is not required for protective earthing conductors. Nevertheless, in order to ensure proper connection to earth, the ground bond test may be performed at 25A (12 V source) or two times the limited fuse/breaker rating depending on the standard. The measured ground resistance must be less than 0.1 ohm for the ground bond test to pass. The duration of the test depends upon the protective current rating of a circuit. For detailed procedure regarding the ground bond test, please refer section 3.36.7.

EN 60950 prescribes test duration as per the following table:

Table 3.15: Ground bond test

Protective Current Rating of the circuit (*I*pc) A	Duration of the test minutes
≤ 30	2
$30 < Ipc \leq 60$	4
$60 < Ipc \leq 100$	6
$100 < Ipc \leq 200$	8
> 200	10

3.12   COMPONENTS AS SAFEGUARDS

Apart from insulation and earthing, protection can also be provided by using safety components. Components may be used as principal safeguard or supplementary safeguard only, if used within their ratings. For components that are used as a principal safeguard the insulation should comply with basic insulation. For components used as a supplementary safeguard the insulation should comply with supplementary insulation. This includes transformers, relays and VDRs (voltage dependant resistors) also called as MOVs (metal oxide varistors).

3.12.1 Capacitors as principal & supplementary safeguard

Capacitors used for safeguard are special type of capacitors called Y capacitors and are usually connected across the live and the earth (and neutral and earth) and as such bridge basic insulation. In such applications, the failure of the capacitor can lead to a dangerous shock. Hence these capacitors are subjected to severe insulation tests and therefore they are of enhanced electrical and mechanical reliability intended to eliminate short–circuits in the capacitor.

Class Y1: This category of Y capacitors is used for bridging reinforced or double insulation. In the past (particularly in Europe) two separate capacitors were required to be used in series to bridge double insulation requirements. Now one class Y1 capacitor is allowed. These capacitors are impulse tested ($1.2 / 50$ µs) with 8 kV and are suitable for line voltages of up to 250 Vac (nominal).

Class Y2: This is the most popular type of Y capacitor. Such a capacitor is allowed to bridge basic and supplementary insulation with line voltages up to 250 Vac (nominal). This is the normal case for power supplies used in data processing equipment. These capacitors are impulse tested with 5 kV.

Class Y3: Class Y3 capacitors are used for bridging basic and supplementary insulation with line voltages up to 250 Vac (nominal) but without an impulse test.

Class Y4: For applications with line voltages up to 150 Vac (nominal) with an impulse test of 2.5 kV and bridging basic and supplementary insulation.

For a capacitor used as principal safeguard, the peak test voltage of the capacitor should be at least equal to the required withstand voltage. Capacitors should comply with the safety requirements of EN 60384-14:1993, subclass Y2 or

Y4. It is permitted to use a higher-grade capacitor than the one specified. For example, subclass Y1 if subclass Y2 is specified OR subclass Y1 or Y2 if subclass Y4 is specified.

Capacitors and RC units used as reinforced insulation should meet one of the following:

a single capacitor or RC unit complying with EN 60384-14, subclass Y1,

OR

a single capacitor or RC unit complying with EN 60384-14, subclass Y2 where the equipment rated voltage is less than 150 V with respect to neutral or earth,

OR

two capacitors in series each complying with EN 60384-14, subclass Y2 or Y4 Where two capacitors are used in series, they should each be rated for the total working voltage across the pair and should be of the same nominal value.

3.12.2   Resistors as principal safeguard and supplementary safeguard

Resistors may be used as principal safeguard. The resistor should comply with the requirements for basic insulation between its terminations for the total r.m.s. working voltage across the insulation. Resistors may be used as supplementary safeguard. The resistor should comply with the requirements for supplementary insulation between its terminations for the total r.m.s. working voltage across the insulation. The resistor should have an adequate stable resistance value under overload. Compliance is checked by inspection and by applying the following test on 10 sample resistors. Before the test, the resistance of each sample is measured and they are then subjected to the damp heat test according to EN 60068-2-3, severity 21 days. Each sample is then subjected to 50 discharges from the 1.2/50 µs impulse test generator (see section 3.36.9) at not more than 12 discharges per minute, with Uc equal to 10 kV. After the test, the resistance of each sample should not have changed by more than 20 %. No failure is allowed.

3.12.3 VDRs (Varistors) as principal and supplementary safeguards

A VDR should comply with EN 61051-2 (Varistors for use in electronic equipment), with the following:

- Preferred climatic categories: -10 °C to +85 °C.

- Duration of damp heat, steady state test ($40\pm2°C$, 95% RH): 21 days.

- Maximum continuous voltage at least 1.25 times the rated voltage of the equipment or at least 1.25 times the upper voltage of the rated voltage range.

Pulse current: For the test, combination wave surge pulses of 6 kV/3 kA of alternating polarity are used, having a combination pulse shape of 1.2 µs rise time and 50 µs pulse-width for voltage and 8 µs rise time and 20 µs pulse-width for current.

3.12.4 Transformers as safeguards

For transformers that are used as a principal safeguard the insulation should comply with basic insulation. For transformers used as a supplementary safeguard the insulation should comply with supplementary insulation. Requirements for transformers to be used as reinforced safeguards are given in section 3.37.5. Precautions should be taken to prevent the reduction below the required minimum values of clearances and creepage distance that provide basic, supplementary or reinforced insulation similar to those discussed in section 3.10.11.2. For acceptable construction method for transformer, requirements of last paragraph of section 3.10.11.2 apply.

3.13 WARNINGS

Even though all of the above considerations are taken into account there are always some residual hazards that cannot be completely eliminated. These can be addressed by providing proper instructions, warnings, and symbols so as to alert users and service personnel to possible hazards (see section 3.35). The type, size, colour, and symbols are specified in the standards and sometimes mentioned in a directive (for example machinery directive). For electrical products, the standards dictate the necessary warnings and where they must be provided (product and/or manual). The standards state that any required safety instructions and markings should be in the language of the country where the product is used and that machine manufacturers must supply user manuals in that language. As specified in the standards, suitable warnings and symbols should be provided for hazards, but warnings should not take the place of a safe design.

3.14   ENERGY HAZARDS

Up till now we have seen hazards due to circuit voltages. But often than not, both voltage and current exist in a circuit and they together can cause energy hazards. It must be stressed that energy hazards can exist even in low voltage circuits where the current is high. They can also exist in outputs of high current power supplies. Injury can result in direct bums or burns caused by ejection of molten metal due to short circuits between adjacent poles (which may be caused, for example, by metal such as ring on finger bridging the supply outputs).

Energy sources can be classified as ES1, ES2 and ES3 depending upon the available voltage and the available current. The voltage and current are the maximum that can be delivered by the source on any resistive load. The classification is as given in table 3.16 for DC and low frequency AC and in table 3.17 for medium and high frequency AC.

Table 3.16: Energy Source Classification For DC And Low Frequency AC

Energy Source Level		ES1 Limit	ES2 Limit	ES3 Limit
DC	Voltage	$\leq$ 60V	>60$\leq$ 120V	> ES2 Limit
	Current	$\leq$ 2 mA	> 2 $\leq$ 25mA	
AC (up to 1 kHz)	Voltage	$\leq$ ES1 Limit	> ES1 Limit	
	Current	$\leq$ 0.5 mA rms	>0.5mA $\leq$10mArms	

Table 3.17: Energy Source Classification For Medium & High Freq. AC

Energy Source Level		ES1 Limit	ES2 Limit	ES3 Limit
AC (1 kHz up to 100kHz)	Voltage	$\leq$ ES1 limit Umax = 50 V	$\geq$ ES1 limit < ES2 limit U max = 120 V	> ES2 Limit
	Current	$\leq$ 0.5 mA x f (kHz) Imax = 50mA	$\leq$ 10 mA x 0.5 f (kHz) Imax = 140mA	

Energy Source Level		ES1 Limit	ES2 Limit	ES3 Limit
AC (above	Voltage	≤ 50V	≤ 120V	
100kHz)	Current	≤ 50mA	≤140mA	

Note: All values in rms. Table to be seen in conjunction with Fig. 3.30

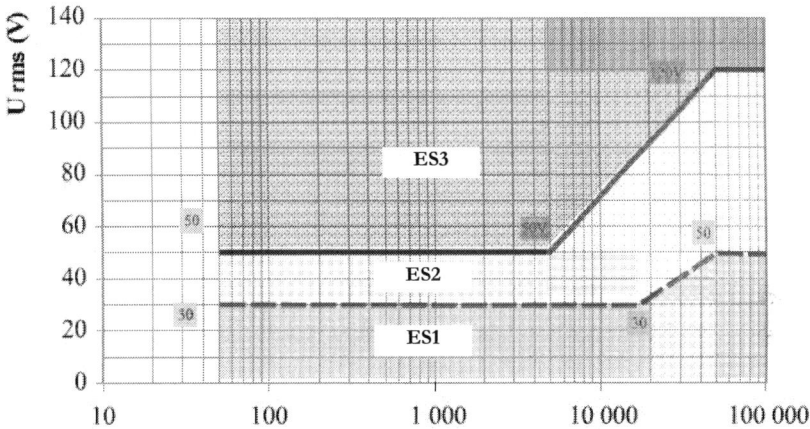

Fig. 3.30: Voltage Limits Depending On Frequency

3.15   PROTECTION AGAINST ENERGY SOURCES

As far as protection between circuits of various energy source levels is concerned ES1 and ES2 circuits should have one level of protection (LOP) in between (basic insulation or earth or protection component) while ES2 and ES3 circuits should have two LOPs (basic insulation + supplementary insulation (or reinforced insulation) or basic insulation + earth or basic insulation + one protection component   in between. As regards to protection of personnel is concerned, standards identify three types of personnel viz. skilled person, instructed person and ordinary person. A skilled person is one who has relevant education and experience to enable him or her to avoid dangers and to prevent risks which may be created by the equipment. An instructed person is one who is adequately advised or supervised by a skilled person to enable him or her to avoid dangers and to prevent risks which may be created by the equipment. While an ordinary person is one who is neither skilled nor instructed.

So far as ordinary person is concerned, ES1 circuits and ES2 circuits (with contact area less than 10mm²) are allowed to be accessible while there should be one LOP between him and ES2 circuit (with contact area greater than 10mm²) and two LOP between ES3 circuit and the person.

An instructed person is allowed access to ES1 and ES2 circuits i.e. no safeguards need be interposed while there should be two LOP between him or her and ES3 circuit. The supplementary safeguard can be accessible but the basic safeguard should not be accessible.

A skilled person is allowed access to ES1 and ES2 circuits while there should be one LOP between him or her and ES3 circuit. Bare parts at energy source ES 3 should be located or guarded so that unintentional contact with such parts is unlikely during service operations involving other parts of the equipment.

The table 3.18 summarises LOPs required

Table 3.18: LOPs Required Between Energy Sources

	LOPs required		
	ES1	ES2	ES3
ES1	0	1	2
ES2	1	0	2
ES3	2	2	0
Ordinary person	0	1	2
Instructed person	0	0	2
Skilled person	0	0	1

3.16   COMPLIANCE IN GENERAL

Compliance is checked by accessibility test (see section 3.9.1) as follows:

For ordinary persons:

No contact to ES 3 and no contact to the principal safeguard of ES 3

No contact to ES 2 for areas >10 mm2.

For instructed persons:

No contact to ES 3 and no contact to the principal safeguard of ES 3

3.17   LIMITED POWER CIRCUITS

Standards like EN 60950 specify requirements for limited power source (LPS) as one of the safeguards against energy hazards. These are used to define power supplies with relatively low maximum voltage, current and power capabilities. Power supplies which qualify as LPS are recognized as unlikely to cause electrocution or a fire due to the limitations on the output current and voltage they can deliver to a load.

The following is a summary of the specifications for power supplies certified as LPS without current limiting devices:

Output (Voc) DC/ sinusoidal AC voltage less than or equal to 30 V dc/Vrms

- – Maximum short circuit current (Isc) of 8 A
- – Maximum VA of 100
- – Maximum marked output power rating of 5 A * Voc
- – Maximum marked output current rating of 5 A

Output (Voc) DC voltage greater than 30 Vdc and less than or equal to 60 Vdc

- – Maximum short circuit current (Isc) of 150 /Voc
- – Maximum VA of 100
- – Maximum marked output power rating of 100 VA
- – Maximum marked output current rating of 100 /Voc

For non-sinusoidal AC or DC with ripple, the output peak voltage not exceeding 42.4 V

Note: Output voltage is open circuit no load voltage

The standard identifies following three methods for meeting the above requirements:

3.17.1   Inherent Power Limiting

This class of circuits does not require additional design considerations to ensure the limited power delivery capability as internal components are incapable of delivering power in excess of the limits. The classic example of a component limiting power delivery capacity is the winding resistance of an isolation transformer. In a well-designed power supply, the components limiting the power delivery capability will not be damaged when they are the limiting factor in the power delivery.

3.17.2 Power Limiting By Linear/Non-Linear Impedance

An impedance in the form of a conventional resistor or PTC resistor can be placed in series with the power conductors to limit the power delivery capability of the power supply. While being simple to implement, conventional resistors are seldom employed for this purpose due to the power dissipation of the resistors causing a reduction of the conversion efficiency of the power supply. The use of PTC resistors maintains the simplicity of the implementation while reducing the associated power losses during normal operation.

3.17.3 Regulating Network Providing Power Limiting

This method is common in modern power supplies due to the low cost and wide availability of the required integrated circuits. However, care must be taken in the power supply design and testing to ensure the required limits are adhered to during both normal and single fault operating conditions.

3.17.4 LPS With Over Current Limiting Devices

Power supplies with external current limiting devices can be classified as LPS even if they do not contain one of the three means of limiting power delivery listed above. A power supply can be certified as conforming to LPS if it employs an over current protection device that limit the current available to be supplied to the load.   The current limiting device must be either a fuse or a non-adjustable, non-auto resetting and electromechanical device (i.e. a circuit breaker). The fuses or circuit breakers must break the circuit within with a current equal to 210% of the current specified in the regulations. The following is a summary of the specifications for power supplies certified as LPS with non-inherent power delivery limits:

Output (Voc) DC voltages /sinusoidal AC voltages ≤ 20 V DC/Vrms

- Short circuit output current must be less than 1,000 /Voc

- Current rating of the over current protection device ≤ 5 A

  - Maximum VA must be ≤ 250
  - Maximum marked output power rating of 5 A * Vmax
  - Maximum marked output current rating of 5 A

Output DC/ sinusoidal AC voltages > 20 Vdc/Vrms and ≤ 60 Vdc/30 Vrms

  - DC voltages with ripple greater than 10% of the peak and non-sinusoidal ac voltages must have peak voltage ≤ 42.4 V.
  - Short circuit current must be less than 1,000 /Voc.
  - Current rating of the overcurrent protection device ≤ 100 Voc.
  - Maximum VA must be ≤ 250.
  - Maximum marked output power rating of 100 VA.
  - Maximum marked output current rating of 100 /Vmax.

3.18   POWER SOURCE CLASSIFICATION

Having seen the limited power sources lets us now see the power source classification. Electric power sources can be classified as PS1, PS2 and PS3. Power source class PS 1 is a power source that does not exceed 100 W until 1 s and 15 W thereafter in normal and single fault conditions. PS2 source is a power source that does not exceed LPS limits while PS3 source is that source which exceeds LPS limits.

3.19   FIRE HAZARDS

Fire or flammability hazards, as the name suggests, are a result of excessive temperature and possibility of fire. Flammability is defined as *ability of a product or material to burn with a flame under specific conditions*. Flammability requirements are intended to minimize the risk of ignition AND the spread of flame (both within the product and outside). Risk of ignition can be reduced by limiting the temperature of components under normal conditions and under single fault conditions and minimizing the spread of flame by the use of appropriate materials, components and by suitable construction that can contain the fire.

Keeping the above facts in mind, safety standards (like EN 60950) accept flammability hazards to have been taken care of if one of the following two methods demonstrate that risk of ignition and spread of flame has been minimised:

METHOD 1: There is no spread of fire after certain single fault conditions have been simulated (that can cause the spread of fire outside the equipment).

OR

METHOD 2: The sources of ignition within the equipment have been reduced or eliminated, fire is contained within the equipment by using flame retardant materials and using a fire enclosure.

Let us take a look at these one by one

3.20   SIMULATING SINGLE FAULT CONDITIONS

This is the preferred test method for an equipment with *small* number of electronic components as per EN 60950. The standard however, does not specify how small is *small*. Single fault conditions (see section 3.36.13), especially those related to heating circuits or cooling mechanisms, are simulated thereby demonstrating that temperature rise is below specified value (as per thermal class) and that there is no spread of fire. Conformity with requirements for protection against the spread of fire is checked by placing the equipment on white tissue-paper and covering the equipment with cheesecloth (100% cotton cloth). Upon simulation of single fault condition, molten metal, burning insulation, flaming particles etc should not fall and there should be no charring, glowing, or flaming of the tissue paper or cheese cloth.

3.21   ELIMINATING/REDUCING IGNITION SOURCES

This is the preferred test method (method 2) for equipment containing *large* number of electronic components as per EN 60950.

Potential ignition source is a location where electrical energy can cause ignition. Such sources can be classified as:

Potential ignition source 1: Location where an arc may occur due to the opening of a conductor or a contact. If the open circuit voltage measured across an interruption or faulty contact exceeds 50 V (peak) AC or DC and the product of the peak of this voltage and the measured rms current under normal condition exceeds 15 W.

Potential ignition source 2: Location where a component may ignite due to excessive power dissipation. Such a source is supplied from a power source above 15 W and under fault condition dissipates 15 W or more and may ignite or ignites when subjected to fault conditions.

The possibility of ignition and occurrence of fire is considered to be reduced to a tolerable level if all the following requirements a), b) c) and d) are met:

a) There are no potential sources of ignition.

b) The voltage, current and power available to the circuit or part of equipment are limited by using PS1 or PS2 circuits. OR insulation between parts at different potentials meets the requirements for basic insulation, or it can be demonstrated that bridging the insulation will not cause ignition.

c) Any ignition hazard related to flammable liquids is reduced to a tolerable level by limiting the liquid temperature by 25°C below the fire point of the liquid or if ignited, the fires are contained within the equipment (by using fire enclosure).

d) In circuits designed to produce heat, no ignition occurs when tested in following single fault condition one at a time:

  - Timers (which limit the heating period) overridden to energize the heating circuit continuously.

  - Temperature controllers overridden to energize the heating circuit continuously.

3.22   CONTAINMENT OF FIRE

In case when PS3 circuits are used or potential ignition sources are present, fires should be contained by using flame retardant materials and using a fire enclosure. Enclosure can be defined as an *external casing protecting the electrical and mechanical parts of an apparatus.* A fire enclosure is defined as *a part of the product intended to minimize the spread of fire or flames within.* A decorative part is a part *outside the enclosure that has no safety function.* Fire enclosures and components and parts inside an enclosure must be so constructed and make use of materials so as to minimize the propagation of fire.

Some materials and components are exempt from the flammability requirements, such as metals, ceramic materials, glass, small components, cable ties, certain wire insulation (PVC/TFE/PTFE/FEP/Neoprene) and small parts mounted on PCBs (I/Cs, transistors, optocouplers, capacitors, etc.), one or more layers of insulation (adhesive tape, etc.), nameplates, mounting feet, keycaps, knobs, lacing tape, twine and components meeting flammability requirements of relevant IEC component standard.

Other than the exempt items, all materials comprising the fire enclosure and components inside the fire enclosure should pass a flammability test or have a flammability rating, such as vertical burning class V-0 or 5V, V-1, V-2 or horizontal burning class HB or for horizontal burning foamed materials, HF-1, HF-2, HBF. The vertical burning class are less flammable and are also known as *self-extinguishing* materials (i.e. they have the ability to self-extinguish once removed from the source of ignition) since they have flame retardants added. The horizontal burning class are most flammable and are known as *slow burning* materials. They are generally polymeric materials with no flame retardants added. The test methodology for HB and VB tests is given in the following sections.

3.23   BURNING TESTS

These tests (as per standards EN 60695-11-10, 20) are used to classify materials according to flammability ratings. During the testing, a rectangular test specimen (120mm length, 13 mm width and 3 to 13mm thick) is supported horizontally or vertically at one end and the free end is exposed to a specified test flame (50watt flame 20mm high as per EN 60695-11-4). The horizontal burning (HB) behaviour is assessed by measuring the linear burning rate (length of material burned per unit time) while the vertical burning (VB) behaviour is assessed by observing whether the materials self-extinguish, the extent of burning and the dripping of flaming particles.

3.23.1 Horizontal Burning (HB) Test

Here the test specimen is clamped horizontally on a stand with a wire gauze kept 10mm below the specimen. The specimen is temperature and humidity conditioned as per the standard before the test. The lateral axis of the specimen should be at angle of 45° to the vertical. Each test specimen is marked with two lines perpendicular to the longitudinal axis of the bar, 25 mm ± 1 mm and 100 mm ± 1 mm from the end that is to be exposed to the test flame. The test flame is then applied to the lower edge of the test specimen's free end for 30 s ± 1 s or removed as soon as the flame front on the test specimen reaches the 25 mm mark. While applying the flame, the central axis of the burner tube is kept at an angle of 45° ± 2° to the horizontal as shown in the Fig. 3.31.

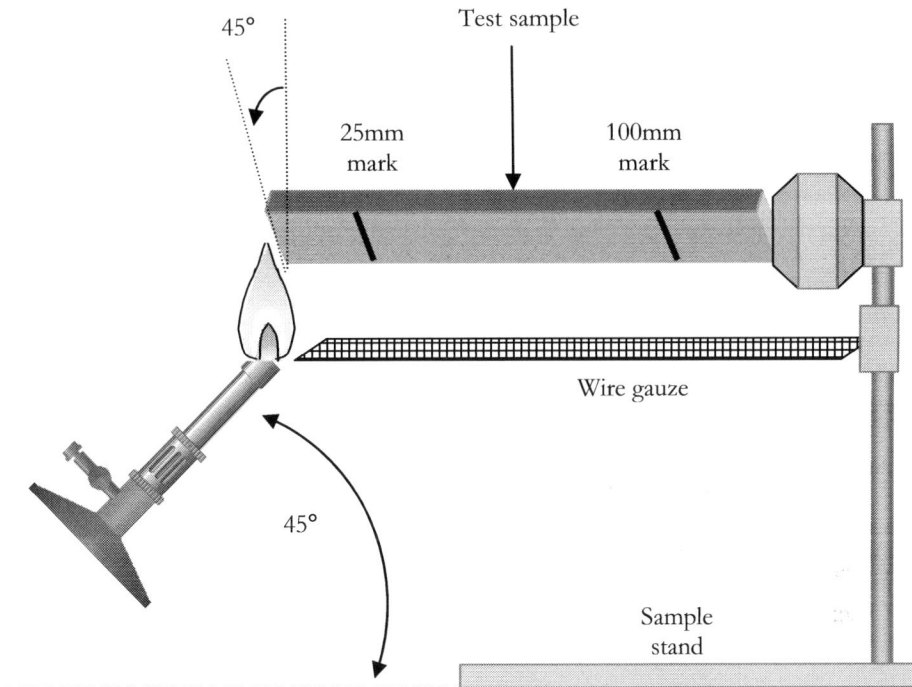

Fig. 3.31: Horizontal Burning Test Set-up

If the test specimen continues to burn with a flame after removal of the test flame, the time taken (in seconds) for the flame front to travel from the 25 mm mark past the 100 mm mark is recorded and the damaged length is recorded as 75 mm. If the flame front passes the 25 mm mark but does not pass the 100 mm mark, the time and the damaged length between the 25 mm mark and the mark where the flame front stops is recorded.

Two sets of three specimens each are tested. If one test specimen from the first set of three test specimens fails to conform to the criteria indicated in section, second set of three test specimens is tested. All test specimens from the second set should conform to all the specified criteria for the relevant classification.

The materials are classified as HB, HB40 or HB75 (HB = horizontal burning) in accordance with the criteria given in table 3.19. It is clear from the above table that, HB is the best rating, followed by HB 40 and then HB 75. According to 2013 version of EN 60695-11-10, HB 75 is likely to be discontinued.

Table 3.19: Horizontal Burning Classification

HB Type	Criteria
HB	The specimen does not burn with a flame after the ignition source is removed. OR The specimen continues to burn with a flame after removal of the ignition source but the flame front does not pass the 100 mm mark; OR If the flame front passes the 100 mm mark: it does not have a linear burning rate exceeding 40 mm/min for a thickness of 3,0 mm to 13,0 mm OR a linear burning rate not exceeding 75 mm/min for a thickness of less than 3,0 mm;
HB 40	Same as HB but with no thickness specification and flame does not have a linear burning rate exceeding 40 mm/min
HB 75	Same as HB but with no thickness specification and flame does not have a linear burning rate exceeding 75 mm/min

Apart from this, foam materials of density less than 250kg/m3 are also subjected to the horizontal burning test. Here the test specimen is 150 x 50 mm marked at 25mm, 60mm and 125mm and is supported horizontally on wire gauze. The end closest to the 25mm mark is exposed to a flame (38 mm, 500W) for 60 seconds. The following are recorded:

– Time taken for the flame front to travel from 25mm mark to the 125 mm mark or when burning ceases.

– After glow time.

– Whether flaming drip ignite cotton wool placed below the specimen.

The material is then classified as follows:

HF-1: The material self-extinguishes in less than 2 seconds and after-glow is less than 30 seconds and cotton wool placed below the sample is not ignited by flaming drips.

HF-2: The material self-extinguishes in less than 2 seconds and after-glow is less than 30 seconds but flaming drip ignites the cotton wool placed below.

HBF: When the burning rate (from 25mm mark to 125mm mark) does not exceed 40mm/minute or specimen ceases to burn before the flaming or afterglow reaches the 125mm mark.

It is therefore clear that HF-1 is best rating for foamed materials followed by HF-2 and HBF.

3.23.2 Vertical Burning Test

The test set-up for vertical burning test is as shown in the Fig. 3.32. Here the sample is clamped vertically on a stand so that its lower end is 300mm above a cotton pad. The samples are temperature and humidity conditioned as per the standard before testing. The flame (20mm, 50 W as per EN 60695-11-4) is applied for 10 seconds to the centre of the lower end of the sample so that the top of the burner is 10mm below the lower end of the sample. The flame is removed after 10s and the following are recorded:

– After-flame time "t1" in seconds. (time for which sample flaming persists after the ignition source has been removed).

– Whether there are any molten drips of the sample.

– Whether these molten drips ignite the cotton pad.

When the flaming of the specimen stops, the flame is applied for second time for 10s and following are noted:

– After-flame time "t2" in seconds.

- After-glow time "t3" (time for which glowing combustion persists after both removal of the flame and flaming of sample) in seconds.

- Whether there are any molten drips of the sample.

- Whether these molten drips ignite the cotton pad.

- Whether the test specimen burned completely up to the clamp.

The vertical burning is then classified as per the table 3.20:

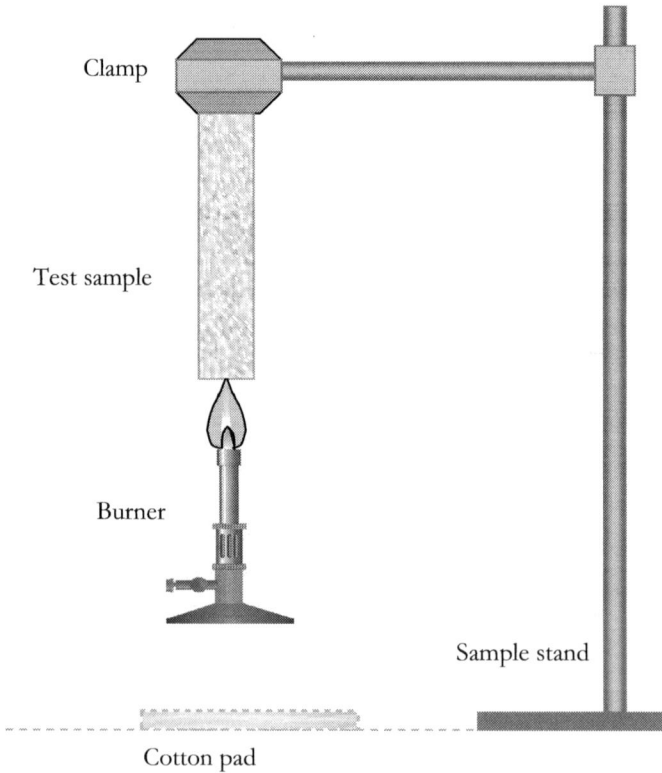

Fig. 3.32: Vertical Burning Test Set-up

Table 3.20: Vertical Burning Classification

Criterion	Classification		
	V-0	V-1	V-2
Individual test specimen after-flame times ($t1$, $t2$)	$\leq$ 10s	$\leq$ 30s	$\leq$ 30s

Individual test specimen after-flame time plus afterglow time after the second flame application ($t2 + t3$)	$\leq$ 30s	$\leq$ 60s	$\leq$ 60s
After-flame and/or after-glow of any specimen burned to the holding clamp	No	No	No
Cotton indicator pad ignited by flaming particles or drops	No	No	Yes

As we can see from the above table, V-0 material is the best followed by V-1 and then V-2. In addition, vertical burning behaviour of materials with density more than 250 kg/m3 is addressed by the standard EN 60695-11-20, using a 500W flame burner and 38mm flame (specified in EN 60695-11-3). This test method requires the use of two test specimens - rectangular bar-shaped test specimens (120mm x 30 mm and preferred thickness of 0,75 mm, 1,5 mm, 3,0 mm, 6,0 mm, and/or 12,0 mm) to assess ignitability and burning behaviour, while square plate test specimens (150mm x 150mm and preferred thickness of 0,75 mm, 1,5 mm, 3,0 mm, 6,0 mm, and/or 12,0 mm) to assess the resistance of the test specimen to burn-through.

During testing, first a bar shaped specimen is taken and the flame is applied for 5s and removed for 5s. This procedure is repeated five times and the following are noted:

− After-flame time t1,

− After-glow time t2,

− Whether any particles or molten drips fall from the bar test specimen. If so, whether they ignite the cotton pad.

If only one test specimen from a set of five bar test specimens does not conform to all of the criteria for a classification, another set of five bars are tested. All the five bar specimens are then categorised as per table 3.21.

Thereafter, the plate shaped sample is subjected to testing wherein the flame is applied to the centre of the specimen in such a way that the inner blue tip of the flame touches the bottom surface of the specimen. The time and repetition are the same as with the bar specimen. Evidence of *burn-through* is `concluded if one of the following occurs:

− visible flame is observed during the test on the opposite surface of the test specimen.

– an opening greater than 3 mm is present in the test specimen after the test, when the test specimen has cooled for at least 30 s.

If only one plate test specimen from a set of three plates does not conform to the burn-through criteria, another set of three plate test specimens are tested. All the three plate specimens are then categorised as per table 3.21.

Table 3.21: Vertical Burning Classification

Criteria	Classification	
	5VA	**5VB**
After flame time plus afterglow time after the fifth flame application ($t1 + t2$) for each individual bar test specimen.	≤60s	≤60s
The cotton pad is ignited by flaming particles or drops from any bar test specimen.	No	No
Specimen burns up to holding clamp	No	No
Burn-through occurs with any of the individual plate test specimens.	No	Yes

It is clear from the table that 5VA classified material is better than 5VB, V-0, V-1, V-2, HB, HB-40 and HB-75 materials, in that order.

Thin film materials are classified as VTM-0, VTM-1 and VTM-3 which are equivalent to V-0, V-2 and V-3. During testing, the materials are wrapped around a metallic cone shaped mandrel as specified by relevant standards.

3.24   FLAMABILITY REQUIREMENTS OF MATERIALS

We will now take a look of some of the common plastic materials and their minimum flammability requirements.

– Fire enclosures for movable equipment not exceeding 18 kg should be of class V-1 while those for movable and stationary equipment exceeding 18 kg should be of material class 5VB.

- Materials for components that fill a fire enclosure (like fuse-holders, switches, pilot lights, connectors and appliance inlets) should be certified as per relevant component standards or should be of material class V-1.

- Plastic materials that comprise the fire enclosure should be located more than 13mm away from arcing parts.

- If the plastic material that comprises the fire enclosure is located at a distance less than 13mm from parts that could attain high temperature sufficient to ignite the material, then they should pass the glow wire test as per EN 60695-2-20. Here a tungsten-chromium wire bent into a U-shape is heated to the specified temperature (usually 650°C) and brought into contact with a sample of the material. If there is an ignition, the flames should extinguish within 30 seconds and the specimen should not be totally consumed.

- Decorative parts, mechanical and electrical enclosures, and parts of such enclosures located outside the fire enclosure should be of class HB 75 if thickness (at thinnest part) is less than 3mm or HB 45 if thickness is greater than or equal to 3mm or else they should be of HBF class.

- Components and parts within a fire enclosure (that do not fall under the exception criteria as prescribed by relevant safety standard) should be certified to relevant component standards or should be of class V-1 or HF-2.

- Air filter assemblies should be of material class V-2/HF-2 with certain exceptions as specified by the relevant safety standard.

- Printed circuit boards should be of material class V-2 (V-1 preferred).

If the manufacturer has used materials as prescribed above, evidence in the form of data sheets or certificates, is normally demanded by the test laboratory doing the safety evaluation for declaring compliance.

3.25   MECHANICAL & OTHER HAZARDS

In the previous chapter we have seen electrical (shock) and fire hazards along with ways and means to comply with them. Apart from electrical (and resultant fire) hazards, safety standards also call for compliance to certain other *non-electrical* hazards that a particular equipment may pose during its normal operation, installation, commissioning, repairs or during movement. In this chapter we will take a look at some other hazards that are addressed by safety standards most notably mechanical, heat, sound, radiation, chemical, material etc. Hazards in each of these categories have the potential to cause injuries or in some extreme cases, even death.

3.26   MECHANICAL HAZARDS

Any electrical equipment should have adequate mechanical strength and its construction should be such that no hazard is created during routine handling of the equipment. Enclosures, handles, knobs, and the like should have adequate strength to withstand rough handling during normal use and pass the relevant tests as described in the standards. Depending on the product and possible hazards, the tests include force tests, impact tests and drop tests.

An equipment, during its normal use or under single fault condition, may give rise to mechanical hazards which include but not limited to:

- Sharp edges which could cause cuts.
- Moving parts that could crush body parts or penetrate the skin.
- Unstable equipment that could topple on a person while in use or while being moved.
- Falling equipment, resulting from breakage of the carrying device (wall mounting bracket or other support part).
- Expelled parts from the equipment.

3.26.1 Cutting Hazard

The edges and corners of enclosures that are easily touched can give rise to cutting hazard. These should be rounded, deburred or smoothened to prevent hazards during normal use. Where sharp edges are needed for functional purposes and access is unavoidable, guarding means should be used to minimise the risk of unintentional contact with such edges. If none of the prevention methods is practical, a clear warning should be reliably affixed (see section 3.35) in a prominent position to warn the user of eminent hazard.

3.26.2 Hazards Due To Moving Parts

Hazardous moving parts should be so arranged or guarded so as to provide adequate protection against personal injury. Protection of the operator is most important, and suitable construction methods should be provided to prevent access to hazardous parts. Permitted methods include locating the moving parts in areas that are not operator-accessible or enclosing the moving parts within an enclosure with mechanical or electrical interlocks that remove the hazard when access is gained. Service personnel should be protected from unintentional contact with hazardous moving parts (and high voltage) during servicing of other parts of the equipment. It must not be possible to touch moving parts with the jointed test finger. In addition, openings preventing the test fingers entry should be tested by a 30 N force from a straight un-jointed version of the test finger.

Where the possibility exists that fingers, jewellery, clothing, hair, and so on may be drawn into the moving parts (like gears or shredder blades) a method must be provided to stop the moving part, which should be placed in a prominent visible position and accessible where the risk is highest.

Equipment that have external moving parts that can be easily touched should be designed in such a way so as to avoid inadvertent touching by providing guards. When it is not possible to make hazardous moving parts directly involved in the process completely inaccessible (like drill or saws or grinders) and where the associated hazard is obvious to the operator, a warning may be considered adequate protection under certain conditions (see section 3.35).

For a hazardous moving or rotating part which will continue to move or rotate through momentum, the removal, opening or withdrawal of a cover, door, etc. should necessitate previous reduction of movement or rotation to a safe level.

3.26.3 Instability Of Equipment

Under conditions of normal use, products should remain *physically stable* to the degree that they cannot present a hazard to operators or service personnel. So as to say, the equipment should not become physically unstable, i.e. overbalance, to a degree that it results in a hazard. Where a means is provided to improve stability when drawers, doors, etc. are opened, it should be permanently in operation when associated with usage by ordinary persons. Where such means is not permanently provided but has to be used during service, warnings should be provided for skilled persons.

The standard EN 60950 gives a good description of the enclosure tests for stability and mechanical hazards. These are discussed in detail in section 3.36.12. The various tilt and tip tests are not applicable for equipment that is intended to be secured to the building before operation, as specified in the installation instructions and to individual units which are designed to be mechanically fixed together on site and are not used individually.

3.26.4 Hazards due to expelled parts

Due to failure or for other reasons, parts might become loose, separated or thrown out from a rotating part. The mechanical enclosure of the equipment should be sufficiently complete to contain or deflect these parts.

3.26.5 Hazards due to explosion

Parts that contain gas at high pressure (like a high-pressure lamp) are potentially exploding parts that can cause injury due to flying debris. The mechanical enclosure of the equipment should be of sufficient strength to contain flying debris in the event of an explosion.

3.26.6 Hazards due to falling parts

Parts should be adequately secured so that should any wire, washer, spring, screw, nut, or similar part fall out of position or come loose, it cannot reduce the distances over the reinforced or supplementary insulation levels specified in the standards. As mentioned, parts fixed in place by screws or nuts with self-locking washers or other means are not liable to come loose, and soldered wires are not considered adequately fixed unless they are held in place near the termination independently of the soldered connection.

Similarly, equipment installed on walls should be fixed with fasteners which should not come loose over time. EN 60950 prescribes that these fasteners should have the capability of bearing four times the weight of the equipment it is expected to support.

3.26.7 Other Considerations

Openings in top, bottom and sides of enclosures should comply with the dimensional requirements of the relevant product or machine standards. The restrictions are necessary to prevent objects from entering the product via top or sides and prevent flaming particles from exiting the bottom of the enclosure in the event of fire. Examples of enclosure opening sizes are as follows:

3.26.7.1 Top and side openings

Top and side openings should not exceed 5 mm in any dimension or should not exceed 1 mm in width regardless of length or top openings should be so constructed that vertical entry of falling object is prevented from reaching bare parts by means of a trap or restriction. Opening on the sides should be provided with louvers to deflect an external vertically falling object or so located that an object entering the enclosure is unlikely to fall on bare parts at hazardous voltages.

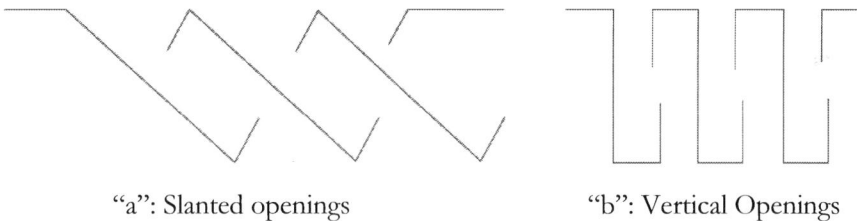

"a": Slanted openings   "b": Vertical Openings

Fig. 3.33: Openings For Preventing Vertical Access

3.26.7.2 Bottom openings

Openings at the bottom of enclosures should be of baffle plate type construction (see Fig 3.34). If this is not possible then metal wire mesh not greater than 2 x 2 mm and a wire diameter of not less than 0.45 mm should be provided. Otherwise holes in metal bottoms should be maximum of 2 mm in diameter while the spacing between holes should be minimum 2 mm. The above dimensions are for metal bottom 1 mm thick. For less thick bottoms. Hole dimensions should be as specified by the standard.

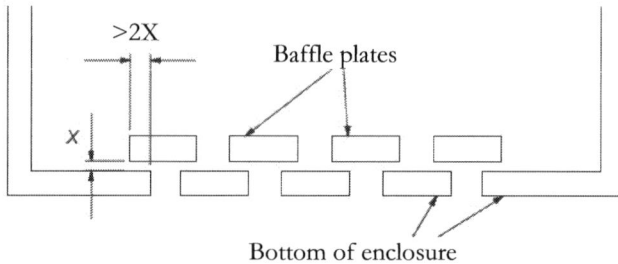

Fig. 3.34: Openings At The Bottom Of Enclosure

3.27   OTHER HAZARDS

Apart from electrical and mechanical hazards, safety standards also address other hazards such as radiation, chemical reaction, gas emission, liquid leakage/spillage, sound pressure etc. that must also be prevented to protect the operator and service personnel. Standards prescribe appropriate methods for limiting, guarding and warning for these hazards. Let us take a look at these one by one.

3.27.1 Radiation Hazards

An electrical equipment should provide adequate measures against radiation including optical, microwave, X-ray, nuclear radiation etc. These are addressed as follows:

  —   For equipment containing ionizing radiation, as per EN 61010, the radiation dose should be measured with a dose-meter at a distance of 50 mm and if it exceeds 5 μSv/h (for non-ionizing the limit is 1μSv/h at 100 mm), the equipment should carry a warning sign for ionizing radiation as given in section 3.35. While taking measurements, the equipment's controls are adjusted to give maximum radiation while maintaining an intelligible picture. Radiation measurements are made after 1 hour.

- If an equipment has a microwave source then the power density of the radiation should not exceed 10 W/m2 (EN 61010) in frequency range of 1 GHz to 100 GHz at a point 50 mm from the equipment under normal and single fault condition.

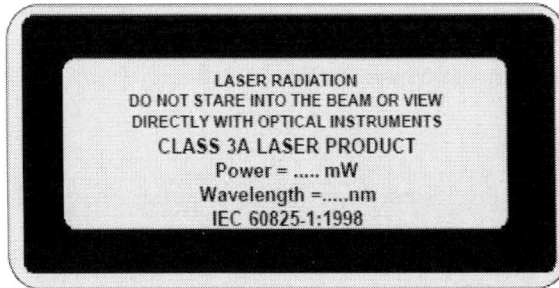

Fig. 3.35: Laser Warning Label

- For equipment incorporating lasers greater than class1, then access without tool is not permitted and the equipment should carry warning label as given in section 3.35 for laser radiation. Class 3A laser products should also bear an explanatory label as per EN 60825-1 as shown in Fig 3.35. The label may also be depicted in the equipment manual.

- For equipment containing UV source, which is not for external illumination, the enclosure should not allow escape of the UV radiation. It is advised that equipment bears appropriate explanatory label, as shown in the Fig 3.36.

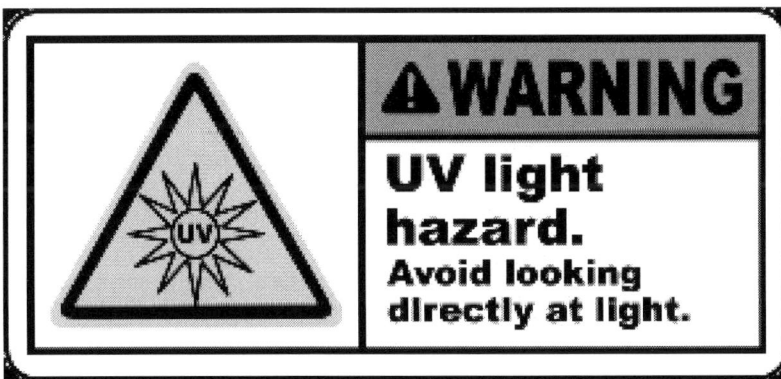

Fig. 3.36: UV Warning Label

3.27.2 Sound Level Hazards

Acoustic sound pressure level including and above 80dBA (A-weighted) is considered hazardous. In such a case the user manual of the equipment should recommend wearing protective earpieces by the operator.

3.27.3 Hazards From Gasses

The equipment enclosure should be such as not to liberate dangerous or poisonous gasses that could be dangerous in normal operating condition. In case there is a possibility of emission of such gasses, the equipment manual should carry appropriate warnings.

3.27.4 Hazards From Liquids

If in normal use, liquid is likely to be spilt into the equipment, the equipment should be designed so that no hazard will occur as a result of the wetting of the insulation or wetting of internal un-insulated parts which are live. Liquid overflowing from containers in the equipment (due to overfilling) or leaking from containers, hoses, couplings, seals, etc., should not cause a hazard during normal use as a result of the wetting of the insulation or wetting of un-insulated live parts. Similarly, equipment that is likely to be moved while a container is full of liquid should be protected against liquid surging out from the container.

3.27.5 Hazards Due To Batteries

- Batteries should be mounted in compartments in such a way that there is no possibility of explosion or fire caused by build-up of flammable gases.
- Batteries should not cause explosion if installed with incorrect polarity or produce a fire hazard as a result of excessive charge or discharge. Suitable protection should be incorporated in the equipment in this regard.
- Appropriate warning on or near the battery compartment or mounting, should be affixed if an explosion or fire hazard could occur through fitting a battery of the wrong type. The equipment manual should also carry appropriate warning instruction.

- If non-rechargeable cells are liable to be fitted, by mistake, in place of rechargeable batteries, there should be a warning marking in or near the compartment. The marking should warn against the charging of non-rechargeable batteries and indicate the type of rechargeable battery that can be used with the recharging circuit.

- Batteries carrying liquid electrolyte should be mounted in such a way that there will not be a safety hazard, should the electrolyte leak.

3.28   COMPONENT APPROVAL

Any item that is used in the composition of, or intended to be built into, end products or machines, is called a component. Products on the other hand, are stand-alone equipment that are comprised of components and are ready to use by an operator. With regard to equipment safety, there are two types of safety sensitive components namely *critical* components and *safety* components.

Table 3.22: Components Requiring Type-Approval Marks

Product and Machine Components	Machinery Specific
Inlets and outlets	Enclosures (for elec. controls)
Plugs and connectors	E-stop switch
Power cords	Service disconnect device
RFI filters, chokes, capacitors	Interlock switches
Circuit breakers	Position sensors
Fuses (user accessible)	Contactors
Fuse holders	Relays
Mains disconnect switch	Circuit breakers
Power supplies	Fuses (internal and external)
External supplies	Function controls/switches
Fans (AC and DC)	Push-buttons and switches
Terminal blocks	Indicator lights and towers
Relays	Heaters and elements
Disk drives (all types)	Solenoid valves
Thermal cut-outs	Alarms (audio and visual)
Current-limiting devices	Robots and controls

Product and Machine Components	Machinery Specific
Voltage select switch	PLCs, drives, and controllers
CRTs, monitors, etc.	Flat panel displays
Transformers	Products like PCs, VDUs, UPS,
Motors	Air conditioners, lighting, etc.
Printed circuit boards	Safety components
Plastics and enclosures	Others
Conductive coatings	
Wire insulation	
Air filters	
Batteries	
Others components >50 Vac/60 V dc	
Components in safety circuit (regardless of voltage, i.e., 12/24 V dc operation) 1	

Critical components are components that may influence the safety of a product, such as those that operate at mains supply (120/230/400 Vac) or hazardous voltages (> 50 V ac or 60 V dc). These components perform functions that protect against electric shock, explosion, mechanical hazards, fire, etc. Examples of critical components are inlets, filters, switches, motors, circuit breakers, opto-couplers, PCBs, power supplies, X capacitors, Y capacitors and transformers. Components that may operate at lower voltages (i.e., 12 or 24 V) and may affect safety are also considered critical components and examples include emergency stop switches, door interlocks, relays, secondary fuses, thermal cut-outs, fans and sensors.

Safety components, as opposed to critical components, fulfil a specific safety function when in use and the failure or malfunctioning of the device places exposed persons in imminent danger. Safety components are regulated by the machinery directive and examples include light curtains, two-hand controls, and sensor mats. The most critical aspect of conformity of the final assembly is that safety and critical components must conform to their EU type-approval standards as given in table 3.22. It is important to distinguish here the difference between type approval and CE marking as people often conclude that critical and safety components that bear the CE marking is equivalent to type approval marking which is not correct. Some components requiring type approval marks are given in the table 3.22.

3.28.1 Approved Vs Components Bearing CE Marking

Under the low-voltage directive, the CE marking is affixed primarily to finished products that (1) are ready to use and (2) can operate with a minimum of connections. Components designed for use within (built-in) products or machines must comply with component safety standards to satisfy the general product safety and product liability directives, but do not require CE marking because they have no autonomous use (i.e. they do not *operate* within the specified voltage range and are not *ready-to-use*) and that their safety depends to a very large extent on how they are integrated into the final product or machine. However, because of the competitive nature of the market, component suppliers often place CE on components for sales and marketing purposes, even when a directive does not specifically allow it. The European commission has disallowed such unqualified CE marking. Manufacturers' CE markings and self-declarations for components, without EU type-approval marks, may make the claim of conformity difficult to prove. Most components do not need CE marking, and if CE marked, it is not an evidence of compliance to a European type approval standard. Products and machines that incorporate safety and critical components that carry only CE marking (using manufacturer's self-declaration) but are not tested to type-approval standards (i.e. lack type-approval markings) are most likely to fail when tested by a test laboratory. Since it is the responsibility of the final equipment manufacturer to ensure that the end product or machine complies with the requirements of CE marking it is always prudent to ensure that the safety and critical components conform with respective type-approval standards.

To ensure safety compliance and limit testing, end-product manufacturers should therefore demand prequalified safety-sensitive components with type-approval marks as shown in Fig. 3.37.

Fig. 3.37: Type Approval Marks

These component marks help to ensure compliance with standards and reduce testing. Components with VDE/TUV marks need only be checked for proper application and use in the product. European-approved components may cost a little more, but they are usually more reliable, and in the long run will save the product designer much aggravation and time. Some of the standards that require components safety compliance are as below:

3.28.1.1 EN 60950

Where safety is involved, components should comply either with the requirements of this standard or with the relevant IEC component standards. Evaluation and testing of components should be carried out as follows:

- A component certified by a recognized testing authority for compliance with a standard harmonized with the relevant IEC component standard should be checked for correct application and use in accordance with its rating.

- A component which is not certified for compliance with the relevant standard as above should be checked for correct application and use in accordance with its specified rating. However, the standard calls for subjecting the component to applicable tests (as mentioned in the standard) while being a part of the equipment. Apart from this, the component should also be separately tested and it should comply to the relevant component standard as well.

- For components for which no relevant IEC standard exists (nowadays we seldom find such components), the standard calls for subjecting the component to applicable tests (as mentioned in the standard) while being a part of the equipment.

3.28.1.2 EN 61010-1

Where safety is involved, components should comply with the applicable safety requirements specified in relevant IEC standards. If components are marked with their operating characteristics, the conditions under which they are used in the equipment should be in accordance with these markings, unless a specific exception is made.

3.28.1.3 EN 60335

Components should comply with safety requirements specified in the relevant IEC standards as far as they reasonably apply.

3.28.1.4 EN 60204-1 —Safety of Machinery, Electrical Equipment of Machines

Electrical components and devices should be suitable for their intended use, e.g., industrial (heavy, light), commercial, leisure, domestic, and should comply with the relevant European standards where such exist. In absence of European standards, compliance should be to available international standards.

Table 3.23: Components Requiring Type-Approval Marks

Component	IEC /EN	UL	CSA
Switches	IEC 61058-1	UL 20 UL917 UL1054	CSA C22.2 No. 55 CSA C22.2 No. 111 CSA C22.2 No. 177
Fuses	IEC 60127-1	UL 248-1, -4, -8, -10, -12, -15	CSA C22.2 Nos. 248, 248.1, 248.4, 248.8, 248.10, 248.12, 248.15
MCBs	IEC 60947		
Attachment Plugs, Receptacles and Connectors	IEC 60083 IEC 60309 IEC 60320	UL 498	CSA C22.2 Nos. 42, 182.1, 182.2, 182.3
Flexible Cords and Cables (PVC and Rubber insulated)	IEC 60227 (PVC) IEC 60245 (Rubber) IEC 60885-1	UL 62	CAN/CSA C22.2 Nos 49, 96
Motors	IEC 60034	UL 1004	CSA C22.2 No. 100
Fans	IEC 60665	UL 507	CSA C22.2 No.113
EMI Suppression Capacitors	IEC 60384-14	UL 1283	CSA C22.2 No. 8
EMI Suppression Varistors	IEC 61051-2		
EMI filters	IEC 60939	UL 1283	CSA 22.2
Power Supplies	IEC 60950	UL 1950	CAN/CSA C22.2 Nos. 234,

Component	IEC /EN	UL	CSA
		UL 1310	950, 223
Transformers	IEC 60742 IEC 60076	UL 1585	CSA C22.2 No. 66
Optical Isolators	IEC 60747-5	UL 1577	CSA C22.2 No. 1
Marking and labelling	IEC 60073	UL 969	CAN/CSA C22.2 No 0.15
Graphical symbols	IEC 60417		
Connectors	IEC 60083, IEC 60309, IEC 60320, IEC 60906-1 or IEC 60906-2		

3.28.2 Component Selection

If a critical or safety component does not have the proper approval mark, the test and certification body may fail the product or require testing of the component. Manufacturers doing a self-assessment of their equipment should use the same criteria as the EU bodies and select only components that they know have been tested, certified, and bear an EU approval mark. Anything less may require additional testing and/or review by a European component expert. Considering the almost universal acceptance of the European standards around the world many suppliers now offer components that are EU type approved, in addition to the U.S. recognition. The U.S. is also moving toward acceptance of the European product standards. Unfortunately, at this time U.S. and EU product and component standards are different. The United States focuses on fire hazards and materials, whereas Europe stresses shock hazards and construction. Because of these and other substantial safety differences, component suppliers realize the need for dual approvals to satisfy both the U.S. and international requirements.

The Fig. 3.37 gives examples of some of the European type-approval marks. The difference between *approval marks* and *self-declarations* (CE, FCC, IEC etc.) should be noted. These self-declaration symbols give the consumer the added assurance they demand and should not be confused with type approval marks of Fig 3.37. As mentioned, there is no such thing as an "IEC" or "CE" approvals and certifications are granted only by accredited testing and certification bodies. The alpha characters IEC and VDE, sometimes found on components and in marketing literature, are a reference to the standard that the component supplier claims to meet and are not independent EU third-party approvals.

Safety and critical components that carry at a minimum, one mark for North America (say UL) and one for Europe (say VDE) should be considered. Many component suppliers now offer components with this dual certification. A *certificate and test report* from the testing and certification body supports the type approval mark. The mark is affixed to the component and is visually recognized by interested parties as positive evidence of compliance.

The table 3.24 gives various evidence of conformity and whether they are acceptable.

Table 3.24: Acceptance of Evidence of Conformity

Rating	Evidence of Conformity	Conformity Status (1) and Actions Required (2)	Verdict
A	EU Type-Approval Mark	Component tested by European accredited testing and certification body and bears the *type-approval mark* as positive evidence on conformity. Test report is on file with EU body are available on request. End-product manufacturer confirms approval by observing *type-approval mark* on component keeps a copy of the approval certificate in technical file/documentation.	Pass
B	Accredited Lab tested	Components are tested by an EU accredited lab but do not bear a type-approval mark. The lab accreditation certificate, test report, and test verification available and copy of test report/certificate placed in the technical file.	Conditional acceptance (case by case)
C	Manufacturer self-declaration	Component supplier claims to meet EU safety standards (i.e., EN, IEC,	Questionable (reject or

Rating	Evidence of Conformity	Conformity Status (1) and Actions Required (2)	Verdict
	(CE marking)	VDE) and offers CE marking in lieu of type approval mark. Since self-declarations and CE are *not* positive evidence, the end-product manufacturer should, in order of preference: a) reject component and select alternative, or (b) test component, if pass document it. or (c) if EU *type-approval* exists on similar components within same series, expert reviews test report and sample component for acceptance /rejection.	test)
D	Evidence not available	No positive evidence available (e.g., no EU approval mark or third-party report). Product manufacturer rejects component, or performs complete testing, according to the relevant standard(s). Testing by EU third-party expert preferred.	Fail

The component manufacturer's declaration or CE marking, if present, should not be used to verify conformity. Lower ratings indicate an increased risk of non-conformity or potential failure when tested. If the component in question falls within ratings C or D the component and test report should be assessed by a European safety expert prior to its acceptance. In the case of testing non-approved components in the end-product, and if successful test results are achieved, the end product manufacturer takes on the conformity responsibility of the component and guarantees its on-going compliance.

For components and products that successfully pass testing, a certificate is issued by the EU body giving the manufacturer the right to affix the approval mark. In addition to the mark on the component a copy of the certificate may be needed to verify the component's ratings and part number and to bring to light any restrictions for use. Some designers and buyers are satisfied if they see the approval mark on the component, but it's a good idea to obtain the approval certificate for review and to place in the technical file. Always request the approval certificate for complex components (such as monitors and power supplies) since restrictions may be written on the certificate or in the installation instructions, which must be checked for suitability in the end application. Some examples of restrictions that may be found on a power supply certificate are:

- For use with external fuse; rating of...
- For use with forced air cooling of...
- For vertical mounting only...
- Isolation system . . . "hazardous energy" outputs.

It's important to request approval certificate to answer all questions. Once someone has a certificate, he/she should also verify that:

- The components part number matches the component number on the certificate;
- The voltage, current, and other ratings are acceptable for the intended use;
- The appropriate standards have been applied; and
- Any restrictions for use.

3.28.3. Critical Components Part List

The critical component parts list, or constructional data form (CDF), is an important tool to aid in component selection, their application criteria, and verification of conformity marks. The table 3.25 shows an example of critical component parts list. It is also extremely useful during the design, purchasing, and manufacturing control of components used in the equipment. The critical component parts list is a listing of all safety-sensitive components with the relevant technical data such as component description, manufacturer's name and part number, rating information, and EU type approval marks found on the component.

Table 3.25: Critical Component Part List

	Certificate No: S98XXXXX	File No: 141-DLAI/E97XXXXX.01	Attachment No: 1
	Critical Component Part List		

Applicant/Manufacturer: XYXCorp.	Rated Voltage: AC 118/230 V, 60/50Hz 1

Type of equipment: Washing machine Rated I/p: 5A

Type or Model No.: RMS-XY (X=0-9, Y=A-2*) Protection Class: Class 1 ~~Class II Class III~~

Built in critical components: (switches, inlets, capacitors, filters, heaters, motors, transformers, power supplies, protection devices, etc.) Complete table below, with EU approval marks) in the right column (CE and IEC not valid). UL/CSA considered for plastics flammability only.

Kind of component	Manufacturer and part number	Information about current and power	Approval mark
Sign Certification Body		Sign Manufacturer	

From the critical component list anyone can see at a glance which are the critical components for safety and most important it identifies the European type-approval mark(s) found on the components (right column of critical component parts list). Once the end product is in conformity with the relevant standards, the critical component parts list can be used by engineering, purchasing, and manufacturing to control the parts and ensure that the company uses only approved replacements or alternates. Alternate or additional components can be easily added during design on the initial critical component parts list or at a later date by simply adding additional pages to the critical component parts list. This product change is known as an *alternate construction*. If the alternate or replacement part is simple and has similar specifications and an EU type-approval mark, the change may involve only paperwork, depending on the component and its application.Substitution of more complex components such as transformers or power supplies may require additional testing in the intended application, again assuming that the component bears an EU approval mark as a minimum. The critical component parts list is especially helpful for factory checks at incoming inspection to verify the components approval marking before accepting them into stock. The critical component parts list can also be referred to during production to verify that manufacturing has assembled the proper components into the product and allows inspectors an easy means of checking for the proper component, number, or approval mark. It is recommended to list the components in order starting from the input (inlet or cord) and working way through the product's circuit diagram.

Components at line voltage (120/230/400 Vac) and at hazardous voltage or energy levels (>50 Vac/60 V dc) must be listed and controlled. Components below these limits are not usually listed unless they have a safety function, such as a door interlock or safety switch operating at any voltage. Components below the safe voltage limits with moving parts or made of hazardous materials, such as DC fans, disk drives, and batteries, should, however, be listed. Do not list any components that are not safety relevant such as I/Cs and most components in SELV circuits. Plastics used in the construction of a product must meet the relevant flammability ratings, such as V-2 for PCBs and V-1 for enclosures. In this case, a plastic that has been tested and rated by a North American agency (UL/CSA) according to the relevant flammability requirements of the standards may be acceptable. Consult an EU testing body for their acceptance and documentation criteria.

3.29   EQUIPMENT CLASSIFICATION

The electric shock classification of equipment determines the extent of insulation needed to protect users and service personnel. Insulation is safer than fusing. As we have seen, insulation is achieved by separating circuits and is required, for example, between user-accessible parts and live parts via insulation layers, thickness, and/or distance (creepage and clearance). As far as safety standards are concerned, equipment are classified according to the operating voltages (including insulation type) or according to mobility.

3.29.1 Classification According To Voltage

There are three classes of equipment namely Class I, II and III, with Class I and II products generally covered by the low voltage directive depending on the type of operating voltage and insulation. Let's us consider these one by one.

3.29.1.1 Class I equipment

This type of equipment utilizes earth as second means of protection in addition to basic insulation. They are typically mains operated (i.e. 115/230/400 Vac) electrical products such as desktop computers, test and measurement devices, machines or stationary appliances. They can be easily identified as having a three-pin plug (for single phase equipment) where the third pin connects the equipment chassis to utility earth. Class I equipment must have their chassis connected to electrical earth by an earth conductor. The failure of equipment's basic insulation may cause a metal chassis to become *live* at full mains voltage. To safeguard against electric shock from metal cased electrical equipment, the metal chassis or case or any exposed metal of equipment (other than double-insulated items) must incorporate a protective earthing (PE) conductor connected to the earth pin of an approved three pin plug incorporating an *earth* terminal. The earth pin is longer than the other two pins so that it is first to make contact when the pin is inserted and the last to break contact when the plug is withdrawn. By connecting to the metal chassis of the equipment, the protective earth wire keeps all this metal at earth potential. What this means is that it is impossible to get an electric shock even when the chassis is connected directly to the live voltage. A fault in the equipment which causes a live conductor to contact the casing will cause a current to flow in the earth conductor. This current should trip either an over current device (fuse or circuit breaker) or a residual current circuit breaker which will cut off the supply of electricity to the equipment.

3.29.1.2 Class II equipment

Class II equipment are ungrounded and rely solely on second insulation layer (or separation) in addition to basic insulation. Examples include TVs, power tools, portable radios, and other handheld or portable appliances. Such equipment are invariably single phase and have a two-pin plug. Class II appliances are also known as double insulated appliances. In Europe, a double insulated appliance must be labelled "Class II", "double insulated" or bear the double insulation symbol consisting of two nested squares (see table 3.26). If one of the basic insulation or the supplementary insulation breaks down, it will not result in an electric shock risk. Protection will be afforded by the other system of insulation. The accessible metal parts will become *live* only in the event of a breakdown of both insulation systems. The probability of this occurring is very remote provided special care is taken when servicing or repairing double insulated electrical appliances to ensure that both insulation barriers remain effective.

Class II electrical products may be any of the following types:

−   Double-insulated electrical products which comprise both basic insulation and supplementary insulation. In other words, there are two layers of insulation between live parts and accessible parts of this type of product. In the case of products with outer casing made from insulating material, the casing will be ranked as one of the required layers of insulation.

−   Reinforced-insulated electrical products which comprise single layer of insulation system to the live parts and provide a degree of protection against electric shock equivalent to double insulation. Electrical products should have durable and substantially continuous enclosure made of insulating material. All metal parts, except small parts such as nameplates, screws and rivets which are separated from the live parts are enclosed by insulation at least equivalent to reinforced insulation. This type of electrical products is called insulation encased Class II products.

−   Electrical products which have substantially continuous metal enclosure in which double insulation is used throughout, except for those reinforced insulated parts where the application of double insulation is obviously impracticable. Such electrical products are called metal-encased Class II products.

3.29.1.3 Class III equipment

These operate on SELV voltages and hence are generally considered electrically safe. Equipment built to the Class III standard is designed to be supplied from a special safety isolating transformer The electrical safety of Class III products is taken care of in the safety isolating transformer design where the separation between the windings of the transformer is equivalent to double insulation. All Class III products are marked with the SELV symbol (see table 3.26 Sr. No. 5156). There is no use of an earth in Class III product. Battery-operated products operating outside the specified range are not covered by the LVD. Products which generate internal high voltages are also excluded from the LVD's scope, provided the high voltages are not accessible via sockets or other accessible parts. The LVD is applicable, however, to battery-operated Class III devices, if they can be operated with a mains-connected power supply or charger (e.g. a laptop computer).

3.29.2 Mobility Classification

The mobility classification of the equipment is another important factor affecting the products testing and construction requirements, such as impact and drop tests, leakage limits, enclosure strength and labelling. The equipment categories are:

1. Handheld equipment : These are the products intended to be held in the user's hand during normal use.

2. Movable equipment : Equipment that is either 18 kg (39.5 pounds) or less in mass and not fixed or that has wheels, castors or other means to allow movement by the operator as required to perform its intended use. Example of such equipment are control rack of C-arm X-ray machine.

3. Stationary equipment: Equipment that is not movable and is designed to operate at a particular location not subject to movement.

4. Fixed equipment: Stationary equipment fastened or secured at a specific location. Such equipment may require grouting to the shop floor or fixing on a wall.

5. Built-in equipment: Equipment that is intended to be installed in a prepared recess, such as a wall. Built-in equipment may not have an enclosure on all sides, as some sides may be protected after installation.

6. Direct plug-in equipment: Products intended to be used without a power supply chord; the mains plug is an integral part of the product's enclosure, and the weight of the product is supported by a socket-outlet (examples are cell phone chargers or adaptors).

3.30   RATING AND MARKING LABELS

Safety standards require every equipment to bear markings and rating labels and in order to properly design these, it is important to know the equipment's actual power requirements. Equipment rating should be decided only after actual measurements as calculating the equipment's input is often unreliable. The equipment's power input under normal operating conditions and with all possible loads applied should be measured at actual. The input is verified by measuring the input current to the product. The product's input power at the desired operating voltage, which is typically 230 Vac (or 400 Vac for three phase) are measured and recorded. The old voltages of 220 to 240 Vac (or 380/415 Vac) may still apply in some countries. It should be noted that 120 Vac, common in the United States, is not generally available in Europe.

For ITE, measurements are taken at the desired frequency (50 Hz) and at $\pm10\%$ for a singular voltage rating (230 Vac or +6 to -10% for a voltage range (220 to 240 Vac). Because input current varies at the extremes of the test range, it may reduce testing if the rated voltage (shown on nameplate) is limited since all testing will be performed at $\pm10\%$ or +6 to -10% of the rating value (see product standard for test ranges). For example, if a computer is rated at 200 to 240 Vac, the safety tests would be performed at +6 to -10% of the rating shown on the label i.e. from 180 to 254 Vac. A printer with a single rating of 230 Vac would require $\pm10\%$ of 230 V for a test range of 207 to 254 Vac.

After taking the input measurements, an input rating for the label slightly higher than the measured value is selected. For example, if measured current is 2.75 A, label marking should be 3.0 A. Do not select an input for the label that is well above the measured value because all product testing would then be performed at that value by adding loads, and so on. The products operating input power should not exceed 110% of the rating shown on the label.

3.30.1 Rating Label

The Figs 3.38, 3.39 and 3.40 show examples of rating labels. All products should be provided with an input rating marking (label or plate) to specify the input voltage, current, and frequency. The label should be durable and legible, since it is subjected to durability tests (see section 3.36.12.6). The label must also indicate the manufacturer's name and type number (model number). It should be located adjacent to the inlet or power entry. In case the manufacturer is based outside the community market, the product should also bear the name, address and contact number of the importer / authorised representative/ distributor /assembler/ installer. Additional information, such as serial number, part number, date codes, and approval marks may be on the same label. The label should be on the exterior of the product and easily recognizable by the user for portable equipment or by the installer for larger equipment. If the rating label is located behind an operator-accessible door (not recommended), for example, within ITE, a visible temporary marking should also be used on the exterior. The words like voltage, amps, hertz or others should not be used when IEC symbols exist. Use proper IEC symbols (as per EN 60417) whenever possible such as, "V" for voltage, "A" for current, and "Hz" for hertz as shown in Fig 3.38.

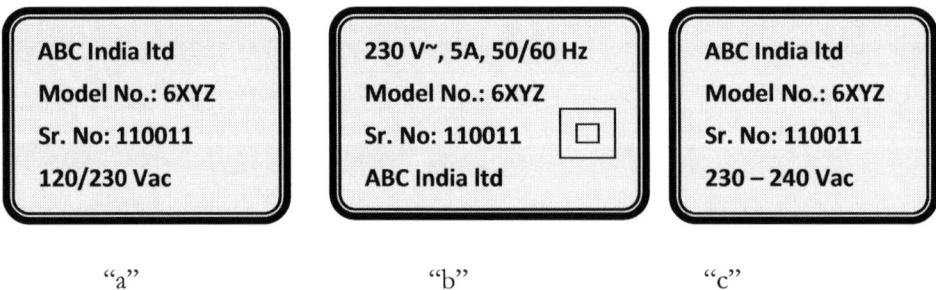

ABC India ltd	230 V~, 5A, 50/60 Hz	ABC India ltd
Model No.: 6XYZ	Model No.: 6XYZ	Model No.: 6XYZ
Sr. No: 110011	Sr. No: 110011 ▫	Sr. No: 110011
120/230 Vac	ABC India ltd	230 – 240 Vac

"a"   "b"   "c"

Fig. 3.38: Rating Labels For Single Phase Product

A dash (-) is used to indicate a range e.g. 220-240 Vac (Fig. 3.38c) and a slash (/) for either or e.g. 120/230 Vac (Fig 3.38a). If the equipment utilizes only two wires for input (line and neutral), as a double insulated product, and there is no ground connection, it is then necessary to mark the product with the Class II symbol

(Fig. 3.38b). Using AC is optional for ITE. The symbol " ~ " may be used instead of AC (Fig 3.40) and "⎓ " for DC(Fig. 3.40).

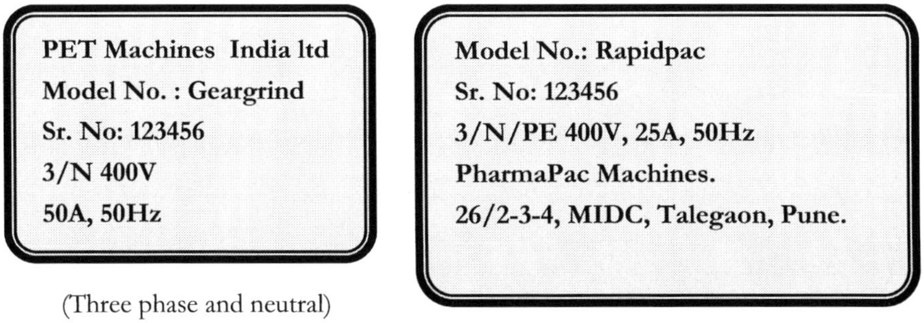

PET Machines India ltd
Model No. : Geargrind
Sr. No: 123456
3/N 400V
50A, 50Hz

(Three phase and neutral)

Model No.: Rapidpac
Sr. No: 123456
3/N/PE 400V, 25A, 50Hz
PharmaPac Machines.
26/2-3-4, MIDC, Talegaon, Pune.

(Three-phase with neutral and Earth)

Fig 3.39: Rating Labels For Three Phase Product

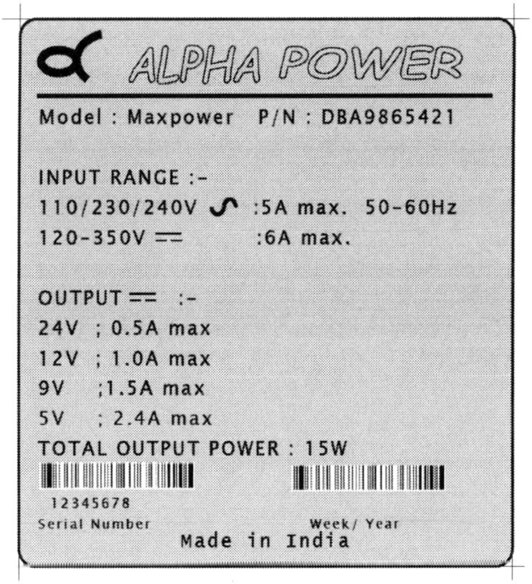

α ALPHA POWER

Model : Maxpower   P/N : DBA9865421

INPUT RANGE :-
110/230/240V ∿ :5A max. 50-60Hz
120-350V ⎓   :6A max.

OUTPUT ⎓ :-
24V ; 0.5A max
12V ; 1.0A max
9V ;1.5A max
5V : 2.4A max
TOTAL OUTPUT POWER : 15W

12345678
Serial Number   Week/ Year
               Made in India

Fig. 3.40: Example of Rating Labels

3.30.2 Markings

Equipment marking, on the outside as well as inside, are needed for proper connection, use and servicing. Safety markings and instructions are required for protection against hazards that remain when other means to eliminate or reduce them are not practical. Where safety markings need to be on the equipment, preference should be given to graphical symbols in accordance with EN 60417. This may be a combination of marking signs and information (explanatory) signs. In the absence of suitable symbols in EN 60417-1 the manufacturer may design specific graphical symbols. Other markings are permitted, provided they do not conflict with required markings or instructions. Where written warnings are given, they should be in a language acceptable to the country where the equipment is intended to be used.

We will now see some of the marking symbols as per in EN 60417-1. The table 3.26 gives some of the frequently used symbols. Manual control devices should be clearly and permanently designated with regard to their functions on or adjacent to the actuator, such as the OFF symbol "O" and ON symbol "I". If only a part of the product is switched off, the standby symbol may be used. Numerous other symbols exist for switch and control indications, such as for push-push or hold-to-run. Fuses, whether internal or external (accessible) should be marked with fuse type and rating adjacent to the fuse or fuse holder or on the fuse holder (see Fig. 3.41). The fuse symbol is also recommended but not mandatory. Only IEC symbols should be used for the fuse ratings and type (F = fast acting, T = time delay, M = medium time delay). A table listing the fuse number and rating may be used where several fuses are found in one location. For connection to 230V mains supply a fuse with breaking capacity (marking 'H') of 1500A is required. Examples of fuse markings are as given in Fig. 3.41. Outlets, (internal or external) also need the voltage and maximum power markings next to the outlet.

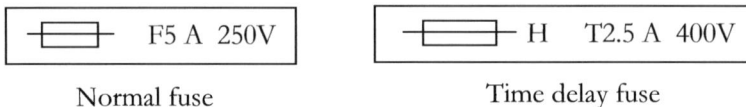

⊶▭⊷ F5 A 250V	⊶▭⊷ H T2.5 A 400V
Normal fuse	Time delay fuse

Fig. 3.41: Fuse markings

Table 3.26: Common Symbols as per EN 60417

Symbol	Sr. No	Description
|	5007	ON (Power) Indicates connection to the mains
○	5008	OFF (Power)

Symbol	Sr. No	Description
		Indicates disconnection from the mains
	5009	Standby To bring equipment in standby condition
	5010	Push On, Push Off Alternatively Each position, "ON" or "OFF", is a stable position.
	5016	Fuse Identifies fuse boxes or their location.
	5017	Identifies Earth Terminal
	5018	Identifies Functional Earth Terminal Earthing to avoid equipment malfunction
	5019	Identifies Protective Earth Terminal For protection against electric shock
	5020	Identifies frame or chassis terminal
	5021	Identifies equipotential terminal
	5031	Direct Current
	5032	Alternating Current
	5104	Identifies start of operation
	5110	Identifies stop of operation
	5156	Identifies connection through a transformer Class III symbol
	5172	Class II Symbol

3.30.3 Marking Requirements And Placement

We will now discuss some of the common requirements of markings and their placement. For additional or specific requirements, the corresponding product specific or product family standards need to be referred. Markings should be permanent, legible and comprehensible (unless specifically permitted to be temporary). They should be easily discernible on the equipment when ready for use (except for marking of internal parts). The marking should preferably be on the exterior of the equipment, excluding the bottom. It is, however, permitted to have it in an area that is easily accessible by hand, for example under a lid, or on the exterior of the bottom of a portable equipment or an equipment with a mass not exceeding 7 kg, provided that the location of the marking is given in the instructions for use. For permanently connected equipment, installation instructions should be provided either as markings on the equipment or in a separate installation instruction document. Markings applying to the equipment as a whole should not be put on parts which can be removed by an ordinary person without the use of a tool. For rack or panel mounted equipment, markings are permitted to be on any surface that becomes visible after removal of the equipment from the rack or panel. Markings which are printed or screened on the equipment should contrast with the background. Markings which are moulded or engraved should have a depth or be raised to a height of minimum 0.5 mm, unless contrasting colours are used. In all cases the meaning of the markings should be explained in the relevant documentation. The marking should be durable in the sense that it should not be possible to erase the marking by rubbing or wiping during normal use of the equipment. Durability is checked by the durability test as per section 3.36.12.6.

3.31   WIRING

Wiring within an equipment is critical to safety and hence wire sizes and insulation must be suitable for their intended use and rating. It should have the proper classification, such as a V-2 flame rating for insulation on internal wiring or <HAR> mark for external fixed power cords. The cross-sectional area of the wires should be adequate for the current they are intended to carry under normal load conditions. All wires used in distribution of primary circuit power should be protected against overcurrent and short circuit by protective devices of suitable rating. Wires should be protected against mechanical damages and it should be ensured that they do not come in contact with moving parts, burrs, sharp points, cutting edges etc. Non-detachable power chords entering through an opening in a metallic enclosure should be through an inlet bushing or chord guard to prevent abrasion. Chord guards or bushing should be made of insulating material, fixed in a reliable manner (requiring a tool to remove) and should extend beyond the inlet opening by at least five times the chord diameter.

Internal wires should be held in place independent (second or double-fixing) of the connection by wire ties or similar methods, to meet single-fault requirement that reduces strain on wire and terminal connections and prevents loosening of terminals. The second fixing should be as close as possible to the initial connection point so that if the wire breaks or comes loose, it will not make contact with metal or live parts. It is assumed that two independent fixings will not come loose. Using proper IEC type terminals as instructed by the terminal manufacturer will reduce any termination problems. Wires are not considered reliably secured to terminals unless there is either an additional fixing provided near the terminal or the terminal has terminators (i.e. ring lugs). For press-on or similar terminators, a double crimp is preferred (e.g. one crimp on the wire and one on the insulation).

Because of the many plug and socket types in Europe, designers often prefer inlets for use with detachable chord-sets instead of fixed cords and plugs. EN 60320 type inlets (Fig 3.42) are available for up to 15 A input, and EN 60309 pin and sleeve types are available through 150 A or higher. The U.S. NEMA twist-lock style inlet does not meet the IEC standards and, therefore, is not allowed for use in operator-accessible areas. Inlets or inlet-filter modules are generally easier to deal with for the equipment manufacturer and the installer than fixed cords. Tabs are typically provided on the internal side of the inlet for solder or press-on connectors. The wire is looped through the hole of the tab, prior to soldering, and tie or sleeve for the second fixing.

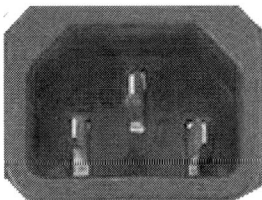

EN 60320 type inlet   EN 60309 pin and sleeve

Fig. 3.42: Power Inlet.

Connection to mains supply should be safe and reliable. The connection to mains can be provided by terminals for permanent connection OR by non-detachable power chord OR by appliance power inlet for detachable power supply chord OR by a mains plug (that is a part of direct plug-in equipment). Rubber insulated power supply chords should comply with EN 60245-1 while PVC insulated ones should comply with EN 60227-1. Appliance inlets for detachable power supply chords should comply with EN 60309 or EN 60320.

Non-detachable power cords are common to machinery and household appliances, but care must be taken to ensure that the cord, strain relief, plug and terminal block are type-approved and that the proper termination methods and makings are utilized. IEC-type terminal blocks (touch safe) are the preferred method for termination of the incoming power wiring. Other methods of termination are possible, such as for connection directly to circuit breakers, mains disconnect switch, PCBs and tabs/ screws on filters. For alternative methods of input terminations, product standards should be referred. Solder connections should not be the only method of fixing and in some cases are not allowed for high current connections.

Non-detachable power chords should be so anchored that the connecting points of the conductors are relieved of strain and outer covering is protected from abrasion. The chord anchor should be of insulating material or should have a lining of insulating material (not applicable for shielded cable). The protective sheath of power supply chord should extend beyond the chord anchor clamp by at least half the diameter of the chord. It is to be ensured that the chord is not clamped by screw fixed directly on the chord or the chord is not fixed by a knot or a string or that when the chord rotates in relation to the equipment, strain is not put on electrical connections.

If protective earthing conductor runs through the cord, it should be anchored in such a way that the PE conductor is the last to take the strain in case the chord slips from its anchorage. Fixed power chords are normally subjected to chord anchorage and pull test wherein the chord is subjected to a steady pull of 30 N (for equipment mass ≤ 1 kg) or 60 N (for equipment mass > 1 and ≤ 4 kg) or 100 N (mass > 4kg) for 1 s and the test being repeated 25 times. The chord should not move more than 2 mm and that creepage and clearances are not reduced below permitted values.

Terminal blocks and stranded wire insertion pose special problems because standards require a loose strand test to determine if a single strand can touch any conductive part upon insertion into a terminal block. Using the proper terminal block and/ or insulation barrier solves this problem. Because of solder cold flow problems, tinning should not be used to consolidate the wire ends for terminal block insertion.

Fixed power cords may require replacement several times over the life of the equipment. For chord replacement, the strain relief must allow for a wide range of chord sizes and stay with the equipment during the replacement. IEC-style strain reliefs are typically plastic (not metal) with ferules/blocks for cord compression. Furthermore, input wiring terminals must be properly marked, such as for protective earth (PE), neutral (N) and line(s) (L), depending on the power distribution system and standard applied as shown in Fig. 3.43.

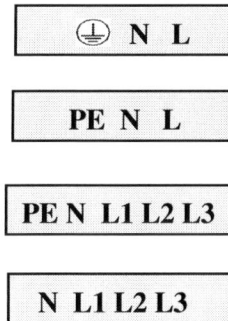

Fig. 3.43: Symbols For Input Wiring.

Proper wire colours and approval markings must be used for fixed power cords. The fixed power cord should have a *<HAR>* mark to ensure it meets the IEC standards for wire size, insulation, and colours (e.g. brown for line, blue for neutral and green-yellow for protective earth [PE]). Green-yellow combination should not be used for purposes other than safety ground. This combination is also not used for SELV, ELV, low-voltage returns or EMI/RFI grounds. Also, light blue signifies the neutral conductor. In general, for some products covered by the low-voltage directive, wires can be any colour with green/ yellow reserved for safety grounds. One should, however, follow the accepted colour conventions throughout.

When dressing the fixed input chord wires, make sure to provide an adequate service loop for the PE wire so it would be the last to break if the strain relief fails to hold when the chord is pulled.

The *HAR* (short for HARmonised) mark is a common mark of conformity according to European EN standards on cables and chords, which is unconditionally recognised by all signatories as equivalent to their own mark. The signatories include about 18 certification bodies in 18 European countries. Considering the stringent rules governing the issuing of HAR licences and the continuous surveillance programme, the HAR scheme provides the manufacturer and the end user with the best possible assurance regarding the quality and compliance of the certified cables and chords. Due to high quality level of the HAR system and its reputation, HAR marked cables and chords enjoy total confidence with European users. The HAR mark is one of the best tools for supporting manufacturers' access to the European market.

Fig. 3.44: The <HAR> Marking.

3.32   POWER DISCONNECT

A hand-operated power disconnect device should be provided to every equipment which should disconnect the whole equipment from main input supply power during servicing. There are several possibilities. A user accessible disconnect switch located on the front of the product is preferred but not the only option. Some of the power disconnect methods are:

   –   Disconnect or isolating switch (ON/OFF) located on the equipment (more so for permanently connected equipment). If not, then suitable instructions should be given in the manual for the provision of switch external to the equipment. Switches located on power chords are not allowed.

- Circuit breaker.

- Plug-on power supply chord (portable or small equipment < 16 A; user warning required).

- Service disconnect switch (lockable for machines).

- Building disconnect device (for permanently connected equipment; warning required).

Mains disconnect devices must meet the following requirements:

- 3 mm minimum contact separation for overvoltage category I, II and III. For non-hazardous DC supply the contact separation should be equal to the clearance distance for basic insulation.

- Connected close to incoming supply, and

- Clearly marked with the symbols "I" for ON and "O" for OFF.

When a plug on the power supply chord is used as a disconnect device, the installation instructions should state that *for permanently connected equipment, a readily accessible disconnect device (switch) should be incorporated external to the equipment* or *for pluggable equipment, the socket-outlet should be installed near the equipment and should be easily accessible.*

For single phase supply a double-pole disconnect device should be used and should disconnect both poles simultaneously. In some products a single-pole device may be allowed to disconnect only the phase conductor when the neutral can be verified, but this is often not possible since plugs are reversible in many EU countries. Parts that remain live after the disconnect device is opened (OFF) should be guarded against accidental contact by service personnel and a warning label is necessary adjacent to the hazard areas. Emergency stop (E-stop) switches, common to machinery and some products, use red mushroom buttons on a yellow background, which is universally recognized by operators and service personnel. E-stop devices have special requirements such as, positive opening, manual reset, fail-safe and should be type-approved as E-stops (see section 3.37.3).

For interconnected equipment, where the interconnections transmit hazardous voltages, a disconnect device should be provided to isolate the parts carrying hazardous voltages during servicing or a suitable warning label indicating accessible hazardous voltage and the removal of all such power before servicing.

3.33   CIRCUIT PROTECTION

Circuit protection devices, such as fuses, breakers or fault-interrupters, may be required in case excessive current is drawn as a result of a short circuit, overcurrent or earth fault. Fuses and circuit breakers are preferred devices for circuit protection. Although several options exist, circuit breakers are normally preferred over fuses. In some of the newer devices several functions are combined into one, thereby reducing the total number of components. An example of this is a combined power switch/ circuit-breaker. The switch/breaker is used as a switch and breaker and senses each line and opens all lines, except the grounded line, simultaneously when a fault current is present. Also, the switch/breaker looks and functions as a standard power ON/OFF rocker switch. In some standards, simultaneous interruption of the phase and neutral is required, thereby, precluding the use of fuses. Fuses must be properly rated for performance in respect to voltage, current characteristics and breaking capacity and they must comply with the European standards. Using fuses and fuse holders that are European approved ensures conformity with the standards. A time-delay fuse, type T, may be used to avoid nuisance tripping, but the delay should not result in a fire or overheat condition during a fault test. Lowest possible value of fuse is selected according to the actual current and not the calculated current.

The rated current of fuses and other protection devices should be selected *as low as possible* based on the actual measured input. In general, for information technology equipment (ITE), user-accessible fuses should be of a European type (IEC; 5 x 20 mm), whereas, internal fuses may be U.S. sized (3 AG; 1/4 x 1-1/4 in) or an IEC type. Other standards, such as those for machinery, require that all fuses, internal or external, be readily available for replacement in the country of use. For Europe this means an IEC type. For smaller fuse types (i.e. 5 x 20 mm) there are fuse holders that can hold either U.S. or IEC fuses in the same fuse holder. For larger fuses it may be difficult to source one fuse holder to meet both the U.S. and IEC fuse sizes. Hence, circuit breakers are the best option in this case.

3.34   THERMAL PROTECTION

Thermal protection devices limit uncontrolled rise in temperature in case of faults and prevent a fire hazard. In addition to fuses or circuit breakers, thermal cut-out and temperature-limiting devices are often required to limit temperature rise during fault tests. Temperature limiters or cut-outs are used where an excess current drawn is not sufficient to open a fuse, thereby causing an overheat condition. Transformers are regarded as temperature sensitive and may be subject to dangerous overheating due to a fault during a secondary short or overload. Motors may also overheat during a locked rotor or running overload test. Transformers and motors are often not protected by the primary fuse and can overheat with a fault in the secondary. Fuses in the secondary or built-in thermal devices or both may be required to limit the hazards as a result of faults. Some standards require the use of temperature limiters in any case. These devices are considered critical to safety. A properly rated device should be specified that is suitable for its application and it should be made sure they are European type-approved to ensure they meet all applicable standards for safety and reliability.

3.35   WARNINGS AND INSTRUCTIONS

In order to avoid hazards when operating, installing, maintaining, transporting or storing equipment, safety information should be provided by the manufacturer in the form of *warning symbols* on the product, along with *instructions* in the product documentation. Maintenance instructions are normally made available only to service personnel. For pluggable equipment intended for user installation, operating and installation instructions should be made available to the user. Further, the instructions should be in a language acceptable to the country in which the equipment is to be installed and used. Warnings are required for all hazards on any enclosure panel or door that does not clearly show that it contains a hazard in order to notify the operator/user. The service persons must also be warned of any possible hazards such as high voltage or energy, moving parts, high temperature or laser radiation before they access a compartment. Additional warning symbols may be required on internal covers and adjacent to hazards within the compartments to protect against accidental or inadvertent contact.

In some cases, a warning may be considered adequate if it is not possible to make hazardous moving parts directly involved in the process completely inaccessible and where the associated hazard is obvious to the operator. In such a case where fingers, jewellery, clothing, etc. can be drawn into the moving parts, a warning should be provided in a visible and prominent position (see table 13.1). An example of where it may not be possible to guard the hazard is the visible moving parts of a paper cutter where hand-feeding is required. Warnings are only permitted when no other means are possible and should not take the place of a safe design. For example, a moving parts warning is allowed only when the hazard is directly involved in the production process and there are no other possible options i.e., guards, interlocks, stop-switch, sensors etc. Therefore, if a guard or other protection means is possible, it must be employed.

3.35.1 Common Warnings And Cautions

Some of the common warnings and caution instructions are discussed in this section and should be a part of instructions or manual supplied along with the equipment. These are given below. Kindly note that warnings in addition to those given below may be required depending upon the safety standard being referred.

CAUTION !! Double-pole/neutral fusing. Disconnect power
        before servicing.
CAUTION !! This unit has two power chords. Remove both chords
        to disconnect power.

For stationary equipment:
WARNING !! High leakage current! Earth connection essential
        before connecting supply.
WARNING !! High touch current! Earth connection essential
        before connecting supply.

For equipment intended for connection to multiple-rated voltages or frequencies, unless means of adjustment is simple:
CAUTION !! See installation instructions before connecting to the supply

If product is provided with a replaceable lithium battery:

CAUTION!! Danger of explosion if battery is incorrectly replaced. Replace only with the same or equivalent type recommended by the manufacturer. Dispose of used batteries according to the manufacturer's instructions,

For pluggable equipment without disconnect device:

CAUTION!! Plug on power cord is disconnect device. The wall's socket outlet should be installed near the product and easily accessible.

For permanently connected equipment without disconnect device:

CAUTION!! A readily accessible disconnect device should be incorporated in the fixed wiring,

For equipment containing hazardous radiation sources:

WARNING !! Turn off UV lamp before opening

The wording of electrical products instructions and products markings related to safety should be in a language that is acceptable in the country in which the equipment is to be installed (ref; EN 60950, EN 61010-1, 73/23/EEC, 89/392/EEC, others). Warning symbols must discourage operator access to compartments containing hazards and must warn service personnel of potential hazards when the hazard is not evident. Safety warnings should be unequivocal by colour, shape, and size and located as close to the hazard as possible. The warning symbols' size and colours are described in the relevant product safety and other supportive standards (ref; EN 60417, EN 60204-1, IEC 1310-1/-2, ISO 3461-1/3864/4196/7000, others). Caution symbols are normally within rectangular box. The preferred method is to use black and yellow warning symbols, without text or signal words (as for as possible), so that the safety warning is unambiguous and negates the need to translate text. A black triangle and pictogram on a yellow background, in accordance with the standards, should be used for warning symbols. The triangle outline and the symbol are black while yellow colour fills the triangle. Other colours may be acceptable depending on the product-specific standard. If signal words and text are used, they should be translated into the appropriate languages. The user instruction or manual must clearly explain the meaning of the various warning symbols that are used on the product or alternatively such information should be marked on the equipment. Some of the more common warning symbols are given in table 3.27:

Table 3.27: Hazard Warning Symbols

	Hazardous voltage
	Hot surface
	Laser Radiation
	Battery Acid
	RF Radiation (Non ionizing)
	Ionizing radiation
	Neutral Fusing
	Stability Hazard
Sharp Edges	Cutting hazard due to sharp edges

	Cutting hazard due to moving objects
	UV light

Except for general warnings, words and text are not required when symbols are present. The warning symbol should be placed as close as possible to the hazard to warn the operator and/or service person before the hazard is accessed.

3.35.2 Documentation/ Manual/Instructions.

Every equipment should be accompanied by a documentation or manual so as to instruct the operator regarding the product operation and safety. Additionally, safety documentation for service personnel, duly authorised by manufacturer must be made available.

Some of the common contents of the manual are as follows:

1) Product description and intended use.

2) Technical specifications like electrical, mechanical etc.

3) Manufacturer profile and his name and address.

4) Equipment ratings including:

 − Supply voltage range, frequency range and power (or current) rating.

 − All input and output connections.

 − Altitude range.

 − Temperature and humidity range.

 − Overvoltage category.

 − Pollution degree.

 − Whether for indoor or outdoor use with IP rating (as per EN 60529).

5) Equipment installation including:

 − Connections to the mains.

- Installation, assembly, location and mounting requirements.
- Instruction for connection to protective earthing.
- Supply wiring requirements.
- Requirements for any external switch or circuit-breaker and external over-current protection devices
- Ventilation requirements.
- Instructions relating to sound level and protection requirement.

6) Equipment operating instructions including:

- Operating controls identification, description and their use in all operating modes.
- Instructions regarding ease of operation of emergency stop / disconnecting device.
- Instructions regarding interconnection of all accessories (if any) and their connection to the equipment.
- In case safety symbols are used, their explanations.
- Instructions for replacement of consumable materials (if any).
- Cleaning and maintenance instructions.
- Risk identification and reduction procedures relating to flammable liquids (if any) and risk of burns from surfaces permitted to exceed the temperature limits.
- A clear warning that if the equipment is used in a manner not specified by the manufacturer, the protection provided by the equipment may be impaired.

7) Equipment maintenance and service instruction including:

- Maintenance, inspection and testing procedure which also ensures safety of the equipment after the maintenance and servicing.
- Repair instruction and verification of the safe state of equipment after repair.
- For equipment using replaceable batteries, the specific battery type should be stated.
- Any parts which are required to be examined or supplied only by the manufacturer.
- The ratings and type of replaceable fuses should be stated.

8) Mitigation of residual risks after risk assessment.

9) An explanation of all the warning symbols used on the equipment.

10) List of potentially poisonous, injurious, hazardous or corrosive substances that are used or that can be liberated from the equipment along with instruction for proper and safe disposal.

The manual can be in electronic form, however the safety instruction should be in hard copy for ready reference. It should be noted that the above contents are general in nature, the reader is advised to consult the standard being referred for additional requirements if any.

3.36   SAFETY TESTING

Once a product meets all of the applicable design, component and construction requirements detailed in the previous sections and the relevant safety standards, the electrical safety testing begins. Type tests are performed on a representative test sample or prototype of the equipment in question. The terms *type* and *model* are interchangeable. When a product complies with the component and construction requirements and successful test results are achieved, then all subsequent samples of a given *type* (e.g., of the same design and identification number) are assumed to comply without testing. The product should be tested at normal operating conditions and the test carried out under the most unfavourable combination of the manufacturer's parameters for supply voltage, frequency and so on and under full load. The number of tests will vary from only a few tests, such as for equipment that uses all EU type-approved critical and safety components, to numerous testing for complex equipment using custom or non-approved components and subassemblies. Using non-approved components, where conformity is not verified, may require considerably more testing on the end-product on the manufacturer's part.

In the sections that follow, we will discuss in detail about the various safety tests prescribed by safety standards. While discussing, an effort has been made to cover the requirements of most safety standards. However, it is always prudent to refer the particular safety standard for specific requirements. Also, for machinery the subsequent sections and the relevant machine safety standards should also be met prior to testing.

3.36.1 Power Consumption.

This test primarily establishes total power (current) consumed by the product for input rating, circuit protection and testing.

3.36.3 Input Test

This test ensures that the equipment draws power as per rating label printed on the equipment or specified in the manual. The equipment is connected to its rated supply voltage while input current is monitored under rated load conditions. The equipment should not draw a current above 110% of rated load.

3.36.4 Voltage Withstand Test.

As discussed earlier, a dielectric barrier is commonly used as one of the means of ensuring separation between hazardous circuits and user accessible circuits or surfaces. The voltage withstand test evaluates the ability of a product's insulation to withstand high voltage between circuits. It is also sometimes referred to as hi-pot, HV test or dielectric strength test. The dielectric strength test is a fundamental method of ensuring that a product is safe before it is placed on the market. Confirming that the proper dielectric barrier (insulation) exists between various areas verifies the existence of a level of protection from electric shock hazards under normal and single fault conditions. It is designed to stress insulation far beyond what it will encounter during normal operation. The assumption is that if the product can withstand higher voltage for a short time, it can function adequately at normal voltage in the long run. Overstressing insulation can also detect possible defects in the design of the insulation barrier, discontinuities like cracks etc. and defects in workmanship like components and conductors spaced too closely. The danger is that air gaps between cracks in insulation, conductors or circuit components may become clogged with dust (both conductive and non-conductive), dirt, and other contaminants over time in typical user environments. If the design spacing is inadequate, a shock hazard can occur after a period of use.

By subjecting the product to a very high voltage, the hi-pot test stresses the product to the point that arcing may occur if the spacing is too close. If the product passes the hi-pot test, it is very unlikely to cause an electrical shock in normal use. A product that can withstand a very high withstand voltage offers a large margin of safety for the consumer. Regulatory agencies usually require a stringent hi-pot test as a product *type test* before releasing the product for sale to the public and another less demanding test to be used on the production line. As a rule, testing laboratories consider the hi-pot test to be the most important safeguard for the consumer. The test is prescribed across circuits that are separated by insulation. For example, EN 60950 calls for testing between primary circuit (carrying hazardous voltages) and body, between primary circuits and secondary circuits and between parts of primary circuits. As we have seen, the test levels depend upon the type of insulation i.e. functional, basic /supplementary and reinforced. It is also required between secondary and body if only functional insulation separates the two. A typical set up for hi-pot test for 2-wire device under test (DUT) is as shown in the Fig. 3.45.

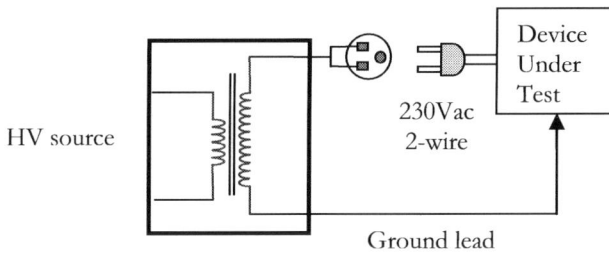

Fig. 3.45: HV Test For 2-Wire DUT

High voltage (HV) generated by a HV source (AC or DC) is applied between live-neutral shorted and DUT body. In case of 2-wire supply, the body of the DUT is normally of non-conductive material. In such a case a metal foil is kept in contact with the insulating body surface. Although care is to be taken that no flashover occurs at the edges of the insulation (especially if the test is done on internal insulating surfaces). The best way is to use adhesive metal foil, with the condition that the adhesive should be conducting. The Fig. 3.46 shows a typical set up for hi-pot test for 3-wire DUT. Here, the HV is applied between live-neutral shorted and earth. The DUT complies if there is no insulation breakdown during the test. Insulation breakdown is considered to have occurred when the current (which flows as a result of the application of the test voltage) rapidly increases in an uncontrolled manner, i.e. insulation does not restrict the flow of the current.

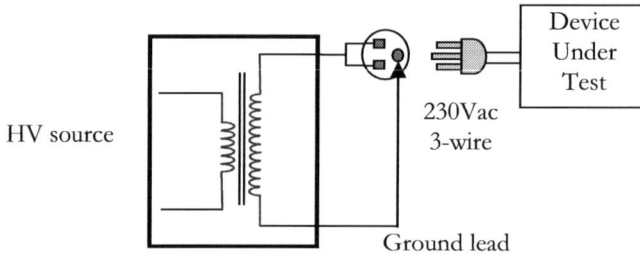

Fig. 3.46: HV Test For Three Wire DUT

A measure of insulation breakdown is arcing or flashover. Flashover is an *electrical breakdown along the surface of solid insulation located in liquid or gas medium.* No arcing or sparking should occur in an insulation stress test. An electrical arc is characterized by very rapid variations in voltage and current. It also produces an audible crackling or zapping sound. Because of these rapid changes, arcing can be detected as soon as it starts to occur by sensing for the presence of high frequency energy. If arcing starts to occur, it is safe to infer that the insulation is about to fail. Nowadays, the available test equipment continuously monitors the current flowing through the DUT (which may be either AC or DC) and checks the magnitude and timing of deviations from normal values. If a high frequency component is found that persists for longer than a specified time (which may be as short as 10 µs) the test equipment interprets this as an arc and immediately sounds an alarm and terminates the test. This indicates that the equipment has failed the test, without actually causing permanent damage to the insulation. Such a test, therefore, can be classed as non-destructive.

3.36.4.1 Choice of DC or AC

The hi-pot test is carried out using either 50 Hz or 60 Hz AC voltage or a DC voltage equivalent to the peak voltage of the AC test voltage. Use of AC or DC depends on the requirements established by the regulatory testing agency. There are advantages and disadvantages of both. The typical rule of thumb used to select an

AC or DC test is if the DUT is powered by AC, then an AC test is used and if it is powered by DC, then a DC test is used. Also, where there are capacitors across the insulation under test (for example, radio-frequency filter capacitors), standards like EN 60950 recommend DC test voltages. Again, if there are components like discharge resistors for filter capacitors, voltage limiting devices or surge suppressors etc which provide a DC path in parallel with the insulation to be tested, then such components should be disconnected.

3.36.4.2 DC hi-pot tests

A typical DC hi-pot test applies a voltage in gradual steps, commonly called ramping, pausing after each increase to allow the capacitance of the DUT to absorb a charge and stabilize. The current increases sharply after each increase in voltage as the capacitance charges, and then decreases to a low steady-state value. The time required for the charging current to decay after each step is called the stabilization time. Current that flows after the stabilization time has passed, represents the leakage current through the insulation. If the voltage steps are too large, the sharp rise in charging current when the step is applied may exceed the high current limit, causing the test to fail prematurely. The magnitude and timing of the steps, therefore, should be carefully matched to the characteristics of the DUT. By monitoring current flow as the applied voltage is increased gradually, waiting for the charging current to decay, and observing the leakage current (if any), a potential insulation breakdown can be detected before it occurs. If the leakage current suddenly starts to increase over the expected value, an insulation breakdown is likely to occur soon. Interrupting the test at this point can save the insulation from breakdown. The test fails but the product is not damaged and may be salvaged by visual inspection or some other means. If the product being tested does not have significant capacitance, there is little or no charging current, and the rate at which the voltage is increased can be much more rapid. Because a DC hi-pot test charges the capacitance of a DUT, the charge itself can present a hazard to testing personnel that must be removed by discharging the DUT to ground after the test is over. Typically, the hi-pot tester automatically discharges the DUT for the same period of time the test voltage was applied.

3.36.4.3 AC hi-pot tests

With an AC hi-pot test, a long ramp time is usually not required (except with certain sensitive devices). AC testing also has the advantages of checking both polarities of voltage and of not needing to discharge the DUT after testing is complete. AC testing, however, has its own limitation and disadvantages. The problem arises because the current an AC test must consider the effects of both real and reactive current. When an AC voltage is applied, the current that flows is equal to the voltage divided by the impedance. The impedance, however, is not purely resistive, it rather is complex since it contains both resistive (real) and capacitive (reactive) components. Now, these two components of AC current are out of phase with each other and the total current is a vector resultant of the two. Since the magnitudes of the two components can be significantly different from each other, the leakage current of a product with large amounts of capacitance can, with some testers, increase significantly without being detected by the test. Hence while doing an AC test, the test equipment must take into account the reactive component before inferring insulation breakdown.

3.36.4.4 Magnitude of the test voltage

The required test voltage for a hi-pot test depends on the type of insulation like basic, supplementary, reinforced or operational and the working voltage across the tested insulation. These have already been discussed in section 3.10.5.

3.36.4.5 Humidity pre-conditioning

As explained earlier, the hi-pot test is used to ascertain the integrity of the insulation and that ingress of moisture into the discontinuities can cause flashover and insulation breakdown. Hence to simulate real life conditions, most of the safety standards like EN 61010 dictate that the DUT should be submitted to humidity pre-conditioning before the voltage tests. For tropical conditions, the pre-conditioning is carried out with the equipment in OFF condition in a humidity chamber containing air with a humidity of 93 ± 3 % RH while the temperature of the air in the chamber is maintained at 40 ± 2 °C. The duration of the pre-conditioning is 120 hours. After removing from the chamber, the equipment is allowed recovery of 2 hours after which the test is carried out within 1 hour.

3.36.5 Insulation Resistance.

This test measures the system's insulation resistance between power circuits and the protective earth circuit. It is done off-line by shorting live and neutral and applying 500-1000V dc between this point and earth. Required IR value is normally specified in $M\Omega$ (e.g. EN 60065 specifies 2 $M\Omega$).

3.36.6 Leakage Current.

This test measures the current that *leaks* onto the equipment enclosure (called enclosure leakage or touch current) and to ground lead (called earth leakage). As shown in the Fig. 3.47, a parallel RC network is introduced between the third pin (or metal enclosure) and earth. The leakage current flowing through this network produces a voltage drop across the 1.5kΩ resistor which is measured and converted to equivalent current. The parameters for the RC network may vary from standard to standard with some standards like the EN 60601 even calling for more than one RC network depending upon where the leakage current is being measured (like patient leads). This test is carried out at rated mains voltage or 110% of rated voltage and under normal conditions and single fault conditions (like neutral open and reversed line and neutral).

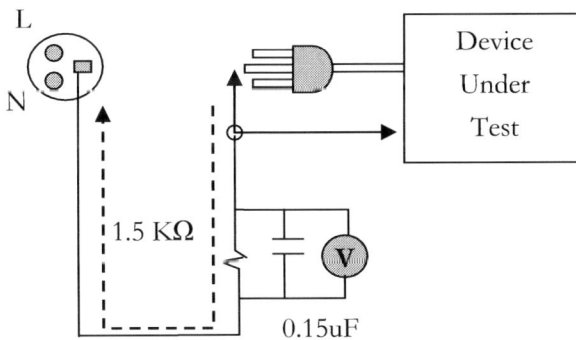

Fig. 3.47: Leakage Current Test Set-up

22222

22222222

2

Safety standards provide certain upper limit for leakage current. For Class I hand held products the limit is normally 0.75 mA, for class I movable and pluggable products it is 3.5 mA, for class III product it is around 0.25 mA. For certain critical products like medical equipment, it may even be lower.

3.36.7 Ground Continuity And Ground Bond Test.

The purpose of a ground continuity test is to verify that all conductive parts of a product that are exposed to user contact are connected to the protective earth (PE). The theory is that if an insulation failure occurs that connects power line voltage to an exposed part and a user then comes into contact with that part, current will flow through the low resistance ground path to the green wire, tripping a circuit breaker or blowing a fuse, rather than flowing through the higher resistance of the user's body. Connecting all exposed conductive parts solidly to PE, safely diverts the current away from the person. The lower the ground resistance, the higher is the fault current and quicker the action of protective devices. Ground continuity tests are normally performed with a low current DC which seeks to ensure that the ground connection has a resistance of less than 1 Ω. Ground continuity testing is not only helpful in determining how well a product will fare during a laboratory investigation, but also is useful in a production line environment to ensure quality and user safety. The purpose of a ground bond test is to protect the user of a product from hazards that could be caused by an inadequate or faulty ground connection. It differs from a ground continuity test in the sense that it tests how much current the ground circuit can safely carry. Ground bond is a high current AC test that measures resistance of the ground path under high current conditions.

For example, a product may pass a ground continuity test with a frayed wire containing only a few strands. The circuit, however, would fail immediately if a high current ground fault should occur - causing an open ground connection. This condition could present a hazard to the user, because part of the product might then have no ground protection at all. If a short occurs between line voltage and an exposed part where no ground exists, users could experience an electrical shock if they touch the part. The ground bond test, therefore, should verify that the ground circuit has a very low resistance and a high current carrying capacity. This ensures that occurrence of a single ground fault on the product will cause the protective circuit breakers or fuses to shut off power to the device automatically.

Ground bond testing is therefore subtly different for ground continuity test in that it requires application of a high current source to a conductive surface of the product and measurement of the voltage drop across the ground connection to determine that bonding is adequate and that the circuit can carry the specified current safely.

Because the resistance to ground is usually a very low value, the resistance of the connecting leads from the tester itself can cause errors in the measurement. Such errors can be corrected either by measuring the resistance of the leads before the test and then subtracting that value from the test value or by using a so-called Kelvin test setup. A Kelvin connection automatically compensates for the lead resistance by bringing an extra lead to the point of measurement. The extra lead is connected so as to balance out the resistance of the test lead. A typical test setup with a Kelvin connection is illustrated in Fig. 3.48. Most standards recommend a ground resistance of 100 mΩ, excluding the power cable.

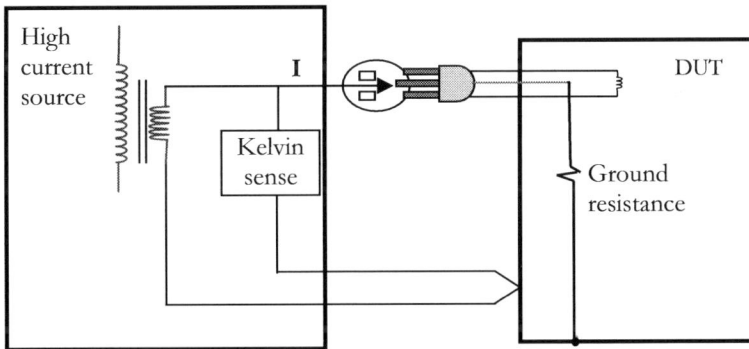

Fig. 3.48: Ground Bond Test Set-up

As shown in the Fig. 3.48, a high current source forces a current of up to 25A through the third pin of the DUT. This current flows through the wire connecting the third pin to the chassis and causes a voltage drop which is measured and converted into resistance value. Safety standards specify the resistance value (upper limit) depending on the gauge of the earth wire. Hence, in order to comply with the test, it is imperative to use earth wire of proper gauge and to ensure that the earth wire is connected to chassis by a low resistance bond.

3.36.8 Capacitor Discharge.

This test is used to determine if the equipment is designed to reduce the risk of electric shock from stored charges in capacitors. Many-a- times high value capacitors are used especially in EMI filters and rectifiers. These retain their charge even after the power supply is disconnected. Thus, there exists a shock hazard, if somebody accidently touches the power supply pins. The capacitor discharge test measures the stored charge (residual voltages) on capacitors after a specified time (1 or 10 s) to determine electric shock potential after power disconnection. If the equipment is pluggable type-A, then the voltage is measured at the power supply pins, one second after disconnection from power supply source. The voltage should reduce to safe value. One may be inclined to think that this test is not required for pluggable type-B equipment, since there is no removal plug. However, if someone happens to touch the bare wire of the power chord (while installing or changing a plug for example) or the pins of the power plug, there may be a potential shock hazard. In case of pluggable type-B equipment, a switch is used to disconnect the equipment from the supply. A storage oscilloscope is then used to monitor the discharge voltage. The voltage should reduce to safe value 10 seconds after disconnection.

In general, laboratories conduct the test using a three-pole switch, which mimics the removal of the plug cap from the mains socket. All three poles are disconnected at approximately the same time in this scenario. Switches that disconnect only the line and neutral, but leave the ground lead connected will not give correct results. The output side of the three-pole switch is connected to the device under test (DUT), and to the 100MΩ, 25pF probe which is connected to a scope channel. The scope is set to read ~±100V/ div to view the entire mains voltage waveform at the maximum voltage generally used with pluggable equipment, 264V rms, which is ±374Vpeak, or 748V from negative peak to positive peak. While the DUT is connected to mains with the three switch contacts closed, the mains waveform is displayed on the scope, and the DUT is placed in the condition which results in the least power drawn, usually a standby mode. Then the three-pole switch is opened. The scope waveform is evaluated to see if the disconnect has happened within ±5% of the peak of the waveform. If not, the setup and switch disconnection are repeated until peak disconnection is seen. Then the waveform is analyzed to see if the voltage is below the standard's pass/fail threshold at the specified time.

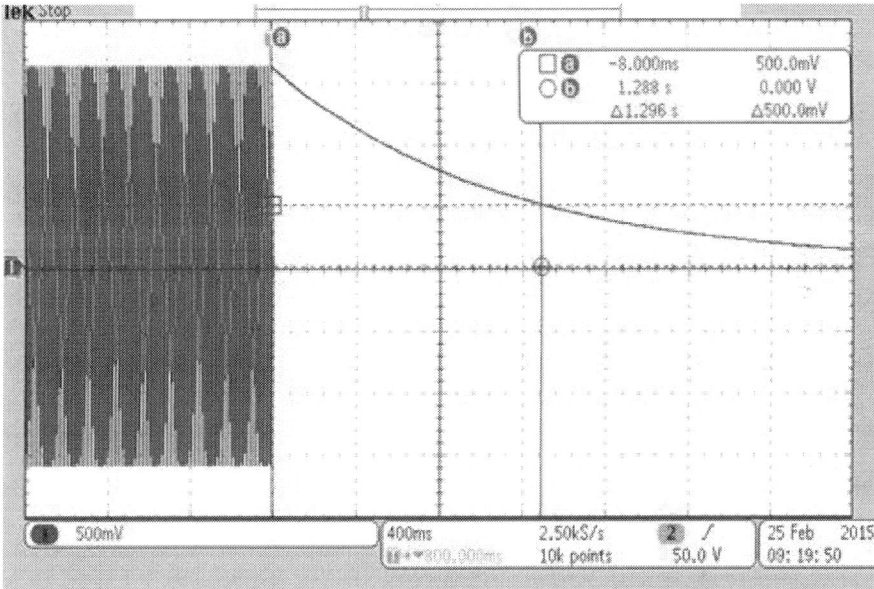

Fig. 3.49: Capacitor Discharge

The following information is presented to show the different pass/fail voltage levels defined per standard:

- EN 61010-1: (For pins) <60V after 5 seconds.
- EN 61010-1: (Terminals for external conductors) <60V after 10 seconds.
- EN 60065: <60V after 2 seconds.
- EN 60335: ≤34V after 1 second.
- EN 60601-1: <60V after 1 second.
- EN 60950: <37% of mains voltage after 1 second (Pluggable Type A).
- EN 60950: <37% of mains voltage after 10 seconds (Pluggable Type B).

Although the pass/fail points of these tests vary, test setups do not contradict each other. EN 60950 requires a 100MΩ input impedance with <25pF capacitance to minimize the influence of the test setup on the results. Other standards do not stipulate input networks. It is therefore clear that, if the equipment is incorporating EMI filters, then the X-capacitors should be provided with suitable discharge resistors in parallel which provides a path for the capacitor to discharge, thus reducing shock hazard.

3.36.9 Impulse Tests.

Safety standards like EN 60950 prescribe two type of impulse tests. The first type simulates lightning transients on telecommunication lines (TNV-1 & 3 circuits) as per ITU-T recommendation k.44. The front time (similar to rise time) of this impulse is 10 µs while time to half value (similar to pulse width) is 700 µs. The second type simulates lightning transients on power lines as per ITU-T recommendation k.44. The front time of this impulse is 1.2 µs while time to half value is 50 µs. The circuit is as shown in the Fig. 3.50. During the tests, there should be no breakdown of insulation. Breakdown is verified in one of two ways, as follows:

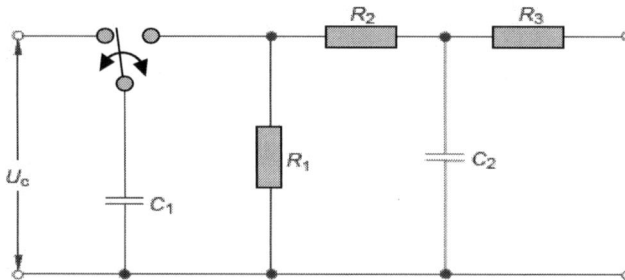

For 10/700µs :-
C1:20µF, C2:0.2µF, R1:50Ω, R2:15Ω, R3:25Ω
For 1.2/50µs:-
C1:1µF, C2:30ηF, R1:76Ω, R2:13Ω, R3:25Ω

Fig. 3.50: Impulse Simulator

— During the application of the impulses, by observation of oscillograms. Surge suppressor operation or breakdown through insulation is judged from the shape of oscillogram. It should be noted that surge suppressor operation during impulse test is permitted in some standards for insulation tests.

OR

−   After application of all the impulses, by an insulation resistance test. Disconnection of surge suppressors is permitted while insulation resistance (IR) is being measured. The test voltage is 500 Vdc or, if surge suppressors are left in place, a dc test voltage that is 10 % less than the surge suppressor operating voltage is used. The IR value should not be less than 2 MΩ.

3.36.10 Temperature Rise (Heating).

This test is used to isolate any components, plastic parts, and touchable surfaces (including handles, knobs, grips etc.) that may attain excessive temperatures or exceed the allowable limits. During the test, the equipment is operated till thermal stabilization. There should be no risk of fire and the temperature rise should be below specified value (see table 3.29). Temperatures higher than these must be accompanied by a warning sign indicating *hot surface*.

Table 3.28: Temperature Rise

Object	Max. temperature °C (Metal)	Max. temperature °C (Plastic)
Handles (Continuous contact)	55	75
Handles (Short period contact)	60	85
External surface (touchable)	70	95
Internal parts	70	95

3.36.11 Ball Pressure Test

Insulating material should have adequate resistance to heat. For this conformity is checked by examination of material data. If the material data is not conclusive, ball pressure test is performed primarily for moulded or plastic enclosures to check warping due to temperature rise. A sample of the insulating material, at least 2,5 mm thick, is subjected to a ball-pressure test (as per EN 60695-10-2) using test apparatus shown in Fig. 3.51.

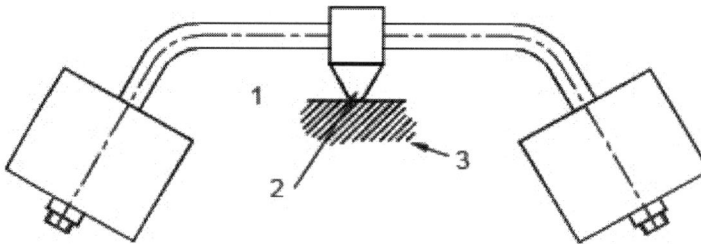

1 Part to be tested

2 Spherical part of the apparatus (diameter 5 mm)

3 Support

Fig. 3.51: Ball-Pressure Test Apparatus

The test is carried out in a heating cabinet at the temperature measured as specified in at 125 ± 2 °C. The part to be tested is supported so that its upper surface is horizontal, and the spherical part of the apparatus is pressed against this surface with a force of 20 N. After 1 h, the apparatus is removed and the sample is cooled within 10 s to approximately room temperature by immersion in cold water. The diameter of the impression caused by the ball should not exceed 2 mm.

3.36.12 Mechanical Tests.

The aim of these tests is that in normal use, equipment must not pose mechanical hazards. The standard EN 60950 gives a good description of the enclosure tests for stability and mechanical hazards and is the basis for this section.

3.36.12.1 Stability test

The following stability tests are performed to assess the stability of an equipment:

- A unit should not overbalance when tilted to an angle of 10° from its normal upright position (the doors and drawers are closed during the test).

- Floor-standing units with a mass of over 25 kg should not tip over when a force of up to 20% of its weight, but not more than 250 N, is applied in any direction except upward, not exceeding 2 m from the floor.

    –   Floor-standing equipment should not overbalance when an 800 N constant downward force is applied at the maximum moment to any horizontal working surface, or surface for obvious foothold, at a height not exceeding 1 m from the floor (the doors and drawers are closed during the test).

3.36.12.2 Steady force tests

To assess mechanical strength of external enclosures, these are subjected to a steady force test of 250 ± 10 N for a period of 5 s. A suitable test tool is used which provides contact over a 30 mm diameter circular plane surface (Fig. 3.52). The force is applied to the top, bottom and sides of the enclosure. Enclosures which are located in operator access area and covered by a door are subjected to a steady force of 30 ± 3 N for a period of 5 s, applied by means of a straight un jointed test finger.

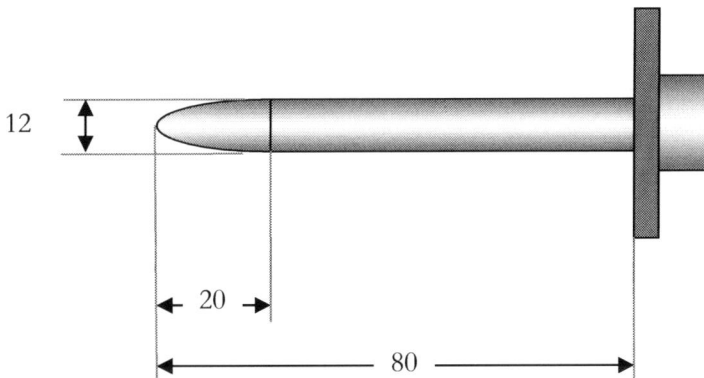

Fig 3.52: Un-jointed Test Finger (IEC 61032) (dimensions in mm)

3.36.12.3 Drop test

Drop tests determine the mechanical strength of products. It is applicable only to those equipment that have a high probability of being dropped during normal use such as hand-held equipment, direct plug-in equipment (equipment designed to be connected directly into mains socket without mains chord, like an adaptor), desktop equipment having a mass of 5 kg or less intended for use with a cord-connected telephone handset and any such equipment that requires lifting by an operator as a part of its normal use. As per EN 60950, the equipment is dropped three times from positions likely to cause the most adverse effects onto a wooden floor (consisting of 13 mm thick hard wood mounted on two layers of plywood each 18 mm thick). Hand-held and direct plug-in equipment is dropped from a height of 1 m, while desktop equipment is dropped from a height of 750 mm. After this test, the product can suffer mechanical damage but it should not be damaged in any way that would cause a hazard. In particular, hazardous parts must not become accessible or clearances should not be reduced below specified values.

3.36.12.4 Impact Test

Equipment other than those mentioned in drop tests, is subjected to the ball impact test (Fig 3.53), wherein a steel ball of diameter 50 mm and weight 500 g is dropped from a height of 1300 mm on to the enclosure of the product.

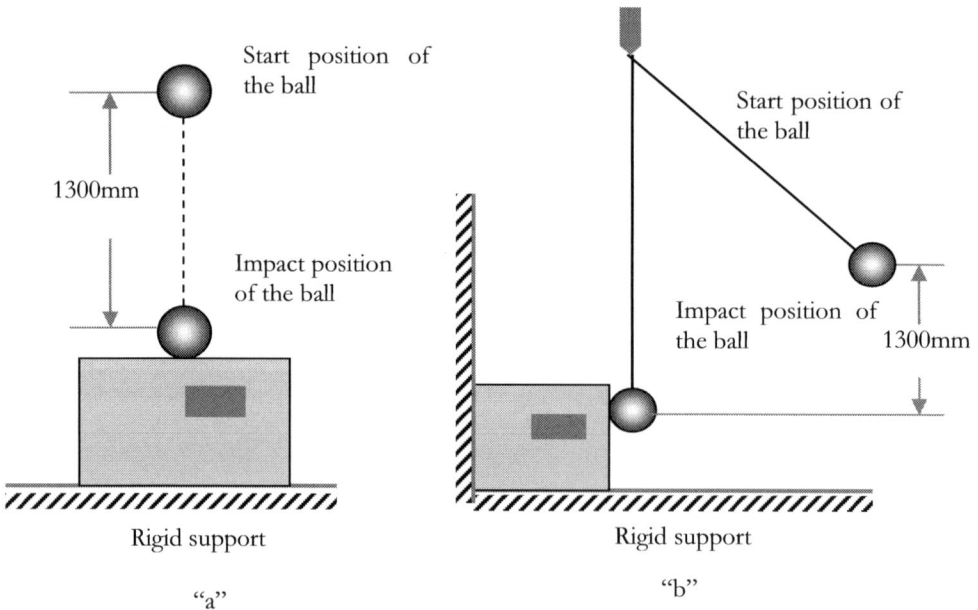

Fig 3.53: Impact test

A horizontal side is subjected to this test either by rotating the product 90° from its normal position. Vertical surfaces are tested by suspending the steel ball from a rope and allowing in to act like a pendulum (Fig 3.53b), falling from a vertical distance of 1300 mm on to the vertical surface of the product. This test is not applicable to CRT, flat panel display, glass platen (of copier machine or scanner). After this test, the product should not be damaged in any way that would cause a hazard. In particular, hazardous parts must not be accessible.

3.36.12.5 Accessibility

The aim of this test is to ascertain that electrical and mechanical hazards must not be accessible to the user. Standards define a test finger and a test pin which are used to probe any openings to see if electrical or mechanical hazards are accessible. The test finger is used with a force of up to 30 N to see whether it can be pushed through openings.

3.36.12.6 Durability of labels

The marking and rating labels (name plate etc.) must be durable and legible. In considering the durability of the marking, the effect of normal use should be taken into account. Different standards call for different tests to assess durability. According to EN 60950, the marking is rubbed by hand for 15 seconds with a piece of cloth soaked with water, and then for a further 15 seconds with a piece of cloth soaked in petroleum spirit. The standard EN 61010 calls for rubbing by a cloth soaked in 70% isopropyl alcohol. After the test, the marking should be legible. There should be no curling and it should not be possible to remove adhesive labels easily.

3.36.13 Abnormal Operation & Fault Conditions.

As we have seen earlier, safety standards rely on the principle of double improbability such that even under a single fault at least one level of protection or insulation is maintained. To ensure this, safety standards call for simulation of single faults wherein various abnormal operation and fault conditions are applied after which the equipment should be rendered safe for the user. It is not required that the equipment should continue to remain ON or maintain normal operation, the only requirement is that it should be rendered safe. The following are some of the commonly referred single fault/ abnormal conditions as applicable:

- Neutral open.

- Reversed live and neutral.

- Short-circuit or disconnection of any components in primary circuits.

- Short-circuit or disconnection of any components where failure could adversely affect supplementary insulation or reinforced insulation.

- Short-circuit, disconnection or overloading (an overload condition is any condition between normal load and maximum current condition up to short-circuit) of all relevant components and parts.

- Faults arising from connection of the most unfavourable load impedance to terminals and connectors that deliver power from the equipment, other than mains power outlets.

- Short-circuits and open circuits of semiconductor devices and capacitors.

- Faults causing continuous dissipation in resistors designed for intermittent dissipation.

- Internal faults in integrated circuits causing excessive dissipation.

- Failure of basic insulation between current-carrying parts of the primary circuit and accessible conductive parts, earthed conductive screens, parts of SELV circuits, parts of limited current circuits.

- Equipment or parts for short-term or intermittent operation operated continuously.

- Motors stopped while fully energized or prevented from starting, whichever is less favourable.

- Interrupting one supply phase of any multi-phase motor while the motor is operating its intended full load.

- Short circuiting of capacitors (except for self-healing capacitors like X-capacitors) in the auxiliary winding circuits of motors.

- Short-circuiting of secondary windings of mains transformers

- Each untapped output winding, and each section of a tapped output winding, which is loaded in normal use, tested in turn, one at a time, to simulate short circuits in the load.

- Each untapped output winding, and each section of a tapped output winding, is overloaded (by connecting a variable resistor across the winding) in turn one at a time.

- Outputs short-circuited one at a time.

- Equipment cooling is restricted by simulating one fault at a time like closing air-holes, stopping forced cooling by motor-driven fans, stopping of cooling by blocking circulation of water or other coolant, simulating loss of cooling liquid.

- In equipment incorporating heating devices, faults are simulated one at a time like overriding of timers which limit the heating period so as to energize the heating continuously, overriding of temperature controllers (except certified ones) so as to energize heating circuit continuously.

- Bridging of insulation between circuits and parts below the level specified for basic insulation to check against the spread of fire.

- Defeating of interlocks one at a time that are in place to prevent access to hazardous parts.

- Voltage selectors which an operator can set for different rated supply voltages set for each possible setting with the equipment connected to each of its rated supply circuits.

And any other single fault condition prescribed by the standard in question.

3.36.13.1 Conformity after application of fault conditions

The product is not required to be in working condition after testing but must remain safe during and after the tests. Conformity with requirements for protection against electric shock after the application of single faults is checked as follows:

a) By making the measurements to check that no accessible conductive parts have become hazardous live.

b) By performing a voltage test on double insulation or reinforced insulation to check that the protection is still at least at the level of basic insulation. The voltage tests are made without humidity preconditioning) with the test voltage for basic insulation.

c) by measuring the temperature of transformer windings if the protection against electrical hazards is achieved by double insulation or reinforced insulation within the transformer. The temperatures of as given in the corresponding standard should not be exceeded.

d) Conformity with requirements for temperature protection is checked by determining the temperature of the outer surface of the enclosure and of parts which can easily be touched.

e) The product should not cause a fire which can propagate beyond the equipment. Conformity with requirements for protection against the spread of fire is checked by placing the equipment on white tissue-paper covering a softwood surface and covering the equipment with cheesecloth. No molten metal, burning insulation, flaming particles, etc. should fall on the surface on which the equipment stands and there should be no charring, glowing, or flaming of the tissue paper or cheesecloth. Melting of insulation material should be ignored if no hazard could arise. The enclosure of the equipment may deform to some extent, but the deformation should not occur in such a way that accessibility and clearance requirements are compromised.

3.36.14 Production Tests

Routine safety tests are tests to which each individual product is subjected during or at the end of the manufacturing process to detect manufacturing variations and unacceptable tolerances in production and materials that could impair safety. These tests are performed to check the insulation between the primary circuits and accessible conductive parts and measure the resistance of the protective earth circuit. There are two production safety tests required for most products:

1. Electric Strength Test
2. Ground Bond Test

The electric strength test (hi-pot) consists of applying to the equipment a sinusoidal AC voltage or an equivalent DC voltage (i.e., 1,500 Vac or 2,121 Vdc) between the primary and ground for 1 to 2 seconds, depending on the product standard. The tolerance of the voltage should be +100 V, -0 V. No insulation breakdown should occur during the test.

The ground bond test is carried out by circulating a test current of not more than 25 A (typical for machines) or 1.5 times the current capacity of the product (i.e., 1.5 times the fuse/breaker rating), for the time required to obtain a meaningful reading through the parts to be tested and the ground (PE) terminal (e.g., between any user-touchable metal parts and the earth ground). The power supply chord should not be included in the measurement. All test results should be kept available with the choice of support and format left to the manufacturer: separate forms or lists of equipment or grouped according to the most suitable parameters are equally acceptable. For all products manufactured and tested, the following data should be filed:

- Date of test.
- Model number.
- Serial number or another identifier permitting unambiguous identification.
- Value of voltage applied during the electric strength test.
- Value of the earthing circuit resistance and corresponding current value.
- Reference information that complete set of tests has/has not been successful.

3.37   SAFETY REQUIREMENTS FOR MACHINERY

This section addresses additional requirements according to EN 60204-1 for the electrical safety of industrial machines so as to comply with the essential requirements of the LVD. The EN 60204-1 is a generic safety standard (type B) used in conjunction with the relevant machine safety standards (type C) which cover machine (mechanical) safety. A machine that complies with EN 60204-1 is presumed to conform to the requirements of the LVD. The requirements presented in the previous sections are applicable to machines and are for the most part contained in EN 60204-1 or other associated standards. We will take a look at some of the key electrical safety items that are often overlooked at the beginning machine of safety evaluation. EN 60204-1 does not cover all machine safety requirements for guarding, interlocks, control etc. that are also applicable and listed in the Machinery directive.

To ensure the machine's conformity with all the electrical safety requirements, a complete assessment according to EN 60204-1 and other applicable standards must be performed by qualified safety persons. Additional requirements may apply depending on the machine's use, environment, and machine safety standards.

The essential factors relating to electrical aspects of machines are to promote the following:

- Safety of persons, property, and machine;

- Consistency of control response; and

- Ease of maintenance.

Safety of the operator and service personnel is most important. High performance should not be obtained at the expense of these essential factors.

3.37.1 Protective Measures

An equipment should employ protective measures to effectively:

1. Protect the operator from shock.

2. Protect the equipment from overload conditions.

The equipment design and overall enclosure should provide protection of persons against electric shock from:

- Direct contact. All measures for protection of personnel that may arise from direct contact with live parts of the electrical equipment.

- Indirect contact. Protection of personnel from hazards that may arise in the event of an insulation failure (single fault) between live parts and exposed conductive parts. Protection against direct contact with live parts is fulfilled in a number of ways, such as enclosures, guards, interlocks, insulation and the use of SELV circuits.

The following measures should be considered to protect the equipment from:

- Overcurrent arising from a short circuit

- Overload currents

- Abnormal temperature

- Loss or reduction in supply voltage

- Over speed of machine elements

The equipment and personnel protection measures and device types, usage, arrangement, and operation of devices are described in EN 60204-1 and other related standards. All fuses used in machines, internal and external, must be of a European IEC type. Circuit breakers may be the best option.

3.37.2 Mains Disconnect Switches

Mains disconnect switches are required on all machines to isolate the entire electrical equipment. The OFF position must be lockable, IEC symbols for ON/OFF positions must be present and meet the relevant EU standards for construction and operation, which is assured if the device is EU type-approved as a mains disconnect or as a combined E-stop/disconnect device (rotary lockable type, red/yellow, etc.). Star-delta, reversing, multi-pole, and U.S. knife-type (pull-down) switches are not acceptable. The switch must be capable of switching the stalled current of the largest motor and at the same time total currents of remaining loads.

3.37.3 Emergency Stop Switches

Emergency stop (E-stop) switches should be provided on every machine to avoid injury to personnel and the surroundings or damage to the machine. The E-stop should stop the dangerous elements of the equipment or stop the entire machine as quickly as practicable. The purpose of the E-stop is to stop the machine in case of danger and disconnect it from the supply voltages. When operating, all loads that may lead to a hazard to personnel or damage to the machine must be disconnected by de-energization so as to de-energize contactors, relays or the under-voltage release of the main disconnect switch. They should have a red button on yellow background, be self-latching, positively opening, readily accessible to operator, manually activated, require manual reset. Resetting the E-stop switch should not restart any part of the machine.

There are several possible E-stops. The palm or *mushroom head* type is the most popular (see table 3.30). The actuator must be readily visible and easily reached by the operator from working and operating positions. Several E-stop devices may be required to cover all the machine's working or operating positions. The mushroom head or other actuator must be red and the background yellow to clearly identify the device as an emergency stop switch. A combined supply disconnect/E-stop device also exists and must meet all the E-stop and colour requirements. The requirements for E-stops are defined in various standards, such as EN 60204-1 and EN 418.

3.37.4 Fault-Tolerant Components And Safety Circuits

Fault-tolerant components and properly designed safety circuits are necessary to ensure safe and reliable machine control. Common terms associated with these special machine components are control reliable, fault tolerant, fail-safe, which means that if a failure occurs (single fault) it will always fail to a safe condition. The machine's circuit design and components must ensure user protection from all safety hazards, such as moving parts or high voltage. The requirements for these highly critical components go beyond the requirements for the typical critical components, both of which must be met for machinery. All components within a safety circuit are of special concern and so as to illustrate some common misconceptions, a few examples of compliant and noncompliant industrial-type components are discussed in the next sections. Fault tolerant components, that have been EU type-approved for proper classification, such as positive opening, guarded actuator, redundancy, cross-monitoring or fault detection are preferred and, in some cases, mandatory. Non approved (CE is not an approval) components, if used, have to be tested to verify their conformity or nonconformity which is a risky alternative and usually costs considerably more time and money.

3.37.5 Transformers

Safety and isolating transformers are frequently used in electrical and electronic products and machines. Transformers are a common cause for non-compliance (e.g., if non-approved). The major reasons for non-compliance are overheating and construction. Tests and construction requirements are detailed in the relevant equipment standards and/or transformer standards EN 60742. Transformers that comply with EN 60742 usually are in compliance with the transformer requirements contained in the equipment standards. The standards have strict construction and testing requirements that are sometimes overlooked by transformer manufacturers or machine builders who source them. For example, the standards require that transformers must be protected against overheating during a short circuit and while under an overload condition. The standards limit the allowed temperature rise in transformers during normal conditions and in the event of an internal or external fault. Some common problems with non-approved transformers could include but not limited to displacement of windings, improper wire terminations, inadequate creepage/ clearance primary-to-secondary, loose parts bridging primary-to-secondary, low-grade or improper class of insulation, thermal limiter/cut-out and secondary fusing missing. Protective devices are recommended, or mandatory in some cases, for effective protection against overloading and short circuits. EN 60204-1 requires the use of a control transformer on machines with more than five electromagnetic coils. The transformer must be installed after the mains switch to ensure isolation. Small transformers are not typically suited for control circuits since they are designed for simple resistive loads. Overcurrent protection is required in accordance with standards EN 60742 and 60204-1. The type and setting of the overcurrent device should also be in accordance with the recommendations of the transformer supplier. The preferred secondary voltages are 24/48/ 115/230 V, 50/60 Hz. Fault-free operation must be verified prior to using lower voltages.

3.37.6 Motors

Motors, like transformers, are another problem facing machine builders. Various standards specify construction and overload protection requirements for motors (EN 60034, 60742, 60664, 60204-1). Motors must meet the requirements for dimensions, construction, insulation class, connection, overload protection, ingress protection, markings, abnormal testing and other parameters.

Effective motor protection is crucial and achieved by using overload protection devices, temperature-sensing devices and current-limiting devices. The protection devices should protect motors against:

- Excessive temperature rise.
- Rapid destruction during start-up or in the locked-rotor condition.

- Unacceptable reduction in service life.

- Nuisance tripping during normal operation.

Overload protection is recommended for all motors, especially coolant pump motors, and should be provided for each motor rated at more than 0.5 kW. Continuously operating motors over 1 kW should be protected against overloads and against a stalled rotor state with built-in thermal sensors typically employed (EN 60034-11). The use of appropriate protection devices for special-duty motors (e.g., rapid traverse, locking, rapid reversal etc.) is recommended. For motors with ratings less than or equal to 2 kW, overload protection may not be required. Fixed terminals for wire termination, within a motor junction box, is the preferred method. If the motor is non approved, as is sometimes the case for larger motors, additional testing and review will be necessary to verify conformity of the motor construction and its protection devices. Motors should meet construction requirements (spacings, materials etc.), employ fixed terminals (no wire-nuts) for wire connections, utilize effective protection devices, and pass the relevant tests.

3.37.7 Wiring

Wiring conductors and cables should run from terminal to terminal without splices (no wire nuts), and terminations of multicore cables should be adequately supported to reduce undue strain on the terminations of the conductors. Internal wiring should be identified by the proper colours and numbers, and identification tags should be legible and permanent. The conductors should be identifiable at each termination and in accordance with the technical documentation. The minimum cross section of single core wire is 1 mm while that for multicore wires is 0.75 mm. The standard provides a table which specifies minimum sizes for wires used for power circuits, control circuits and data communication circuits. Insulation should be flame-retardant V-2 grade minimum. If the cables are identified by colour alone, the colour coding should be as per EN 60204-1 with the following colours specified:

- Neutral: Light blue.
- Safety and protective earth (PE): Green-and-yellow

- AC or DC power circuits: Black

- AC control circuits: Red

- DC control circuits: Blue

- Interlock circuits (external source): Orange

Pure green or pure yellow should not be used since there is a possibility of confusion with bicolour green-and-yellow (PE). All live power conductors, except the earthed neutral, should employ over-current protection selected as low as possible but adequate for the anticipated start-up currents. Control circuits conductors connected directly to supply voltages and circuits feeding control circuit transformers should be protected.

3.37.8 Protective Earth

Protective earth (PE) terminal for external ground (earth) conductor should be marked with the letters PE. The PE designation is restricted to the terminal for bonding of the external protective conductor (green and yellow) of the incoming supply. Other grounds must not be placed on the same ground terminal as the PE conductor. To avoid confusion, light blue is reserved for neutral. Green-and-yellow is for safety and PE ground only and should not be used for low-voltage (< 50V) or EMI/RFI grounds. The wiring *terminals* should be appropriately and plainly marked and correspond with the circuit diagrams.

Terminals should be of IEC type, which means touch-safe. All connections must be double fixed to secure them from accidental loosening, especially the protective earth conductor. Placing two or more wires into one terminal is permitted only in the cases where the terminal is designed and tested for that purpose. Terminal blocks should be so mounted that wiring does not cross over the terminals. Selecting an EU type-approved terminal block is the best option to ensure conformity without further testing. Wire splicing is prohibited.

3.37.9 Access Areas

Operator access area is any area to which, under normal operating conditions, access can be gained without the aid of a tool (e.g., by a person's hand or fingers alone). Opening a hinged door or removing a cover by hand, without a tool, makes the area behind the door/panel an operator access area, and all hazards should be adequately guarded or interlocked to remove the hazards before access. Fixing the door or cover in place with screws is one way of protecting persons, but if door or cover interlock switches are used to reduce a hazard (high voltage, moving parts, etc.), they must be of a fail-safe type and non-overridable by the standard test finger.

The non-safety type that may be activated (ON/OFF) by a person's finger through a push action and/or pulled into a locked 'ON' position is not permitted. In addition, the restoration of the interlock should not initiate machine motion or operation if this could give rise to a hazardous condition.   Machine builders sometimes forget that tops and bottoms may be necessary on the equipment to keep persons and foreign objects out, and to minimize the spread of fire from within.

3.37.10 Enclosures

Selecting an electrical control enclosure that meets all of the relevant requirements is one of the problems facing machine manufacturers. Enclosures must be mounted to facilitate accessibility and maintenance and protected against external influences under normal use conditions. This can be accomplished only if the enclosure meets the EU standards and is used properly in the equipment. EN 60204-1, EN 60529, and other standards are very specific concerning the enclosures' location, construction, component types and placement, grounding and ingress protection (IP). EN 60529 outlines an international classification for sealing effectiveness of enclosures of electrical equipment against the intrusion of foreign bodies (i.e. dust, tools, fingers) and water (moisture) as shown in Table 3.29.

This classification system utilizes the letters "IP" (Ingress Protection) followed by two digits. The first digit refers to the dust rating and the second to the water. For example, in table 3.29, protection against limited ingress of dust (first numeral "5") and splashing water (second numeral "4") is designated as IP 54. An "X" is used for one of the digits if there is only one class of protection.

Table 3.29: Ingress Protection (IP) Levels per IEC/ EN 60529

IP.	First digit: Ingress of solid objects		Second digit: Ingress of liquids
0	No protection		No protection
1		Protected against solid objects over 50mm e.g. hands, large tools.	Protected against vertically falling drops of water or condensation.
2		Protected against solid objects over 12.5mm e.g. fingers, tools.	Protected against falling drops of water, if the enclosure is tilted up to 15 from vertical.
3		Protected against solid objects over 2.5mm e.g. wire, small tools.	Protected against sprays of water from any direction, if the enclosure is tilted up to 60 from vertical.
4		Protected against solid objects over 1.0mm e.g. wires.	Protected against splash water from any direction.
5		Limited protection against dust ingress. (no harmful deposit)	Protected against low pressure water jets from any direction.
6		Totally protected against dust ingress.	Protected against high pressure water jets from any direction
7	N/A		Protected against short periods of immersion in water between 15cm and 1m for 30 minutes.
8	N/A		Protected against long, durable periods of immersion in water.

Example is IP X4, which addresses splashing water resistance only. EN 60529 does not specify sealing effectiveness against risk of explosions, certain types of moisture conditions, e.g., those that are produced by condensation, corrosive vapours, fungus and vermin. IP 54 is the minimum degree of protection for most machine enclosures with a higher or lower degree allowed depending on the installation conditions or environment.

In addition to the IP rating, other construction requirements exist. All openings, including those in the bottom or those for mounting purposes, should be closed in a manner ensuring the specified degree of ingress protection. Components and devices must not be placed on doors except those for operating/ indicating. One should not forget that the door must have a ground wire or strap since the hinges are not considered adequate for grounding purposes. Enclosure doors should have vertical hinges (lift off type) and captive fasteners. In general, if the enclosure bears an EU type approval mark it may be considered to comply with the IP and construction requirements without further testing.

Also, enclosure manufacturers whose product lines are EU type-approved usually offer complimentary products with EU approvals, such as air conditioners, lights, power strips, door interlocks, and grounding kits. Both the dust and water tests must be performed according to EN 60529 with passing results. Enclosures often fail the IP requirements, especially the dust tests. It is best to use a type-approved and marked (VDE/TUV) enclosure to ensure conformity.

3.37.11 Functional Markings

Functional markings should be used for all controls, switches and indicators whenever possible, unless their function is obvious. Proper markings and colours are essential if the control or indicator in question is related to safety. Red, green and yellow are important push-button and indicator colours and strictly controlled by the standards for safe operation of the equipment. A red light is used to indicate an emergency condition and instructs the user to operate the emergency-stop (E-stop), thereby disconnecting power to the equipment. A green light shows a normal or run condition and is commonly used as a power ON indicator. Yellow is used as a warning to indicate an impending hazardous condition may occur and instructs the user to take appropriate action.

Most safety standards will cover specific information required for safety functions and markings for the product in question with reference to other standards for details. Tables 3.30 and 3.31 give push-button type and colour, while Table 3.32

shows indicator light colours and their meanings. Markings for machinery are the best defined, and various standards detail the accepted conventions (e.g., EN 60073 and EN 60204-1 for colours and meanings, IEC 601310-1/-2 and ISO 7000 for actuator movements, hazard symbol types and sizes.

3.37.12 Item Designations

An item designation is a distinctive code that serves to identify an item in a diagram, list, chart and on the equipment. Item designation markings are required within the equipment for machine parts and terminals to ensure ease of identification and maintenance. These markings help service personnel to correlate the parts in the equipment to the different diagrams, parts lists, circuit descriptions and instructions. All items, such as basic parts, components, terminals, sub-assemblies and assemblies should be plainly identified with the same symbols as shown in the technical documentation. The item designations are generally assigned when the circuit diagram is drawn. The designations are located adjacent to the items and on or adjacent to terminals (not on replaceable components). The item designation markings (labels, etc.) should be permanent and visible after the components and wiring are in place. The item letter codes and methods for establishing these designations are recommended in EN 60750 and EN 61346-1/-2.

A basic part is one piece (or several pieces joined together) that cannot normally be disassembled without destroying its function, such as an integrated circuit or resistor. A subassembly is two or more basic parts that form a portion of an assembly, replaceable as a whole, but having a part or parts that are individually replaceable, such as over-current protective device, filter unit, terminal board etc. An assembly is a number of basic parts or subassemblies or any combination thereof joined together to perform a specific function, such as electrical generator, audio amplifier, power supply, or switchgear assembly. Items are defined as a basic part, component, equipment, or functional unit and are usually denoted by graphical symbols on diagrams.

Resistors, relays, generators, amplifiers, power supply units, and switchgear assemblies may all be described as items for designation purposes. Terminal designations are applied to the conducting parts of an apparatus (screw terminal, terminal block, quick-connect tab) provided for electrical connection to the circuits and conductors. Aspect is the specific way of selecting information on or describing a system or an object of a system, such as:

-   What the system or object is doing (function viewpoint);
-   How the system or object is constructed (product viewpoint); or
-   Where the system or object is located (location viewpoint).

The item designations and symbols may be placed in a block format with each block consisting of Latin letters or Arabic numerals or both, uppercase letters being preferred. The blocks may be preceded by a prefix sign. Prefix signs are used to distinguish the various designation blocks and enable the blocks to be combined in any suitable manner:

Prefix Sign	Location	Type/Number/Function	Example
=	Block 1	Higher level (function)	=T2
+	Block 2	Location -	+D126
-	Block 3	Item (aspect)	-K5
:	Block 4	Terminal	:14

3.37.13 Push-Buttons And Actuators

Actuators in the form of push buttons are used in machines to initiate or control various operations. These are available in many shapes and sizes based on their functionality and application. Some common types are as given in table 3.30.

Table 3.30: Type Of Pushbutton Switches

Type	Description	Indicative photo
Flush Pushbutton	Require direct pressure on the operator surface, making it less likely to be accidentally activated.	
Extended Pushbutton	Usually used for most prominent commands on a control panel, as it is easily accessible because of its height.	

Type	Description	Indicative photo
Selector Switch	Used for multiple commands in one control device, or often when maintained control is desired.	
Mushroom Pushbutton	Large push area makes an easy activation; often used for emergency controls.	
Emergency Stop	Demands a more active decision on behalf of the worker to reset a control function.	
Keyed Selector Switch	Used for safety-related issues, i.e., locking down a machine for maintenance.	
Double Pushbutton	Saves space and cost; combines two operators (and an indicator in the illuminated version) into one 22 mm position on a panel.	

3.37.13.1 Push-button colour codes

The colour coding for pushbutton control devices used on machines can be defined according to standard EN 60204-1(table 3.31). There are other regional standards like VDE which may call for slightly different coding more so in case of colours such as white, blue, grey or black. For example, grey and white colours are not allowed for stop or off function.

Table 3.31: Pushbutton Colour Codes

Colour	Meaning	Description	Typical Application
Green	Safe	Actuate in case of safe or normal condition.	Reserved for functions indicating safe or normal condition.
Amber (Yellow)	Abnormal	Actuate in case of unsafe or abnormal condition.	– Intervention to suppress abnormal condition. – Intervention to prevent undesired changes. – Intervention to restart an interrupted automation cycle.
Red	Emergency	Actuate in case of	– Reserved for emergency

Colour	Meaning	Description	Typical Application
		hazardous condition or emergency.	condition where there is immediate danger to human life/property. – Initiation of emergency condition. – For actuating emergency stop (E-stop). Examples are stopping of drives and other units of a machine, halt of an operational cycle etc.
Blue	Mandatory	Actuated for a condition requiring mandatory action	Reset function
White, grey or Black	No specific meaning assigned.	All other operations except safe, abnormal or emergency operations.	– For push-buttons actuators that alternately act as START/ON and STOP/OFF. – STOP/OFF button (main preference being for black colour). – Actuators that cause operation while they are actuated and cease the operation when they are released (e.g. hold-to-run) – Selector switches for multiple commands in one control device. Examples are : Any switching on or starting of one or more motors, starting of machine units, unlocking, auxiliary functions that are not directly related to the operating cycle etc.

Push buttons can also be illuminated and used for indication and acknowledgement. In the former case, the button lights up to prompt the operator to take some action, for example, to press the button. The colour coding is the same as for un-illuminated push buttons. In the latter case, the colour can be white or the button can be colourless and the button lights up after having been pressed to acknowledge that the instruction has been accepted or to indicate the new operating status. The use of flashing light for intermediate acknowledgements is generally acceptable. As far as possible, an illuminated pushbutton should be avoided for emergency stop function and if used, the colour red should be visible even when the light is not illuminated.

3.37.14 Indicator Lights

Indicator lights are also sometimes referred to as tower lights or stack lights. These are normally installed on top of the machine or its control panel so as to be visible from a fair distance. Although there are displays, pilot lights, panel meters etc. on a typical machine, these are useful when the operator is working in front of the control panel. In a typical factory situation, operator may have to move away from the control panel to attend other functions at times. In such a scenario a high visibility stack light provides information about machine status across a greater area. In addition to operator, other plant personnel like maintenance technicians, fork lift drivers supplying raw materials or production supervisors may also need machine status information which can be provided by the stack light.

Indicator lights can be used for indication (with or without alarm) and confirmation. In the former case, the light is used to attract operator's attention or to indicate to him that a certain task needs to be performed while in the latter case it is used to confirm a command or a condition or to confirm the termination of a change. For further distinction or information and especially, to give additional emphasis, flashing lights can be provided to request immediate action or to indicate a discrepancy between the command and actual state or to indicate a change in process (for example flashing during transition). It is recommended that higher frequency flashing lights or display be used for higher priority information (see EN 60073 for recommended flashing rates and pulse/pause ratios). Where flashing lights or displays are used to provide higher priority information, audible warning devices should also be provided.

For indication, the colours RED, YELLOW, BLUE and GREEN are normally used, while for confirmation, blue or white is recommended. Indicator lights should be colour-coded with respect to the condition (status) of the machine in accordance with the table 3.32.

Table 3.32: Type Of Indicator Lights

Colour	Meaning	Description	Action by operator
RED	Emergency	Hazardous condition Like Machine Down, E-Stop Activated, Out of Material, Jam Detected, Servo Fault etc.	Immediate action to deal with hazardous Condition. For example — Switching off the machine supply. — Being alert to the hazardous condition. — Staying clear of the machine.
YELLOW	Abnormal	Abnormal condition or Impending critical condition like low material, temperature out of range, process nearing time out, manual bypass active, cycle interrupted etc.	Monitoring and/or intervention for example by re-establishing the intended function.
BLUE	Mandatory	Indication of a condition that requires action by the operator like machine down, waiting for service call or raw materials. out of work orders to process.	Mandatory action
GREEN	Normal	Normal condition like machine running, process in cycle, part passed inspection, automatic mode active	Optional
WHITE	Neutral	other conditions; may be used whenever doubt exists about the application of red, yellow, green, blue.	Monitoring

3.37.15 Documentation

Machine manufacturers should deliver a machine that is safe to operate and, therefore, the operator's instructions should be in the language of the country the machine is intended for. For machinery, the complete operator's manual, with safety instructions, should be in a language acceptable in the country for its operation. Translation of the installation and maintenance instructions may also be necessary depending on the language of the installer and service personnel. All manuals requiring a country-specific language must be translated by the equipment manufacturer/supplier and made available prior to putting the product or machine into service. The equipment manufacturer or supplier is responsible for translations. Translation of the operator's manual is a must! Translation of service and installation manuals may also be required if requested by the customer. The language and translation requirements are mandatory and the equipment buyer or user may not accept liability for language translations.

3.37.16 Manual

Technical manual necessary for the installation, operation, and maintenance of the machine should be supplied. The information must be provided by the machine supplier to the buyer before delivery of the equipment. For non-European manufacturer the documentation should be supplied by the manufacturer to the importer. Circuit documentation (diagrams, tables, descriptions) serve to explain the function of circuits, power connections, and process-oriented functions. Documentation should include, in addition to that discussed in section 3.35.2:

- Description of machine, installation and mounting, and the connection to the electrical supply.
- Electrical supply requirements.
- Physical environmental information (i.e., lighting, vibration, noise levels etc.)
- Atmospheric contaminants where appropriate.
- Circuit diagram(s).
- System or block diagram(s) where appropriate.
- Information (where appropriate) on programming, sequence of operation and inspection, frequency and method of functional testing, guidance on adjustments and maintenance (especially on the protective devices and circuits), parts list, in particular the spare parts.

- Description of the safeguards, interacting functions and interlocking of guards for hazardous movements and

- Description of safeguarding means and methods where the safeguards are suspended.

3.38   RISK ASSESSMENT

Electrical equipment, during its normal operation and use, will always pose a certain amount of risk. The manufacturer of the electrical equipment should endeavour to reduce the risk to such an extent that the residual risk (i.e. the risk remaining after all possible protective measures have been taken) is tolerable (i.e. which is accepted in a given context based on current values of society). The best way the manufacturer can do this is to get his product evaluated to the relevant electrical safety standard. The standardization committee, while preparing the standard, normally takes into consideration almost all the risks that can be posed by the equipment. There are certain documents (like the CENELEC guide 32) that guide the standard making committee regarding risk assessment. In other words, risk assessment is ingrained in the basic structure of the standard and compliance to the standard means that the residual risk posed by the equipment is tolerable. Thus, compliance to a relevant safety standard will give the *manufacturer presumption of conformity* with the essential requirements of the LVD.

Well, at-least that is what we thought up till now! Then comes the new LVD which makes risk assessment compulsory even if the equipment complies with all the requirements of the relevant product standards. This has left manufacturers confused and often than not there has been a query as to why a separate risk assessment is required when the standards itself are created on the basis of risk assessment and compliance with the standard means that all the possible risks have been addressed and complied with. Not that the risk assessment is new to manufacturers, especially those making measurement, control and laboratory use equipment where the product standard EN 61326-1 has a clause on risk assessment (Clause 17). However, in many a report this clause has simply been addressed (or should we say dismissed?) by such statement as *complied with all previous clauses and the equipment possesses no risk in addition to these clauses.*

While answering the query regarding a separate risk assessment (in addition to compliance to a standard), clarifications from the authorities suggest that even though the standards are made on the basis of risk assessment, a particular product may give rise to certain unforeseeable risks. Again, not all products may present the same risks hence, the new LVD requires that the conformity assessment procedure requires the manufacturer to carry out a risk assessment of the specific risks posed by the product and to address them in order to comply with all the essential health and safety requirements of the directive. Once these risks have been identified and addressed, the manufacturer can choose to apply the relevant harmonized standards. This seems to be in line with the machinery directive, which calls for separate mechanical risk assessment in addition to the conformity with relevant harmonized standard. It should be noted that the risk assessment of a product only complements the product compliance evaluation and does not replace the requirements laid down in the directive and the relevant harmonised standards. Product compliance or non-compliance as per harmonised standards or the directives remains the basis on which authorities decide whether corrective action is needed or not.

3.38.1 Some Definitions

The following are some of the key terms used in the risk assessment based on the ISO/IEC guide 51:

- **Risk**: A combination of the probability of occurrence of a hazard that generates a harm in a given scenario and the severity of that harm.
- **Harm**: Physical injury or damage to the health of persons or damage to property and domestic animals.
- **Hazard**: Potential source of harm. The hazard is intrinsic to the product.
- **Probability of occurrence of that harm**: The likelihood of the harm occurring.
-

3.38.2 Risk Assessment

We will now look at the procedure for risk assessment. This is in line with what is done for machinery directive and is based on CENELEC guidance document (CENELEC guide 32), which is actually a guide for the standardization committee responsible for developing a particular safety standard. The risk assessment procedure needs to be documented and has to be a part of technical documentation in support of CE marking. We will consider a representative product –an LCD projector working on 230V, 50 Hz AC supply. This is a class –I product with a three-pin plug i.e. with external protective earth connection. Its power rating is 300 W with a 200 W bulb rating. The major modules in the projector are the SMPS power unit which takes the 230 Vac input and generates DC voltages (15V,16.5V, 6V and 4V) required for operating other modules like the lamp ballast, image processor, LCD module (for driving the LCD display) and fan driver module for the cooling fan. The Fig. 3.54 shows a typical risk assessment procedure to be used to reduce risks to a tolerable level.

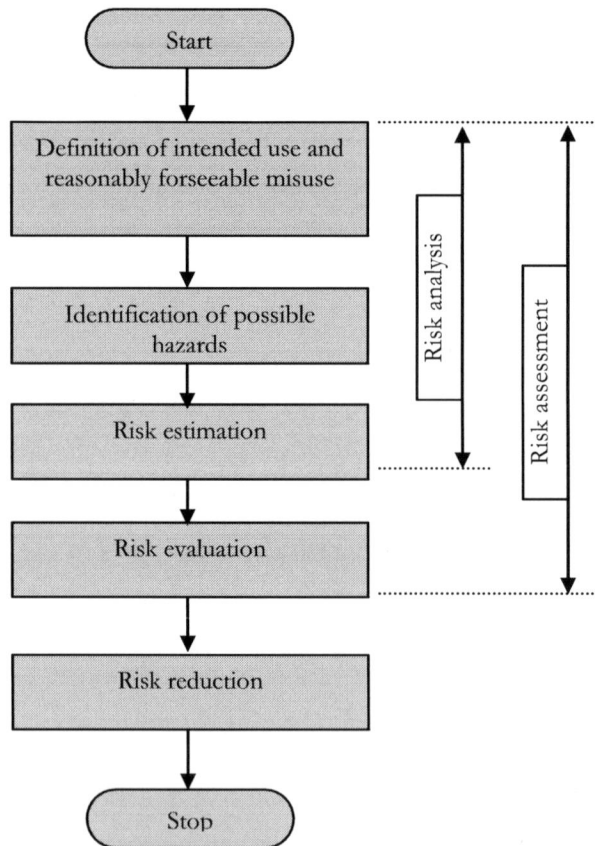

Fig. 3.54: Risk Assessment Flow

The first step is to identify the likely user group or groups for the product (including those with special needs and the elderly), and any known contact group (e.g. use or contact by children) and then to identify the intended use and assess the reasonably foreseeable misuse of the product. After that one needs to identify each hazard (including any hazardous situation and harmful event) arising in all stages and conditions for the use of the product including transportation, installation, maintenance, repair and disposal. This is followed by estimation and evaluation of the risk to each identified user/contact group arising from each hazard identified. The risk is then judged whether it is tolerable (e.g. by comparison with similar products, processes or services). If the risk is not tolerable, it has to be reduced until it becomes tolerable.

3.38.3 Intended Use And Reasonably Forseeable Misuse

Here the manufacturer has to provide details regarding the product, its application and intended use (i.e. use of the equipment as per information for use provided by the manufacturer). The manufacturer needs also to provide the circumstances in which the product can be possibly misused (i.e. used in a way not intended by the designer but which may result from readily predictable human behaviour) which can be reasonably foreseen. The description for the product in question i.e. the LCD projector could be in tabular form as below:

3.37.3.1 Limits of the product and intended use:

Intended use	This product is designed for use in a typical residential/ commercial environment. It is for indoor use only and works on 230 Vac supply from normal utility socket. It can be mounted on a table or fixed permanently on the ceiling (requires special mounting). It can be used for making presentations to a large audience and generally is connected to a PC/Laptop.
Misuse of the product	Mounting the product outdoors and exposing it to environmental conditions beyond specifications that may cause ingress of dust and moisture leading to a hazard. Using the equipment beyond its environmental specifications may give rise to temperature hazard. Keeping the product ON continuously beyond recommended

	time which may give rise to temperature hazard. Not following mounting instructions, when mounted on the ceiling, may give rise to mechanical hazard. EMI hazards can occur in the equipment is used in places with high electromagnetic fields or high static charges.
Use limits	Installation environment: Commercial, indoors only Environmental limitations: Temperature range: 0 °C to 40 °C. Humidity range: 10% to 80% humidity Altitude limitations: Ground to 2700m EMI precautions: Cell phones should not be used within one meter of the periphery. Should not be used in locations prone to static.
Space limits	The product manual provided details about space requirements for the product as regards to table mounting and ceiling mounting. The height of the table or distance below the ceiling and distance from the screen for different aspect ratios is also provided.
Time limits	The equipment should not be used continuously for over twelve hours. During use frequent ON/OFF of lamp should be avoided.

3.38.3.2 Environment of use

Private	NO
Commercial	Yes

3.38.3.3 User groups

Qualified staff	Yes
Non-professional users	Yes
Apprentices/students	Yes
Children	No
Elderly persons	No
People with limited physical abilities	Blind: No Deaf/Mute: Yes (with monitoring) Wheelchair users: Yes (table mount only with

	monitoring).
People with limited mental abilities	No

3.38.3.4 Hazardous materials

Mercury	In lamp

3.38.3.5 Product life-cycles

Transport.	Transport of the product for internal or external relocation.
Installation	Assembly, setting, testing, programming, start-up, all operating modes, switching ON, switching OFF, resetting, fault-finding and troubleshooting.
Operation	Switching ON, setting, programming, wireless setting, LAN settings, switching OFF, resetting, all operating modes.
Maintenance	Cleaning and housekeeping, changing lamp, Fault-finding and troubleshooting (operator intervention).
Disposal	By authorized dealers/distributors, authorized collectors of e-waste.

3.38.4 Identification Of Possible Hazards

There may be a considerable number of hazards arising from any given product. The correct identification and estimation of the relevance of these hazards is very much dependent upon the essential requirements of the directive. The hazards have to be identified at every stage of the product lifecycle.

	Hazards				
Life cycle	**Mechanical**	**Electrical**	**Thermal**	**Radiation**	**Materials**
Transport.	Yes (Falling)	No	No	No	Yes (Mercury)
Installation	Yes (Falling) Explosion (of lamp)	Yes Shock Leakage current	Yes High temp. Fire	Yes High Optical radiation	Yes (Mercury)
Operation	Yes (Falling) Explosion (of	Yes Shock	Yes High temp.	Yes High	Yes (Mercury)

	lamp)	Leakage current	Fire	Optical radiation	
Maintenance	Yes (Falling) Explosion (of lamp)	Yes Shock Leakage current	Yes High temp. Fire	Yes High Optical radiation	Yes (Mercury)
Disposal	Yes (Falling) Explosion (of lamp)	No	No	No	Yes (Mercury) Environme ntal contaminati on

Other hazards like noise, vibration, ergonomics may be applicable for other equipment types.

3.38.5 Risk Estimation

After having identified the possible hazards in various equipment lifecycles, the risk posed by the hazards can be estimated on the basis of the severity of the harm and the probability of the harm. The following table can serve as guidance for the severity of the harm. Although the severity has been classified as severe, moderate and minor, other intermediate severities may also be applicable for other equipment types.

Severity group	People	Equipment / Facility	Environment
Severe	Disabling injury/illness.	Major subsystem loss or facility damage.	Chemical release with temporary environmental or public health impact.
Moderate	Medical treatment or restricted work activity.	Minor subsystem loss or facility damage.	Chemical release triggering external reporting requirements.
Minor	First aid only.	Non-serious equipment or facility damage.	Chemical release requiring only routine clean-up without reporting.

While the probability of the harm can be summarized in the following table. Although the probability has been classified as likely, possible and rare, other intermediate probabilities may also be applicable for other equipment types.

Likelihood	Likelihood expected rate of occurrence
Likely	More than once per year, but not more than five times a year.
Possible	More than once in five years, but not more than one a year.
Rare	More than once in ten years, but not more than one in five years.

3.38.5 Risk Evaluation

It involves combining the severity of the harm and the probability of the harm. Each of the hazards as identified in previous section can be evaluated as per the following table:

Hazard	Risk Evaluation
Shock.	During routine operation and maintenance of our representative equipment, the likelihood of this hazard can be categorized as "possible" meaning there can be no more than one instance of shock per year. While the severity of the harm posed by this hazard is "moderate" meaning that the person could require medical treatment or activity may be hampered temporarily. There could be minor subsystem loss but no chemical hazard.
Leakage Current.	The risk posed by this hazard is the same as that posed by the shock hazard.
Stored Charges.	The posed by this hazard is the same as that posed by the shock hazard.
Thermal.	Since the equipment uses a high wattage lamp. Heat is generated during operation and the surface may get hot and the ventilation opening can be a source of hot air. The user may touch the hot surface or accidently put his hand over ventilation vent. Therefore, the severity of the harm posed by this hazard can be categorised as "minor". The likelihood of this harm could be can be categorized as "likely" more than once per year.
Burn.	The risk posed by this hazard is the same as that posed by the thermal hazard.
Radiation (Optical).	Since our representative equipment uses a high lumen projector lamp and that the user may, accidently or otherwise, look directly in the lamp from close proximity, the severity of the harm posed by this hazard can be categorised as "severe" that may result in blindness. The likelihood of this harm can be categorized as "likely" could be more than once per year.

Electromagnetic Interference (EMI)	Since our representative equipment uses SMPS and digital circuits, EMI hazard may result. The likelihood of this hazard can be categorized as "likely". However, the severity of the harm posed by this hazard can be categorised as "minor" that may result in blindness. The likelihood of this harm could be once per year.
Explosion.	Since our representative equipment uses a projector lamp containing mercury vapour under high pressure, there is a "rare" possibility of explosion. This may result in severity of hazard that can be categorized a "minor" causing minor subsystem loss or facility damage.
Fire	The risk posed by this hazard is the same as that posed by the explosion hazard.
Chemical	Since our representative equipment uses a projector lamp containing mercury, a "rare" explosion may result in emission mercury vapour. This chemical release can be categorised as "minor" requiring only routine clean up without reporting.
Mechanical	When installed on a table, mechanical hazard can result in the event of the equipment falling off the table. The likelihood of the harm in this case can be categorized as "likely". The severity of the harm can be categorized as "minor".
	When installed on the ceiling, mechanical hazard can result in the event of the equipment falling down. The likelihood of the harm can be categorized as "rare". However, the severity of the harm, in this case, can be categorized as "moderate".
	If the lamp explodes from the shock due to the fall, the damage to the facility may be "minor". The likelihood of the harm can be categorized as "rare". This resulting chemical release may require routine clean up without reporting.

The following table summarises the risk evaluation:

Hazard	Probability / Severity	Likely	Possible	Rare
Shock,	Severe			
Leakage Current,	Moderate		✔	
Stored Charges.	Minor			
Thermal, Burn	Severe			
	Moderate			
	Minor	✔		
Radiation (Optical)	Severe	✔		
	Moderate			

Hazard	Probability \ Severity	Likely	Possible	Rare
	Minor			
EMI	Severe			
	Moderate			
	Minor	✓		
Explosion, Fire	Severe			
	Moderate			
	Minor			✓
Chemical	Severe			
	Moderate			
	Minor			✓
Mechanical (Falling from table)	Severe			
	Moderate			
	Minor	✓		
Mechanical (Falling from ceiling)	Severe			
	Moderate			✓
	Minor			
Mechanical (Explosion due to fall)	Severe			
	Moderate			
	Minor			✓

3.38.6. Risk Reduction

After having identified the hazards and subsequent risk estimation and evaluation, the final step towards designing a safe product is risk reduction to tolerable levels. This can be done in three steps as shown in the Fig. 3.55.

3.38.6.1 Risk reduction by inherent design measures.

This is the risk reduction purely based on good engineering practices without resorting to additional components or protective measures per se. Measures include, but not restricted to separation of circuits carrying hazardous voltage, proper ingress protection measures, certified (approved) components, selection of proper and low flammability materials, selection of suitable technology, using SELV, geometrical design of enclosures to avoid injury, limitation of actuating forces, using ergonomic principles to reduce stress and many others.

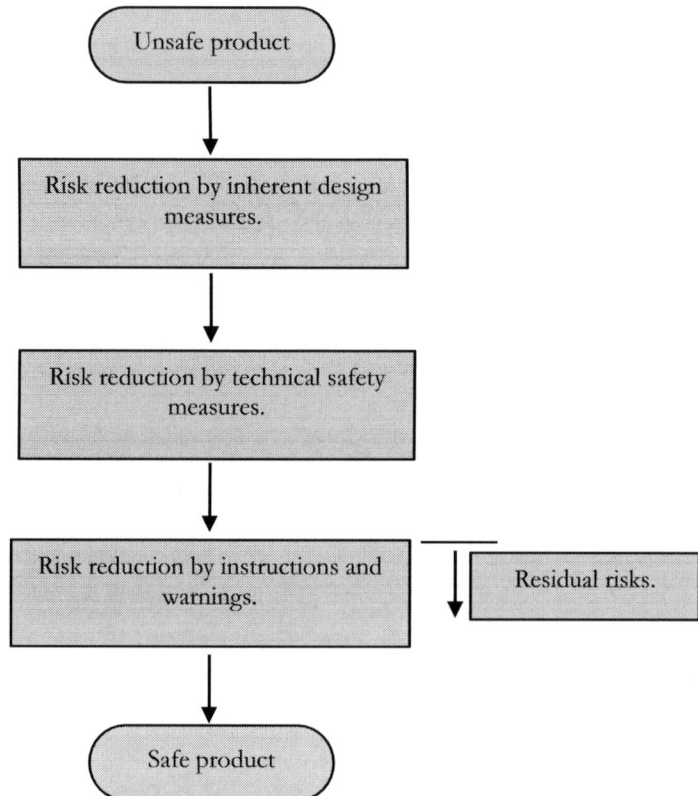

Fig. 3.55: Risk Reduction

3.38.6.2 Risk reduction by technical safety measures.

This is achieved by the application and use of additional safety components (that have no circuit function) which adequately reduce risk for the intended use and are appropriate for the application such as fuses and electrical cut-outs, thermal cut-outs, limit switches to limit movement, proximity switches, warning lamps, guards, fire enclosures, access control devices etc.

3.38.6.3. Risk reduction by instructions and warnings.

The risks that cannot be reduced even after implementing all possible design and technical measures are called residual risks. For example, for our LCD projector we cannot reduce the light intensity below a certain level since the device will become useless or for a drilling machine or a grinding machine, one cannot possibly cover the entire drilling head or grinding tool. These risks can be reduced by providing adequate information the user through appropriate instructions in the product literature and by using relevant warning signs. Instruction to the user should clearly indicate residual risks and the measures to be taken by the user to take care of these risks, about shortcomings of the protective devices, whether specialized training is required and specify appropriate personnel protective equipment.

Warning signs can be effectively used to warn the user about eminent danger. These can be located on the equipment and placed strategically so as to be easily seen. The signs can also accompany safety instructions in equipment literature. At the end of the risk reduction procedure, it should be checked that all operating conditions and all intervention procedures have been taken into account. The protective devices or the safety measures are compatible to each other and themselves do not generate new hazards and that the user's working conditions and the usability of the equipment are not jeopardized by the protective measures taken. The users are sufficiently informed and warned about the residual risks and sufficient consideration has been given to the consequences that can arise from the use of equipment designed for professional / industrial use when it is used in a non-professional / non-industrial context.

The section that follows provides a format of documentation of the risk reduction that has to be included in the technical documentation as per the requirements of the new LVD.

3.38.6.4 Risk assessment documentation

The result of the risk assessment can be documented as given in section 8.12 chapter 8.

CONCLUSION

In this chapter we have seen the effect of electric shock on the human body and ways and means to protect the operator/user and service personnel from this hazard. It is usually assumed that equipment that do not use hazardous voltages are safe, however such equipment may pose energy hazards that are required to be taken of. Hazards other than electrical and fire also need to be addressed so that the equipment complies with the requirements of various safety standards particularly mechanical hazards. Components, especially those influencing safety are critical so far as the equipment safety is concerned. All critical components should be certified and must carry a European type approval mark in order to reduce testing requirements of the equipment. After taking all recommended safety precautions like insulation, separation, earthing etc. some risks may still remain. These are called residual risks for which warning signs are required so that the user is appropriately warned of a possible danger. With the advent of strict liability in Europe the *duty-to-warn* viewpoint has recently been expanded in the courts to a considerable extent i.e. it is the duty of the manufacturer to sufficiently warn the user of hazards posed by the equipment. Further, the equipment should also be accompanied by operating instruction or manual in order to make the operator aware of correct operating procedure and use of the equipment.

Risk assessment has now become an integral part of compliance to the LVD. The risk assessment needs to be documented as a part of the technical documentation/file for CE marking. A proper risk assessment will be a valuable aid to the manufacturer and will prove that he has done his due diligence as far as product conformity is concerned. It should be reiterated that risk assessment is not a substitute for product safety evaluation as per harmonized standard as it is only compliance to the standard that gives presumption of conformity to the directive.

References

IEC 61051-2: Varistors for use in electronic equipment.

IEC/EN 61010-1: Safety Requirements for Electrical Equipment for Measurement, Control, and Laboratory Use.

IEC/EN 60065: Safety Requirements for mains operated electronic and related apparatus for household and similar general use.

IEC/EN 60601-1: General Safety Requirements for Medical Electrical Equipment.

IEC/EN 60950: Safety of Information Technology Equipment, Including Electrical Business Equipment.

Infinion Application Note AN 2012-10: Electrical Safety And Isolation

IEC 60664:1980 "Insulation Co-ordination within Low-Voltage Systems Including Clearances and Creepage Distances for Equipment,"

EN/IEC 61032: Protection of persons and equipment by enclosures - Probes for verification ECMA Std 287: Safety of electronic equipment

CUIINC Application note: Overview of LPS requirements

IEC/EN 60950: Safety of Information Technology Equipment, Including Electrical Business Equipment.

IEC 61695-11-10, 20: Test flames – 50 W horizontal and vertical flame test methods

IEC 61695-11- 20: Test flames – 500 W horizontal and vertical flame test methods

IEC 60695-2-12: Glow-wire flammability index (GWFI) test method for materials.

IEC 60695-11-3: Test flames – 500 W flame – Apparatus and confirmational test method.

IEC 60695-11-4: Test flames – 50 W flame – Apparatus and confirmational test method.

Safety Considerations in Power Supply Design: Bob Mammano, Texas Instruments Lal Bahra, Underwriters Laboratories

IEC/EN 61010-1: Safety Requirements for Electrical Equipment for Measurement, Control, and Laboratory Use.

IEC/EN 60065: Safety Requirements for mains operated electronic and related apparatus for household and similar general use.

IEC/EN 60601-1: General Safety Requirements for Medical Electrical Equipment.

IEC/EN 60950: Safety of Information Technology Equipment, Including Electrical Business Equipment.

IEC/EN 60950: Safety of Information Technology Equipment, Including Electrical Business Equipment.

ECMA Std 287: Safety of electronic equipment

Application note AN1102 Exelsys Technologies Ltd

IEC 60417: Graphical symbols for use on equipment

IEC 60320: Appliance couplers for household and similar general purposes - Part 1: General requirements.

IEC 60309: Plugs, socket-outlets and couplers for industrial purposes - Part 1: General requirements.

IEC 60245-1: Rubber insulated cables - Rated voltages up to and including 450/750 V.

IEC 60227-1: PVC insulated cables - Rated voltages up to and including 450/750 V

• • •

4

THE EMC DIRECTIVE

4.1   INTRODUCTION

The Electromagnetic Compatibility (EMC) directive 2014/30/EU was published on 26th February 2014. There was a transition period of almost two years and the directive came into force on 20th April 2016 i.e. it is legally binding for products falling under the scope of this directive and placed into market (or put into use) on or after this date to meet the requirements of this directive. The directive aims to ensure EMC in community market by ensuring that an equipment does not emit electromagnetic disturbance (see section 4.4) above a certain limit (as prescribed by EMC standards) and at the same time is not affected (i.e. its operation is not disturbed) by the electromagnetic disturbance generated by equipment in its vicinity or by any electromagnetic phenomenon for that matter. We will now take a look at the various provisions of the directive which are given in the form of articles.

4.2   ARTICLE 1: SUBJECT MATTER

This directive aims at regulating electromagnetic compatibility (see section 4.4). Its basic goal is to ensure functioning of the EU market by requiring an equipment to comply with *adequate level* of electromagnetic compatibility. The directive has used the words *adequate level* since it is practically impossible to ensure that the emission of electromagnetic disturbance is zero or that it should tolerate infinite level of electromagnetic disturbance. Thus, only *reasonable* electromagnetic compatibility is expected out of the equipment depending on its operational environment.

4.3   ARTICLE 2: SCOPE

Within the scope of this directive are only those electrical/electronic equipment capable of generating EMI and/or the performance of which is liable to be affected by such disturbance. Equipment excluded from scope are as follows:

4.3.1 Equipment which do not contain electrical and/or electronic parts, since these will not generate electromagnetic disturbances and its normal operation and will not be affected by such disturbances.

4.3.2 Equipment which are inherently *benign* in terms of EMI. Such equipment are those whose inherent physical characteristics are such that they are incapable of generating or contributing to electromagnetic emissions and which will operate without unacceptable degradation in the presence of the electromagnetic disturbance. Examples of such equipment are:

- Cables and cabling, cables accessories, considered separately.
- Equipment containing only resistive loads without any automatic switching device; e.g. simple domestic heaters with no controls, thermostat, or fan.
- Batteries and accumulators (without active electronic circuits).
- Headphones, loudspeakers without amplification.
- Pocket lamps without active electronic circuits.

- Protection equipment such as fuses and circuit breakers without active electronic parts or active components which only produces transitory disturbances of short duration during the clearing of a fault.

- High voltage types of equipment in which possible sources of disturbances are due only to localised insulation stresses which may be the result of the ageing process and are under the control of other technical measures included in non-EMC product standards, and which do not include active electronic components such as high voltage inductors, transformers etc.

- Other benign products like capacitor, induction motors, quartz watches, filament lamps (bulbs), home and building switches plugs, sockets, terminal blocks, etc which do not contain any active electronic components and passive antennas used for TV and radio broadcast reception.

4.3.3. Equipment covered by other specific community directives which contain their own EMC requirements from the date of mandatory application of those directives. Examples are:

- Radio equipment covered by radio equipment directive (RED) 2014/53/EU
- Active implantable medical devices directive 90/385/EEC.
- Medical devices directive (MDD) 93/42/EC.
- In vitro diagnostic medical devices directive 98/79/EC.
- Marine equipment directive 96/98/EU.
- Agricultural and forestry tractors covered by the directive 2003/37/EC.
- Two or three-wheel motor vehicles within scope of directive 2009/67/EU.

4.3.4. Equipment covered by regulations other than CE, examples are:

- Aeronautical products, parts and appliances as referred to in regulation (EC) No 1321/2014,
- Radio equipment used by radio amateurs as defined in international telecommunication union (ITU) radio regulations.
- Defence equipment

4.3.5. Also excluded are custom built evaluation kits destined for professionals to be used solely at research and development facilities for such purposes.

4.3.6. Exclusions only for immunity for measuring instruments falling under the scope of measuring instruments directive 2014/32/EU.

4.4   ARTICLE 3: DEFINITIONS

This article gives some basic definitions used in the EMC vocabulary. These are as follows:

Electromagnetic disturbance is defined as *any electromagnetic phenomenon (noise or unwanted signal) that may degrade the performance of a device, equipment or a system.*

The electromagnetic phenomenon can be natural or man–made and includes electromagnetic noise or unwanted signals or even change in propagation medium itself.

Electromagnetic interference (EMI) is defined by the international electrotechnical commission (IEC) as *degradation in performance of a device, equipment or system caused by electromagnetic disturbance.* Although, the definition of EMI is not given in the directive, EMI occurs only when the electromagnetic disturbance actually disturbs an equipment.

Electromagnetic environment is *all electromagnetic phenomenon observable at a given location.*

Immunity is defined as the *ability of equipment to perform as intended w/o degradation in the presence of EMI.*

Electromagnetic compatibility (EMC) is defined as *"the ability of an equipment to perform satisfactorily in its intended environment without introducing intolerable interference (EMI) into anything in that environment.* The first part of the definition says that the equipment shall *perform satisfactorily* in its intended environment, meaning that it should tolerate certain level of EMI OR should have certain level of *immunity* OR should *not be susceptible* to EMI already present in the environment. And while doing so, says the second part of the definition, the equipment shall not introduce *intolerable* EMI into the environment, meaning that the EMI it emits shall be *below* a certain limit. See next chapter for details on EMC, emission and immunity.

4.5   EQUIPMENT CLASSIFICATION

The classification of equipment is given in article 3 itself. According to the directive, *equipment* is defined as any apparatus or fixed installation.

Apparatus is defined as:

- Any finished appliance (any device or unit that delivers a function and has its own enclosure) or combination thereof.

- Made commercially available (placed in the EU market) as a single unit.

- For end user.

- Liable to generate electromagnetic disturbance or liable to be affected by it.

It should be noted that any product is deemed as an apparatus only when it is *commercially available*. This means products for personally use, otherwise not available commercially (like products brought in by EU citizens from outside the EU for own consumption) are not deemed as equipment and hence is outside the scope of the directive. Also combining two or more finished appliances bearing CE marking does not automatically produce a *compliant* system e.g. a combination of CE marked PLC and motor drives may fail to meet the protection requirements.

Again, the *end user* is any natural person (e.g. consumer) or legal entity (e.g. enterprise) using or intending to use the apparatus for its intended purpose. Generally, the end user is deemed to have no qualifications in the field of electromagnetic compatibility.

Another caveat is that the apparatus should be liable to cause electromagnetic disturbances or its normal operation should be affected by such disturbances. If both of these conditions are not fulfilled due to inherent characteristics of the apparatus, then the apparatus may be considered as inherently benign in terms of electromagnetic compatibility and hence, the EMC directive may not apply.

Apparatus also includes:

- Components or subassemblies for incorporation into apparatus by end-users, otherwise also available commercially to end-users and which are liable to generate electromagnetic disturbance or affected by it. Some components /sub-assemblies that may not have a proper enclosure but still could be incorporated into a bigger apparatus with a minimum of connections. Examples could be add-on cards for computers, expansion modules for PLC etc.

- Mobile installation which is a combination of apparatus/other devices intended to be moved and operated in a range of locations. Examples are portable broadcast studio, OB vans and the like.

- *Fixed installation* is a particular combination of several types of apparatus and other devices, which are assembled, installed and intended to be used permanently at a predefined location. Examples of fixed installations are industrial plants, power plants, power supply networks, telecommunication networks, cable TV networks, computer networks, airport luggage handling installations, airport runway lighting installations, automatic warehouses, skating hall ice rink machinery installations, storm surge barrier installations (with the control room etc), wind turbine stations, car assembly plants, water pumping stations, water treatment plants, railway infrastructures, air conditioning installations. These are however, subject to a different compliance procedure (see section 4.13).

Certain components may be benign in nature or their function may depend on how they are connected or they cannot be readily incorporated or *plugged-in* into an apparatus. Such components are excluded e.g. resistors, capacitors, inductors, filters, diodes, transistors, thyristors, triacs, integrated circuits, simple electromagnetic relays, LEDs, simple thermostats etc.

4.6   ARTICLE 4: MAKING AVAILABLE IN THE MARKET

This article states that member states should take all measures so that the equipment is made available (placed) in the market/put into service *only* if it meets requirement of this directive (when properly installed, maintained and used for intended purpose).

4.7   ARTICLE 5 : FREE MOVEMENT

This article states that member state cannot impede the placement of compliant equipment in the market. However, it also empowers member states to take special measures:

- To overcome an existing EMI problem at a specified site.
- To protect telecom networks/transmitting/receiving stations for safety purposes.

Also, member state cannot create obstacles for *display* of non-compliant equipment at trade fairs provided a visible sign clearly indicates so. The directives further states that demonstration (powering ON) may take place provided adequate measure are taken to avoid electromagnetic disturbance.

4.8   ARTICLE 6: ESSENTIAL REQUIREMENTS

As discussed, directives only give broad legal guidelines called essential requirements. As per this directive, the essential requirement is that an equipment should be so designed and manufactured, so as to ensure that:

– the electromagnetic disturbance generated does not exceed   the level above which radio, telecom or other equipment cannot operate as intended;

– it has a level of immunity to the electromagnetic disturbance to be expected in its intended use which allows it to operate without unacceptable degradation of its intended use.

Thus, there are basically two requirements. Firstly, under normal operational environment, the electro-magnetic disturbance generated by an equipment (i.e. emission) is below a certain limit. Secondly, the equipment has an adequate level of immunity to electro-magnetic disturbance expected in its normal operational environment. Obviously, the goal of the essential requirements is not to guarantee absolute protection of equipment (e.g. zero emission level or total immunity). These requirements accommodate both physical facts and practical reasons. See chapter 5 for details on EMC, emission and immunity.

Then there are specific requirements for fixed installations. Accordingly, a fixed installation should be installed applying good engineering practices with a view to meeting the above essential requirements. The good engineering practices should be documented and the documentation should be held by the responsible person(s) (i.e. manufacturer or if the manufacturer is non-European then by the authorized representative/importer/distributor/assembler/employer), at the disposal of the relevant national authorities for inspection purposes for as long as the fixed installation is in operation.

4.9   ARTICLES 7, 8, 9, 10.

These define obligations for manufacturers, authorised representatives, importers and distributors. These are the same as explained in the chapter on low voltage directive.

4.10   ARTICLES 11, 12.

Article 11 deals with cases where obligations of manufacturers are applicable to importers and distributors. Article 12 deals with economic operators. These are the same as explained in the chapter on low voltage directive.

4.11   ARTICLE 13: PRESUMPTION OF CONFORMITY

This article states that equipment complying with harmonised EN standards (published in official journal of European union) are presumed to comply with the essential requirements of this directive. This has been dealt in detail in section 1.9 of the first chapter.

4.12   ARTICLE   14:   CONFORMITY   ASSESSMENT PROCEDURES

The directive prescribes different procedures for apparatus and fixed installations. The   conformity assessment procedures for apparatus are given in this article while those for fixed installations are given in article 19 (section   4.13). As far as the the conformity assessment procedures for apparatus are concerned, full responsibility is on the manufacturer for all aspects of conformity assessment. If the manufacturer is non-European then the person who places the product in the EU market (the importer/distributor/installer or authorized representative) should ensure that the manufacturer has carried out all the procedures as per the directive. There is no mandatory 3rd party (notified body) involvement whatsoever. Where a manufacturer assembles a final apparatus using components from other manufacturers, the manufacturer must retain overall control. The manufacturer at all times, is responsible for the compliance of the final apparatus.

4.12.1 Conformity Assessment Procedures For Apparatus.

The Fig 4.1 describes the conformity assessment procedure for apparatus as given in annexure II of the directive. We will discuss all steps in detail.

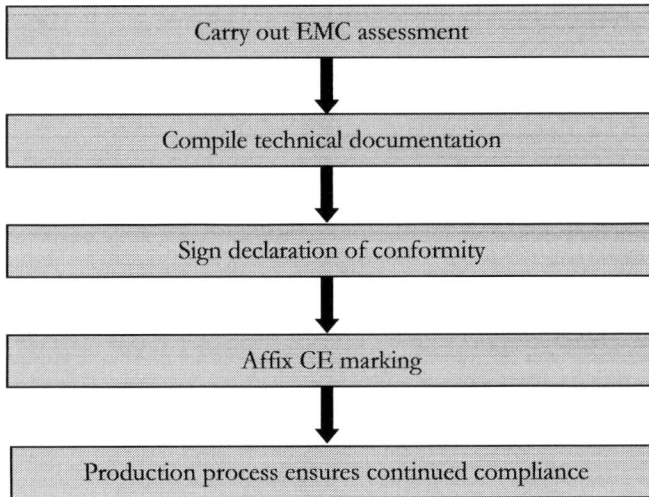

Fig. 4.1: Conformity assessment procedure for apparatus

4.12.1.1 EMC assessment

An EMC assessment must be performed, on the basis of relevant phenomena, to ensure that the essential requirements of the directive are met. The electromagnetic compatibility assessment should take into account all normal intended operating conditions Where the apparatus is capable of taking different configurations (or in case of products *variants*), the electromagnetic compatibility assessment should confirm whether the apparatus meets the protection requirements. In this case the *worst-case* approach can be taken meaning that the EMC assessment can be done in that configuration which is most likely to cause maximum *emissions* or in that configuration which is most like to be *susceptible*. This worst-case configuration can then be treated as a representative for the entire series of variants. The basis of selection of worst-case configuration and the justification that other configurations/variants meet EMC requirements (despite not testing) should be included in technical documentation (refer chapter 8).

The EMC assessment can be done in either of the following ways:

- **Using harmonised standards.**

  The correct application of the relevant European harmonised standards (whose *references* have been published in the official journal of the EU) covering all the essential requirements of the EMC directive is equivalent to carrying out detailed technical EMC assessment i.e. no separate assessment is required. *Correct application* means ensuring compliance with all applicable tests as prescribed in the standard. The tests can be carried out by the manufacturer himself (if he has the requisite test facilities as per standards) or he can get the apparatus tested from a third party (such as EMC test laboratory or a notified body). If latter is the case, it must be reiterated that it is still the manufacturer who is responsible for conformity of the apparatus. As we have seen, complying with all the requirements of the standard gives presumption of conformity with essential requirements of the directive. Hence the standards method is the most frequently used and recommended way to demonstrate EMC compliance. While choosing a standard, product or product family-standards take all relevant phenomena into account, and hence these should be selected if available for the product of product family. If not, then generic standards can be applied. A detailed information about EMC standards and their selection has already been discussed chapter 1.

- **Using other methods.**

  These methods are required in cases where the apparatus is physically too large to be tested in a proper test laboratory. In such cases, a thorough investigation of EMC phenomena that are relevant to the apparatus must be done considering intended use and location of use. For example, if the apparatus is for use in residential area it means a controlled environment and close distance between source and victim but if the apparatus is used in non-residential area, it is exposed to uncontrolled EMC environment with longer distance between apparatus. Emission limits and severity levels can then be chosen by referring to basic standard.

- **Combination of methods**

  In cases where the apparatus is physically too large to be tested in a proper test laboratory, EMC tests may have to be carried out at site and in case, it is not possible to carry out certain EMC tests, then a technical justification can be given for the tests not carried out. This justification can be based on the use of certified components on which EMC tests have already been carried out by the respective manufacturers. It is prudent to carry out at least the emission tests (like conducted and radiated emission) and give technical justification for immunity tests. EMC testing has been dealt in detail in chapter 5.

4.12.1.2 Compile technical documentation

After carrying out EMC assessment, the manufacturer has to compile a technical documentation in support of his conformity claim. The contents of technical documentation are elaborated in annexure-II of the directive and are discussed in detail in chapter 8. Where the manufacturer has not applied harmonised standards, or applied them only in part, explanation of the steps taken to meet the essential requirements of the directive like:

-The EMC assessment.

-Results of design calculations made.

-Examinations carried out, test reports, etc.

The documentation should also carry information on specific precautions to be taken during assembling, installing, maintaining and operating along with information enabling use for intended purpose. The manufacturer or his authorised representative or the importer based inside the EU is required to hold the technical documentation at the disposal of the competent authorities for at least ten years after the date on which such apparatus was made available in the market.

4.12.1.3 Article 15: EC declaration of conformity (DOC).

After compiling technical documentation, the manufacturer signs and issues the EC declaration of conformity. The format of the DOC is given in Annexure IV of the directive and is explained in chapter 1. The manufacturer or his authorised representative or the importer based inside the EU is required to hold the DOC at the disposal of the competent authorities for at least ten years after the date on which such apparatus was made available in the market.

4.12.1.4 Article 16,17: Affixing the CE Marking

After compiling documentation and issuing the DOC, the manufacturer has to affix the CE marking on the apparatus. The directive calls for CE marking only on apparatus and not on fixed installations. The affixing requirements have been already dealt with in section 1.16 chapter 1. In addition to CE marking, each apparatus should carry a label indicating type, batch, serial number, the name and address of the manufacturer and authorized representative (or in his absence, the importer). For products used in industrial environment, the label should clearly indicate restriction of use in residential premises.

4.12.1.4 Continued compliance

Since testing is usually carried on one representative sample of the apparatus, the manufacturer has to demonstrate that the manufacturing process ensures that the samples produced afterwards are identical to the sample that was tested, to ensure continued compliance. In this regard, a description as to how this is achieved has to be included in the technical documentation. This has been explained in the chapter 8 of technical documentation.

4.12.1.5 Article 18: Information concerning use of apparatus.

The apparatus should be accompanied by instructions (manual) regarding use in accordance with the intended purpose. The apparatus should also be accompanied by information on any specific precautions that must be taken when the apparatus is assembled, installed, maintained or used, in order to ensure that, when put into service, the apparatus always stays in compliance. Refer chapter 8.

4.13   ARTICLE 19: FIXED INSTALLATIONS

As the name suggests, fixed installations are intended to be used permanently at a predetermined location and hence the term *free movement* is not applicable to them. Hence CE marking (which is to ensure free movement) need not be affixed on fixed installations and certain provisions such as DOC are not applicable. Now, as per definition, fixed installation is an assembly of several types of apparatus. Accordingly, apparatus incorporated into a given fixed installation and otherwise also commercially made available in the community market, should bear the CE marking and must meet all requirements of the directive. On the other hand, apparatus made specifically to be incorporated in a given fixed installation and otherwise is NOT available commercially, may not bear the CE marking.

As a whole (whether using CE marked apparatus or not), fixed installation must comply with essential requirements of the directive concerning emission and immunity. Now, this appears a bit confusing as one may think that if CE marking is not applicable then how come fixed installation should meet essential requirements? Well, essential requirements are required as a safeguard so that internal market remains protected while CE marking is not required since the installation is fixed and that there is no movement. Thus, for the fixed installation, in order to prove that the essential requirements of emission and immunity are met, it becomes imperative to specify EMC characteristics of all apparatus in the technical documentation like copies of test reports and DOC or technical justification (for non-CE apparatus). Needless to say, CE marked apparatus should be preferred over *non-CE* apparatus since their EMC characteristics are specified in their DOC and test reports.

Furthermore, fixed installations are also subjected to *specific requirements* as below:

– *Documented good engineering practices.*

– Information regarding intended use of components that make up the installation.

Good engineering practices mean those installation practices that are good for EMC purposes with regards to environment at the specific installation site. This includes:

– Precautions to control conducted EMI emissions and improving immunity like EMI filters, absorption devices, surge suppression components etc in appropriate places like connections of the installation to mains or wired interfaces to other equipment.

– Controlling radiated emissions and improving radiated immunity by electromagnetic shielding of control panels that house apparatus known to be prolific emitters (like motor drives) or shielding of susceptible apparatus (like controllers).

- Avoiding coupling of EMI by giving proper attention to positioning and segregation of apparatus, selection and routing of interface cables especially to noisy components like drives, SMPS etc.

- Proper grounding strategies (including dedicated earthing) which is essential for good performance of filters and shields and proper bonding of apparatus to avoid corrosion and ensure proper low resistance grounds.

All these practices should be documented in detail and hence the installation should be accompanied by technical documentation.

4.14   OTHER ARTICLES

The requirements of conformity assessment bodies, notified bodies and notifying authorities (authorities that are responsible for notification, assessment and monitoring of conformity assessment bodies and notified bodies are given in articles 21 to 36. Articles 37 to 38 are regarding market surveillance and these have already been discussed in chapter 1. Article 39 is regarding safeguard procedure and has been discussed in section 2.19. Article 40 deals with non-compliances of economic operators (manufacturers, importers, authorised representatives) and these have been explained in chapter 1. Article 42 talks about penalties for substantial non-compliance and this has been explained in chapter 1. Articles 41 and 43 to 46 deal with electromagnetic committees, transitional provisions, transposition (of directive into their law), repeal (of old directive) and entry into force (of new directive).

CONCLUSION

In this chapter we have seen the legal requirements of the EMC directive. The technical requirements are given in the various EMC standards and are discussed in the next chapter which also introduces the reader to the fundamentals of EMC and EMC test methodologies.

Reference

Directive 2014/30/EC relating to electromagnetic compatibility.

• • •

5

EMC FUNDAMENTALS, STANDARDS AND TESTING

5.1   INTRODUCTION TO EMI

Pollution is the bane of modern society. It is the undesirable by–product of mankind's scientific and technological progress. Air pollution, for example, is the introduction of certain undesirable gasses like sulphur–dioxide or carbon–mono oxide, to name a few, which otherwise are not part of the standard atmosphere and which can have adverse effect on human health. The most recent form of pollution is electromagnetic pollution (if we can call it that) i.e. the generation of electromagnetic energy that can have an adverse effect on the *health* of an electrical/electronic equipment. We have already seen electromagnetic disturbance and the definition of EMI. Electromagnetic disturbance can occur across the entire range of the electromagnetic spectrum and some of it may fall within the   radio frequency range of 3kHz to 300GHz (which is normally used for radio communication) in which case it is referred to as radio frequency disturbance. We can then define RFI or radio frequency interference as *the degradation in reception of a wanted signal caused by radio frequency disturbance.* In simpler terms, EMI/RFI is an electromagnetic pollution caused by the generation of electromagnetic energy that is unwanted or unwarranted. This can interfere with the normal operation of an electronic equipment resulting in the degradation of its health. The electromagnetic phenomenon is the source of EMI, which travels through the intervening medium and affects the operation of nearby electronic equipment that receives the EMI and becomes a victim.

5.2   THE PROBLEM OF EMI

In recent years the market has been flooded with a myriad of electronic gadgets that are potential generators of electromagnetic energy. It is not their business to generate such energy, but because of the very nature of circuitry and technology they do so inadvertently or *unintentionally* and interfere with radio communication and other electronic gadgets. The problems have compounded with the advent of digital devices with high clock rates and low rise-times, which pour electromagnetic energy over entire range of the radio spectrum. The situation is worsened by the use of switching power supplies, thyristor converters and similar equipment that switch voltages at high rates causing unwanted emission of electromagnetic noise.

5.3   NEED FOR EMC

We have already seen the definition of electromagnetic compatibility in the previous chapter. Now, if we have a group of equipment, each emitting EMI below a certain level and each having sufficient immunity to this EMI, then we can say that each equipment is *electro–magnetically compatible* with other equipment and hence the name electromagnetic compatibility. EMC/EMI is used many-a-times as a combined term since we are trying to achieve EMC by limiting EMI. EMI is what we are trying to avoid and if we succeed in doing so, we achieve EMC.

The Fig. 5.1 demonstrates the effect of EMC laws or directives introduced in the European union. On the upper side we have a situation where miniaturized structures as well as increasing complexity, integration and interconnection are making electronic installations and components more susceptible, decreasing their immunity to EMI. The lower portion of the graph shows that high currents, voltages and power levels of power installations as well as proliferation of personal communication devices increase the level and potential for EMI in the environment. If the two graphs were to meet, then the situation would become hopeless. The EMC directive avoids such a situation by making it mandatory for all equipment to have some level of immunity and restrict emissions below a certain limit, thus seeking to maintain a *guard band* by ensuring electromagnetic compatibility.

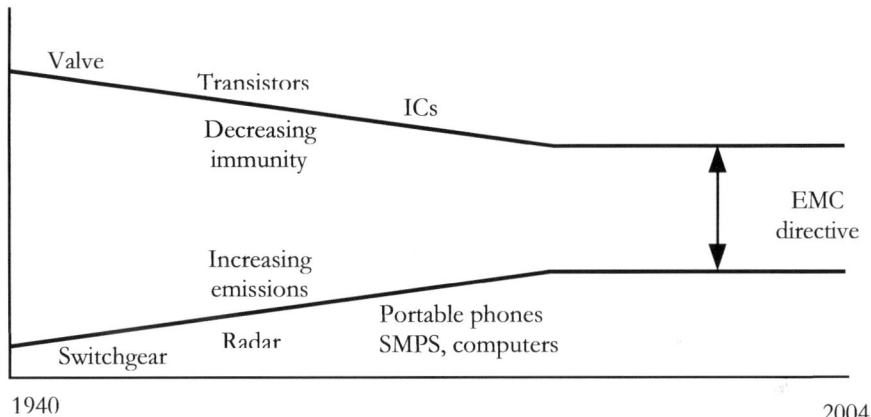

Fig. 5.1: Effect Of The EMC Directive

5.4   REALISATION OF EMC

Consider a typical electronic equipment as shown in Fig. 5.2. Such equipment normally has, among other things, a mains chord that draws power from the utility socket, cables for input /output (I/O) and control signals, openings for keypad / display, ventilation etc. For this equipment to be declared electromagnetically compatible, firstly the EMI generated by the equipment i.e. *emission* should be below a certain limit. Electromagnetic emission is defined as *the phenomenon by which electromagnetic energy emanates from a source*. This emission can be conducted out of the equipment in the form of a current, via conducting wires like the mains chord or I/O cables. This is called *conducted emission (CE)* and is defined as *electromagnetic energy emitted via one or more conductors.*

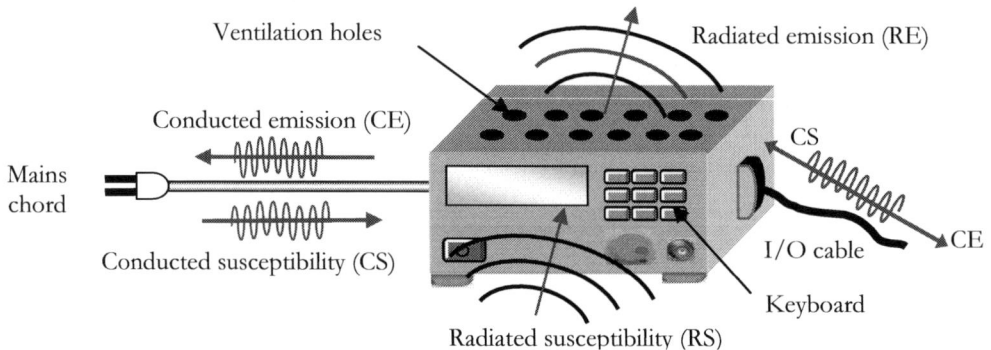

Fig. 5.2: Realization Of EMC

The emission can also be in the form of radiation and can escape through openings like displays, ventilation holes, cable entries, door seams etc. This is called as *radiated emission (RE)* and is defined as *electromagnetic energy emitted through space in the form of electromagnetic waves.* Secondly, radiations from other equipment can enter through these very openings and can degrade performance of the equipment. This is called as *radiated susceptibility (RS).* External EMI can also enter through mains cables or I/O lines and cause performance degradation. This is called as *conducted susceptibility (CS).* Many-a-times, the word *immunity* is used instead of *susceptibility.* Now immunity is the ability of an equipment to perform without degradation in the presence of an electromagnetic disturbance and as such, is exactly the reverse of susceptibility (which is the *inability* of an equipment to perform without degradation). In simpler words, if an equipment does not have sufficient immunity it will be susceptible. Optimistic people generally use the term immunity, although a lot depends on the EMC standards and what the standardization body thinks is a better term.

5.5   EMC TESTS AND MEASUREMENT

As seen in the previous section, there are four broad categories of tests for EMC viz. conducted emission, radiated emission, conducted susceptibility/immunity and radiated susceptibility/immunity. The chart in Fig. 5.3 gives some of the EMC test and measurements as prescribed by civilian EMC standards like the IEC/CISPR and EN. Emission measurement involves measuring conducted and radiated emissions from equipment and comparing them with prescribed limits. Conducted emission is prescribed mainly on power cables (AC/DC) and under some conditions on the signal cables as well. Conducted emission on power cables can be further categorized as high frequency (above 9kHz) which includes continuous emission and discontinuous emission (called clicks) and as low frequency (below 9kHz) which includes harmonics and flicker measurements. Radiated emission test has to be carried on the equipment as a whole and includes measurement of electric and magnetic fields, while some standards (especially for household appliances) call for measurement of radiated power from mains cables and is called disturbance power measurement.

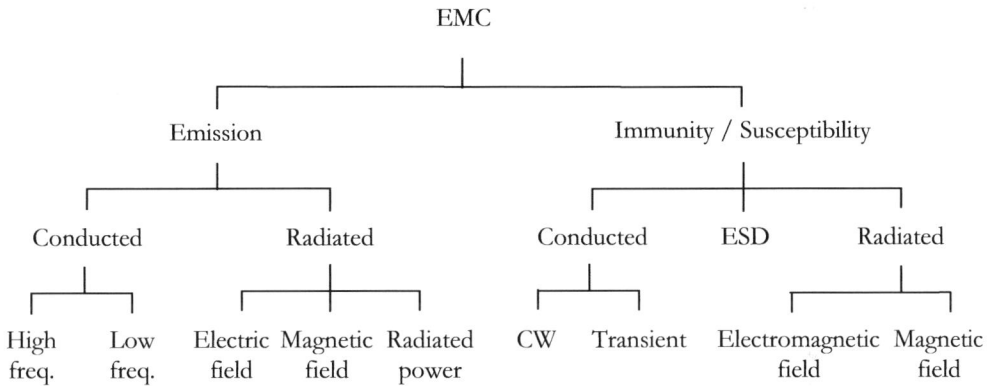

Fig. 5.3: EMC Tests

In the case of susceptibility tests, interference of a certain level or severity is deliberately generated under controlled conditions. Equipment performance is then monitored for signs of degradation when exposed to this *simulated* EMI. Conducted susceptibility tests includes transients as well as continuous wave (CW). Transient tests like electrical fast transients (EFT), surges and oscillatory waves involve injection of transients or pulses on mains and I/O chords. While in continuous tests, a current at audio/radio frequencies is injected into mains and I/O lines. Radiated susceptibility test involves exposing the equipment to simulated electric or magnetic or electromagnetic fields. Electrostatic discharge (ESD) is a unique test involving both conducted and radiated susceptibility. It basically simulates the effect on equipment when an operator, charged to static voltage, happens to touch it. The charge is simulated using an ESD gun (which has a discharge *tip* simulating the human finger) and is injected into the equipment either by actually touching it (conducted susceptibility) or discharging the tip through air–gap ionization by bringing the tip very close to the equipment (radiated susceptibility).

5.6   ELEMENTS OF EMI

As shown in the Fig. 5.4, any interference event comprises of three elements viz the source, the coupling media and the victim. The source is where the interference originates, the coupling medium is that which carries or couples the interference (it can be a conductor or free space) to the victim which is affected by the interference.

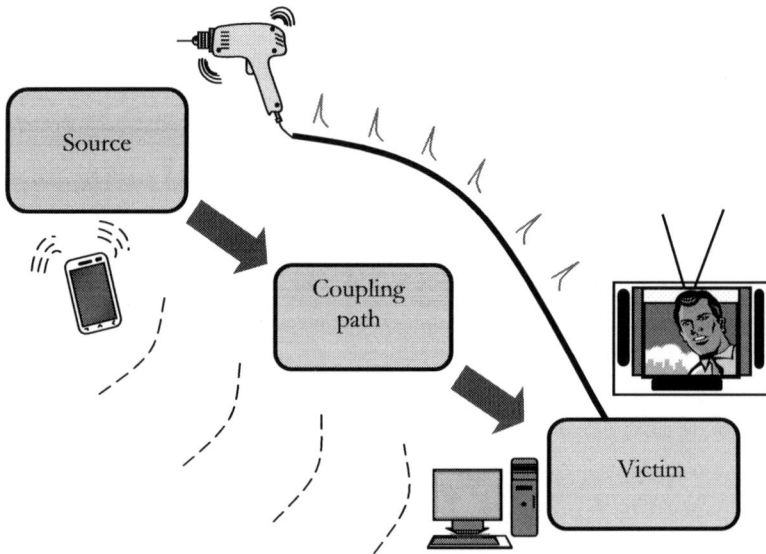

Fig. 5.4: Elements Of EMI

In the Fig. 5.4, the source is the drill machine and the mobile. Now, drill machines generally use AC universal motors which have brush contacts energizing the rotor. The contacts usually wear off after repeated use and the air gap thus created leads to arcing. An arc in frequency domain, translates into spurious frequency generation over a large band which is nothing but EMI. This EMI is coupled via the mains chord to the TV which is sharing the same mains as the drill machine. The coupling path in this case is *conducted*. While the EMI generated by the cell phone is coupled to the T.V. antenna, and the coupling path is *radiated* i.e. through space.

5.6.1   Sources of EMI

EMI sources can be classified according to their origin i.e. natural and man–made. Natural source is the electromagnetic noise generated as a result of natural phenomena and can be classified as terrestrial or extra-terrestrial. Terrestrial sources have their origins in the phenomenon on our planet. The most common source of terrestrial EMI is lightning, which is a huge discharge of static electricity. The most prolific of extra-terrestrial sources is our Sun which at times emits very powerful alpha particles that can penetrate the Earth's atmosphere and cause catastrophic failures in microwave communication, aircraft navigation system etc.

Manmade source is any electromagnetic noise having its source in man–made devices and can be further classified as intentional or unintentional. Intentional sources are those wherein electromagnetic energy is deliberately generated for communication and include all type of transmitters and transceivers. They are narrowband in nature i.e. consists of only one frequency (and some sidebands). Unintentional sources are those where generation of electromagnetic energy is not intended but is an undesired product of the equipment operation. Unintentional sources are broad–band in nature i.e. have many frequency components spread over a large part of the spectrum. The best example is a switch mode power supply (SMPS) where electromagnetic energy is unintentionally generated due to the switching. Other examples are DC–DC converters, motor drives, as well as out–of–band emission from RF sources like transmitters and microwave ovens. Portable tools like drills and household appliances are a source of repetitive transients of low energy normally referred to as electrical fast transients (EFT) or bursts. Fast transients are also generated in industrial environment during inductive load switching and also by vehicle ignition and spark plugs. Arc welding is carried out at very high currents, and the arcing is a source of high energy transients. Such transients (called surges) are also generated by lightning strike or switching of capacitive loads. Business machines like a P.C or photocopiers normally use SMPS that generate harmonics (of the switching frequency) which spread over a large region of the spectrum.

Sources can also be classified according to their type i.e. electric, magnetic and electromagnetic (plane waves). Electric fields are produced by high impedance circuits such as monopole /dipole antennas, where voltage is a dominant factor. Magnetic fields are produced by low- impedance circuits such as loop antennas where current is a dominant factor. Electric and magnetic fields are called near fields as they exist near the source only. As one moves away from the source, the electric field creates a magnetic field (and vice-versa), so that at a distance equal to $\lambda/2\pi$ (where λ is the wavelength) the ratio of electric and magnetic field amplitudes equals 377 Ω in which case it is called as electromagnetic or plane wave of far field.

5.6.2   Coupling Mechanisms

Coupling is the process by which the energy from an EMI source is transferred to the victim. Basically, there are two modes of EMI namely, the differential or symmetric or normal or balanced mode and the common or asymmetrical or unbalanced mode. In a two-wire line, differential mode currents flow in opposite directions while common mode currents flow in the same direction (and return via earth). EMI generated by the source reaches the victim via any or all of the following coupling mechanisms:
-   Inductive coupling.
-   Capacitive coupling.
-   Galvanic coupling.
-   Electromagnetic coupling.
-   Field-to-cable common mode coupling.
-   Field-to-cable differential mode coupling.
-   Cable–cable coupling.

5.6.2.1 Inductive coupling

In low impedance, high current systems, coupling of EMI is usually inductive. This is also referred to as magnetic field coupling. A changing current in one system produces a magnetic field that couples to another system and induces a current in it. A common example of inductive coupling is that of 50Hz power currents through a transformer.

5.6.2.2 Capacitive coupling

This is also referred to as electric field coupling and occurs in high impedance circuits where voltage plays a dominant role. A changing voltage in one system couples to another system through a stray capacitance and induces a voltage in it. The most common example of this type is the coupling of EMI through a transformer by a stray capacitance between primary and secondary.

5.6.2.3 Radiation or electromagnetic coupling

Capacitive and inductive coupling occur in the near field region where either the electric field or magnetic field is dominant. But if the victim lies far away from the source, the coupling will be via electro–magnetic field and as such is referred to as electromagnetic or radiation coupling. Such coupling occurs between a radio receiver and the transmitter.

5.6.2.4 Galvanic coupling

This type of coupling occurs when there exists a direct conducting path between two circuits for example when two circuits share common mains or earth.

5.6.2.5 Field-to-cable common mode coupling (ground loop coupling)

This type of coupling occurs when wires or traces are routed away from the ground forming a loop. This loop acts as an antenna for stray fields causing common mode EMI to be coupled into the system. The loop can also cause radiation.

5.6.2.6 Field- to- cable differential mode coupling

This type of coupling occurs when forward and return wires (or traces) are routed at a large distance from each other forming a loop. This loop acts as an antenna for stray fields causing external differential EMI to be coupled into the cables. The loop can also cause radiation.

5.6.2.7 Cable–cable coupling (cross – talk)

Cross talk is said to occur when high frequency signals on one cable couple (or *talk*) to adjacent cable electrically, magnetically or even electro–magnetically. This is just a special case of inductive or capacitive coupling. It presents some of its unique problems and hence has to be considered separately.

5.6.3   EMI Victims

EMI emitted by the source is transported to the victim by various coupling mechanisms that we have discussed. Mostly EMI victims are such equipment which have high sensitivity. The Fig. 5.5 illustrates some common victims of EMI, although there can be several scenarios where EMI victims can cause undesirable and unexpected events.

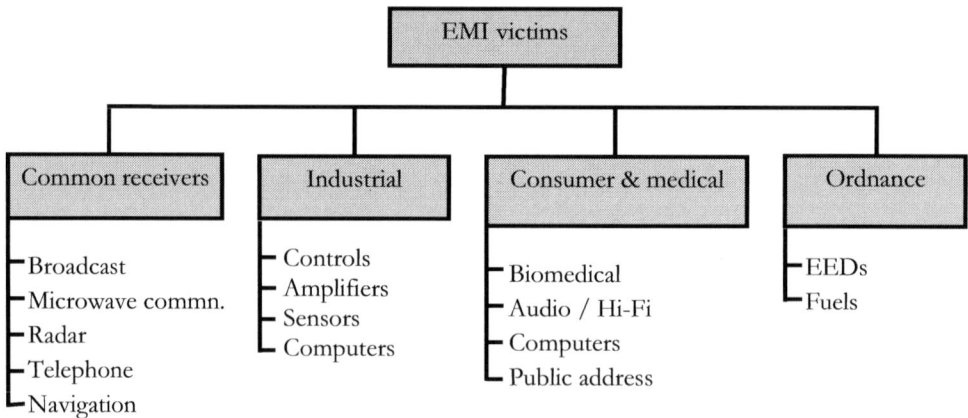

Fig. 5.5: Common EMI Victims

5.7   EMC STANDARDS

EMC standard is defined as *a technical specification adopted by a standardization body for the purpose of establishing rules, guidelines, methods or characteristics for assessing the*

EMC performance of a product. The objective of EMC standards is to ensure reasonable electromagnetic compatibility between electrical, electronic, electromechanical and RF communication systems for trouble-free co-existence by limiting *emission* of EMI and ensuring that these systems have adequate level of *immunity* to EMI generated by other equipment. EMC standards also provide a means of comparing the EMC characteristics of various equipment so that the user may choose an equipment appropriate to his requirement.

5.8   CONTENTS OF EMC STANDARDS

EMC standards are structured documents and a majority of them are given in a format that is more or less the same. They contain, at the minimum, the following:

5.8.1   Scope

This section defines the scope of the standard as regards to the phenomenon, tests, test methods and the type of equipment being considered.

5.8.2   Definitions

This part gives definitions of the various terms used in the standard.

5.8.3   EMC Test Environment

All EMC standards call for testing under controlled environmental conditions for the tests results to be accurate and reproducible. Therefore, they give the limits and tolerances for temperature, pressure, humidity, RF ambient and other environmental parameters of an EMC lab.

5.8.4   Test Set-up

This includes placement of the equipment under test (EUT) is be placed i.e. distances from the floor, walls and other metallic surfaces, placement of EMC test equipment/ EMI receiver, the EUT wiring and energizing.

5.8.5   Limits, Severity Levels And Performance Requirements

An EMC standard may contain either emission or immunity requirements or sometimes both. For emission measurement, standards provide a maximum allowable level or *limit to* emission under a particular test environment. For immunity/susceptibility tests, standards dictate *severity level* i.e. what level of simulated EMI the EUT has to be exposed to, to decide compliance for a particular test environment. In addition to this, immunity/susceptibility tests also lay down certain conditions on the extent of performance degradation acceptable after the equipment under test is exposed to simulated EMI (i.e. EMI which deliberately generated to mimic certain phenomenon) of a certain severity. This is normally referred to as *performance criterion* and is based upon the nature of simulated EMI (continuous or transient) and equipment reliability requirements. For example, for transient nature of the simulated EMI (like a short duration pulse) and less critical operational requirements, EMI standard may define a criterion that allows temporary degradation of product performance, provided it recovers automatically or by operator intervention. For continuous nature of simulated EMI (like continuous wave) and high operational reliability requirement, the standard may define a criterion that does not allow degradation in performance, temporary or otherwise.

5.8.6   Characteristics Of EMC Test Equipment

This includes the design criteria and circuit diagrams of EMI simulators/generators (so as to generate the specified transients/RF), EMI receivers (for EMI measurement) and associated ancillary equipment like antennas and transducers. Some standards also include procedures for performance verification and calibration of EMC test equipment.

5.8.7   Test Method & Results

This includes detailed procedures as to how the test is to be carried out, what are the settings for the EMI simulators and receivers, how the EUT should be conditioned, how to decide on compliance and method of presentation /documentation of the results to be reported via test reports.

5.9   EUROPEAN EMC STANDARDS

We shall now take a look at some of the important European EMC standards. All European EMC standards are based on IEC standards since the IEC has been on the forefront of developing and publishing EMC standards. The IEC was founded in 1906 and is headquartered in Geneva. Work on EMC standards started in 1934 when the IEC formed the *International Special Committee on Radio Interference* or *CISPR* (which in French actually stands for *Comite International Special des Perturbation Radioelectrique* and many pronounce it as *SISPER*) to address quantification and measurement of interference from 150kHz – 1,605kHz. CISPR worked on the design of measuring receivers, artificial mains networks, field strength measurement procedures and techniques for measuring radio noise. From 1934 – 1939 CISPR issued reports RI–1 through 8 that specified EMC receivers, bandwidths and detectors, field strength measurements, artificial mains networks and the techniques for measuring EMI. As we have seen in chapter 1, there are three European standardisation bodies namely CEN, CENELEC and ETSI. Out of these the CENELEC and ETSI are responsible for making EMC standards. CENELEC provides EMC standards for EMC directive and all other directives requiring EMC tests, except the radio equipment directive (for which ETSI makes EMC standards). The procedure for making an EMC standard is the same as discussed in chapter 1. Further, EN EMC standards can also be grouped into three broad categories viz. basic standards, product standards and generic standards.

5.11.1 Basic Standards

As the name suggests, these standards deal with basic EMC phenomenon and methods to test equipment for such phenomenon. Basic standards are general and hence are not dedicated to specific product families or products; they relate to general information, to the disturbing phenomena and to the measurement or testing techniques. CENELEC produces the *EN 61000* series of basic standards which generally cover the *safety aspects* of electromagnetic compatibility. A majority of EN 61000 series embody immunity aspect of EMC testing and prescribe methods to assess immunity of equipment to various EMC phenomenon which may otherwise cause impairment of equipment safety. For example, the basic standard EN 61000-4-

2 deals with methods to test equipment's immunity towards the phenomenon of electrostatic discharge (ESD). Some basic standards (notably the 61000-3 series) also deal with low frequency emission phenomenon (like harmonics) which, if unchecked, can cause safety impairment. Some basic standards are also based on CISPR standards, notably the EN 55016 series which give detailed requirements for EMC instrumentation for emission measurement like transducers and EMI receivers.

5.11.2   Product Standards

These standards provide EMC requirements for a particular product (called product specific) or to a particular group of products depending upon end use (called product family standards). Product specific standards are prepared by product committees while many of product family standards are based on CISPR standards. Here basic standards are taken as reference and limits/severity levels/performance criterion are specified for the type and severity of EMC phenomenon applicable for that environment in which the product operates.

5.11.3   Generic Standards

These are generally applicable to those products or product family for which no product standard exists. Here again, basic standards are taken as reference for immunity and low frequency (below 9kHz) emission aspects and product family standards for high frequency aspects above 9kHz.

5.11.4   More On Basic Standards

This series consists of six parts EN 61000-1 to -6. EN 61000-1 is introductory in nature and explains fundamental principles, functional safety, definitions, terminology etc. and consists of subparts 61000-1-1 to 61000-1-6. EN 61000-2 deals with the description of test and operational environment, classification of the environment and compatibility levels and consists of subparts 61000-2-1 to 61000-2-14. EN 61000-3 with subparts 61000-3-1 to 61000-3-12, deals with low frequency phenomenon of harmonics and flicker emission and prescribes appropriate limits. EN 61000-4 series with subparts 61000-4-1 to 61000-4-29 titled *Testing and measurement techniques* for gauging *immunity* of equipment to various EMC phenomenon and is by far the most frequently referred series of basic standards. They give a basic introduction to the phenomenon, a description of the various transients/RF to be simulated, the severity levels to which the equipment under test should be exposed to, the design of generators and coupling networks to generate and couple transients onto power line and on equipment enclosure, the method of presentation of results and calibration of the generators and couplers. EN 61000-5 deals with installation and mitigation guidelines while EN 61000-6 are the generic standards (EN 61000-6-1 to 61000-6-4).

5.12   OVERVIEW OF EN BASIC STANDARDS

Given below is a list of some basic standards that have been published until now. Of immediate use for a majority of products and product manufacturers, are the EN 61000-3, 61000-4 and 61000-6 series. We will now take a brief look at some of the frequently referred basic standards. The tests are discussed in detail in the next chapter.

5.12.1   EN 61000-3 series : The following table gives the standards included in this series:

Table 5.1: EN 61000-3 series

Standard	Regarding
EN 61000-3-2	Measurement of 50Hz *harmonic current* emissions up to 2kHz for electrical and electronic equipment having an input current up to and including 16A per phase, and intended to be connected to public low-voltage distribution systems (nominal voltage 220V or higher).
EN 61000-3-3	Measurement of *voltage fluctuations and flicker* for electrical and electronic equipment as defined in EN61000—3-2.
EN 61000-3-11:	Measurement of *voltage fluctuations and flicker* for electrical and electronic equipment having an input current greater than

	16A per phase and up to and including 75A per phase.
EN 61000-3-12	Measurement of **harmonic current** emissions for electrical and electronic equipment having an input current greater than 16A per phase and up to and including 75A per phase.

5.12.2 EN 61000-4 series of basic standards

This is the most important series of EN standards and gives test and measurement techniques for most EMC phenomenon. A cursory look at these standards is presented in table 5.2. Detailed description of the tests is available in the next chapter on EMC testing. These are immunity standards i.e. they prescribe various tests to ascertain immunity of an equipment to various EMC phenomenon.

Table 5.2: EN 61000-4 series

Standard	Regarding
EN 61000-4-1	Reference to the technical committees of IEC or other bodies, users and manufactures of electrical and electronic equipment on EMC immunity specifications and tests.
EN 61000-4-2	Simulates immunity against electrostatic discharge when a person charged to static voltage touches an equipment.
EN 61000-4-3	Test method for gauging equipment immunity against radio waves generated by radio and TV transmitters, cell phones, wireless sets etc. (called *radiated immunity*/susceptibility).
EN 61000-4-4	Effect on an equipment of *electrical fast transients* (EFT) which are repetitive transients of fast (short) rise times generated in "bursts" due to effects like relay contact bouncing and switching of inductive loads.
EN 61000-4-5	Effect on an equipment of *surges* generated by indirect lightning strikes, switching of capacitor banks, electrical faults etc which can cause high energy surges.
EN 61000-4-6	Effect on equipment operation because of disturbance currents induced in cables (mains or data) by high power RF transmitters and other RF sources (called *conducted RF immunity*/susceptibility).
EN 61000-4-8	Effect on equipment of *power frequency magnetic fields* generated in the vicinity of transmission lines.

EN 61000-4-9	Effect of *pulsed magnetic field* generated in electrical plants.
EN 61000-4-10	Effect of *damped oscillatory magnetic fields* generated due to faults on equipment installed in electricity generating stations and switchyards.
EN 61000-4-11	Simulation of effect of v*oltage dips, short interruptions and voltage variations* (also called sags or brown outs) on electrical and electronic equipment fed by low-voltage power supply networks and having an input current not exceeding 16A per phase.
EN 61000-4-18	Effect of *damped oscillatory surges* generated due to faults on equipment installed in electricity generating stations and switchyards.

5.13   GENERIC STANDARDS

5.13.1 EN 61000-6-1

This standard gives immunity requirements for all apparatus used in residential, commercial and light industry environment for which no dedicated product or product family immunity standards exist. It calls for tests like electrostatic discharge, radiated immunity, fast transients, surges, conducted RF immunity, power frequency magnetic field and voltage dips and interrupts. It specifies the severity levels and performance criterion and refers to basic standards for test methods.

5.13.2 EN 61000-6-2

Here immunity requirements for all apparatus used in industrial environment are prescribed, for which no dedicated product or product-family immunity standard exists, but excluding radio transmitters. Tests are the same as 61000-6-1 except at severities one level higher.

5.13.3 EN 61000-6-3

This is a standard which gives emission requirements for all apparatus used in residential, commercial and light industrial environment for which no dedicated product or product family standards exist. It prescribes two tests namely conducted and radiated emission to be carried out according to EN 55011, Class 'B'.

5.13.4 EN 61000-6-4

This standard gives emission requirements for all apparatus used in industrial environment, for which no dedicated product or product-family immunity standard exists. It prescribes two tests namely conducted and radiated emission to be carried out according to EN 55011, Class 'A'.

5.14   "EN 550 " SERIES OF STANDARDS

As seen earlier, CISPR was established as an independent committee of IEC in 1934. In 1950, CISPR came under the umbrella of IEC as one of its committees. Presently CISPR develops EMI emission standards which are essentially product family standards. When adopted by CENELEC, the letters CISPR are replaced by *"EN 550"*. Given below is the scope and a detailed list of equipment within the scope of each of this series.

5.14.1 EN 55011: Industrial, Scientific And Medical (ISM) Equipment-Radio Disturbance Characteristics-Limits & Methods Of Measurement

It gives emission requirements for equipment designed to intentionally generate RF energy for industrial, scientific and medical (ISM) purposes, including spark erosion. It classifies equipment into classes and groups. Equipment suitable for use in domestic establishments are classified as Class 'B' while those used all establishments other than domestic (read industrial) are classified as Class 'A'. Equipment which generate conductively coupled RF energy for its internal functioning is classified into Group 1 while those that generate RF in the form of radiation for material inspection, analysis/treatment and processing are classified as Group 2. It also includes spark erosion equipment. Tests specified include conducted emissions (referred to as mains terminal disturbance voltage) from 150kHz to 30 MHz and radiated emission of electromagnetic fields (called electromagnetic radiation disturbance), outside ISM bands, from 30MHz to

1000MHz and from 1GHz to 18 GHz for equipment using frequencies above 400MHz. Group 2 Class A equipment are subject to relaxed limits as compared to Group1 Class A, while limits for Class B equipment are the most stringent.

5.14.2 EN 55014: Requirements For Household Appliances, Electric Tools And Similar Apparatus.

Part 1 (14-1): Emission Requirements

Part 1 of this standard calls for conducted emission (called terminal voltage) from 148.5kHz to 30MHz. Discontinuous interference (clicks) must also be measured as a special case of conducted emissions for appliances which generate such interference through switching operations. Disturbance power from 30MHz to 300MHz on mains lead is specified to be measured by means of the absorbing clamp. It also specifies radiated emission measurement from 30 MHz to 1 GHz if disturbance power measurements are above limits reduced by 10 dB and equipment clock frequency is above 30MHz. It gives less stringent limits for electric tools and the load terminal of regulating controls.

Part 2 (14-2): Immunity Requirements.

This standard gives immunity requirements. It calls for tests like electrostatic discharge, radiated RF field, fast transients, surges, radio frequency common mode, power frequency magnetic field and voltage dips and interrupts on the AC power input port. It specifies the severity levels and performance criterion and refers to basic standards for test methods.

5.14.3 EN 55015: Limits And Methods Of Measurement Of Radio Disturbance Characteristics Of Electrical Lighting Equipment.

This standard includes all lighting equipment with a primary function of generating and/or distributing light intended for illumination purposes, and intended either for connection to the low voltage electricity supply or battery operation. It includes neon signs and flood lighting equipment but excludes lighting equipment operating in ISM bands, lighting equipment in automobiles, aircraft and runway lighting. It specifies tests like conducted emissions (called disturbance voltage) at mains (from 9kHz to 30 MHz), load and control terminals (from 150kHz to 30 MHz), radiated emissions magnetic fields (measured with a Van-Veen loop antenna) from 9kHz to 30 MHz and radiated emissions electromagnetic fields from 30MHz to 300MHz in open area test site or anechoic chamber.

5.14.4 EN 55016-1: Specification For Radio Disturbance And Immunity Measuring Apparatus And Methods.

EN 55016-1-1: Measuring apparatus.

This part of EN 55016 is designated as a basic standard, which specifies the characteristics and performance of EMI receivers, spectrum analyzers or other measuring equipment used for the measurement of radio disturbance voltages and currents (conducted emission) and fields (radiated emission) in the frequency range 9 kHz to 18 GHz. In addition, requirements are specified for specialized equipment for discontinuous disturbance measurements i.e. the *click analyzer*. The requirements include the measurement of broadband and narrowband types of radio disturbance using different detectors. The detectors covered include the quasi-peak, the peak, the average and the r.m.s type.

EN 55016-1-2: Ancillary equipment - Conducted disturbances.

This part specifies the characteristics and performance of ancillary equipment for the measurement of radio disturbance voltages and currents in the frequency range 9 kHz to 1 GHz and includes LISNs or AMNs (see section 5.15), current and voltage probes and coupling networks for current injection on cables.

EN 55016-1-3: Ancillary equipment - Disturbance power

This part defines the characteristics and calibration of the absorbing clamp for the measurement of radio disturbance power in the frequency range 30 MHz to 1 GHz. A procedure for the validation of the absorbing clamp test site (ACTS) is also included in the document.

EN 55016-1-4: Ancillary equipment - Radiated disturbances

This part gives characteristics and performance of equipment for the measurement of radiated emission in the frequency range 9 kHz to 18 GHz. Specifications for ancillary equipment like antennas, TEM cells and reverberating chambers are also given.

EN 55016-1-5: Antenna calibration test sites for 30 MHz to 1000 MHz

This part outlines the requirements for calibration test sites, used to perform antenna calibrations, as well as the test antenna characteristics, calibration site verification procedure and site compliance criteria.

5.14.5 EN 55016-2: Specification For Radio Disturbance And Immunity Measuring Apparatus And Methods

EN 55016-2-1: Conducted disturbance measurements

This standard specifies the methods of measurement of disturbance phenomena in general in the frequency range 9 kHz to 18 GHz and especially of conducted disturbance phenomena in the frequency range 9 kHz to 30 MHz. Guidelines are given as to how resonances can be avoided while connecting the LISN/ AMN to reference ground, how to avoid ground loops, and avoidance of ambiguities in the test setup.

EN 55016-2-2: Methods of measurement of disturbances and immunity - Measurement of disturbance power

This part of specifies the methods of measurement of disturbance power using the absorbing clamp in the frequency range 30 MHz to 1 000 MHz. EN 55016-2-2 has been reorganised into 4 parts, to accommodate growth and easier maintenance (see Fig. 2.4). This edition of EN 55016-2-2, together with EN 55016-2-1, EN 55016-2-3 and EN 55016-2-4, cancels and replaces the second edition of EN 55016-2, published in 2003. It contains the relevant clauses of EN 55016-2 without technical changes.

EN 55016-2-3: Radiated disturbance measurements

This part specifies the methods of measurement of radiated disturbance phenomena in the frequency range 9 kHz to 18 GHz.

EN 55016-2-4: Immunity measurements.

This part specifies the methods of measurement of immunity to EMC phenomena in the frequency range 9 kHz to 18 GHz.

5.14.6 EN 55016-3: Technical Reports

This part of EN 55016 contains specific technical reports and information on the history of CISPR. With the reorganization of EN 55016 in 2003, a significant portion of 16-3 has been moved into the EN 55016-4-1, EN 55016-4-3 and EN 55016-4-4.

5.14.7 EN 55016-4: Uncertainties, Statistics And Limit Modelling.

This standard has following sub-parts:

EN 55016-4-1: Uncertainties in standardized EMC tests

The objectives of this part are to identify the parameters or sources governing the uncertainty (called *standards compliance uncertainty*) during measurement and to provide guidance on the estimation of the magnitude of the standards compliance uncertainty and to implement it into the compliance criterion of an EMC test.

EN 55016-4-2: Uncertainty in EMC measurement instrumentation.

This part specifies the manner in which measurement uncertainty is to be taken into account in determining compliance with limits.

EN 55016-4-3: Statistical considerations in the determination of EMC compliance of mass-produced products

This part of EN 55016 deals with statistical considerations in the determination of EMC compliance of mass-produced products.

EN 55016-4-4: Statistics of complaints & model for calculation of limits.

This part contains information on how to deal with statistics of radio interference complaints.

5.14.8 EN 55022: Information technology equipment – Radio disturbance characteristics- Limits and methods of measurement.

This standard deals with information technology equipment (ITE). It has been withdrawn and replaced by EN55032 –EMC of multimedia equipment (MME).

5.14.9 EN 55024: Information technology equipment – Immunity characteristics

It gives immunity requirements for IT equipment. Tests are the same as those called for EN 61000-6-1, with the difference that test set-ups, severity levels and performance criteria are as per typical environment in which IT equipment operate.

5.14.10 EN 55032: Electromagnetic Compatibility of Multimedia Equipment (MME)–Emission requirements

This standard replaces EN 55022. Multimedia equipment is defined as any equipment which is:

I.T. equipment: Equipment whose primary function is either (or a combination of) data entry, storage, display, retrieval, transmission, processing, switching or control and which may be equipped with one or more terminal ports typically operated for information transfer.

OR

Audio Equipment: Equipment which has a primary function of either (or a combination of) generation, input, storage, play, retrieval, transmission, reception, amplification, processing, switching or control of audio signals.

OR

Video equipment: Equipment which has a primary function of either (or a combination of) generation, input, storage, display, play, retrieval, transmission, reception, amplification, processing, switching or control of video signals

OR

Broadcast receiver equipment: Equipment containing a tuner that is intended for the reception of broadcast services.

OR

Entertainment lighting control equipment: Equipment generating or processing electrical signals for controlling the intensity, colour, nature or direction of the light from a luminaire, where the intention is to create artistic effects in theatrical, televisual or musical productions and visual presentations.

OR

A combination of the above equipment

Equipment is classified as Class 'A' and Class 'B' similar to CISPR11. It specifies conducted emissions (called interference voltage) from 150kHz to 30MHz on mains and on telecommunications port, radiated emission (called radiated disturbance) from 30MHz to 1GHz measured at 10m or 3m in an open area or semi anechoic chamber (SAC) or fully anechoic room (FAR). It also specifies additional radiated emission measurement from 1GHz to 6GHz if the equipment has internal frequency sources greater than 108 MHz.

5.15   EMC TESTING

To ascertain whether a particular product meets EMC requirements and to form a basis of comparison of EMC performances of different products, it is required to test products in a controlled environment like an EMC test lab. In the chapters that follow, we are going to take a look at various EMC test methodologies, test setups and instrumentation used for EMC tests. For an equipment to be declared as EMC compliant, it is required to carry out two basic type of tests. One category checks whether the emissions are below the limit and these tests are called emission tests. The other category checks the immunity of the equipment towards EMI (generated by other equipment) and such tests are called as immunity or susceptibility tests. The Fig. 5.6 gives some of the basic EMC tests as prescribed by EN standards.

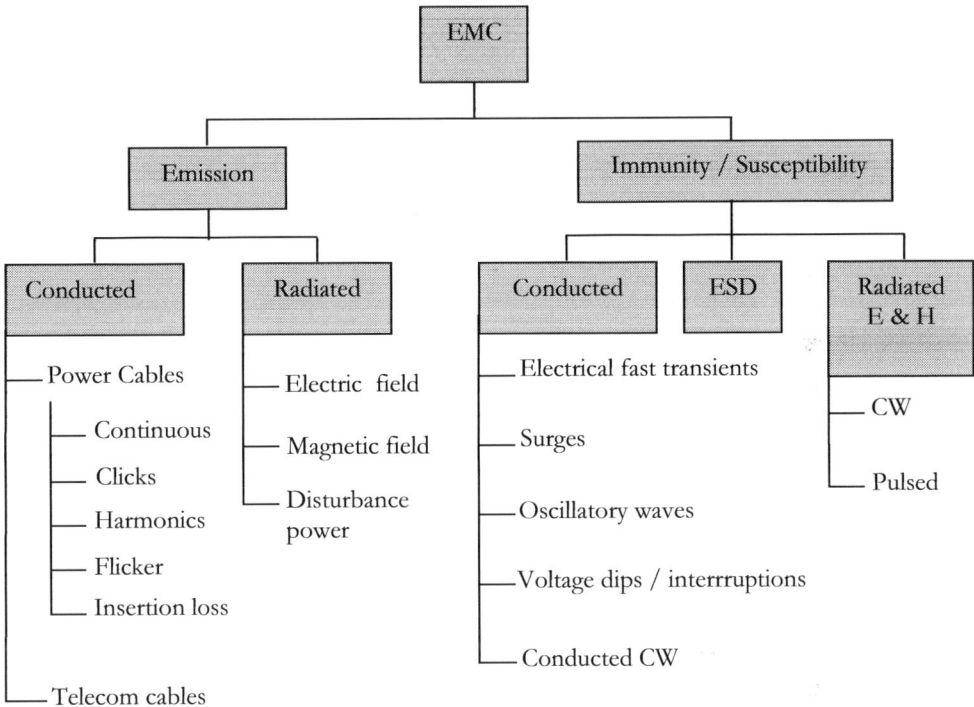

Fig. 5.6: EMC Tests

5.18.1: Emission Tests

Emission tests measure RF noise emitted by the equipment onto power mains and into space. There are two broad categories of emission tests namely conducted emission which is the measurement of emission on the power lines and radiated emission which measures the EMI generated in the form of radiation.

5.18.2: Immunity Tests

Immunity tests are exact opposite to emission tests wherein the EMI is deliberately generated in a controlled manner, by a dedicated test equipment. The equipment under test (EUT) is then exposed to this simulated EMI and its behaviour is studied. Like emission, immunity tests can also be classified broadly into conducted immunity and radiated immunity. The former studies the effect of EMI simulated in the form of a current or voltage (which can be transient or continuous) and superimposed on the lines that interface the EUT with the outside world which can be mains or I/O lines. In the latter category, the EUT is exposed to simulated radiations of EMI (usually in a shielded enclosure) and its performance monitored for degradation.

5.19   CONDUCTED EMISSION

Conducted emission (CE) is the EMI emitted by equipment in the form of a current through any conductor that interfaces the equipment with the outside environment. The three most commonly referenced standards like EN 55011, 55014–1 and 55032 specify measurement in the frequency range from 150 kHz to 30 MHz on the mains cable. EN55013 (for broadcast receivers) and EN 55015 (for lighting equipment) also require a similar test, although EN 55015 calls for measurement from 9 kHz to 30 MHz. In addition to mains cable, the standard EN 55032 also calls for measurement on telecom cables. There are certain standards which call for CE measurement below 9kHz. Such measurement includes power frequency harmonics (EN 61000–3–2) which is measured from fundamental mains frequency to its 40th harmonic.

5.20   CE TEST SETUP

The basic test set up for high frequency CE measurement, as prescribed by a majority of EMC standards, is shown in Fig. 5.7. The EUT is connected to the mains through a line impedance stabilisation network (LISN), also called as artificial mains network (AMN). The LISN/AMN separates the conducted emission from the mains, which is then measured by a scanning EMI receiver that scans the spectrum from the lowest frequency of measurement (called *start* frequency) to the highest (called *stop* frequency) in discreet steps. In simpler terms, it tunes to a frequency much like a radio receiver, stays there for some time (called *dwell time*), plots the amplitude (in dBμV) and moves to the next frequency. The measurement result is thus a plot of emission amplitude (Y axis) against frequency (X axis) as shown in the Fig. 5.7. The emissions are compared with recommended limits to decide compliance.

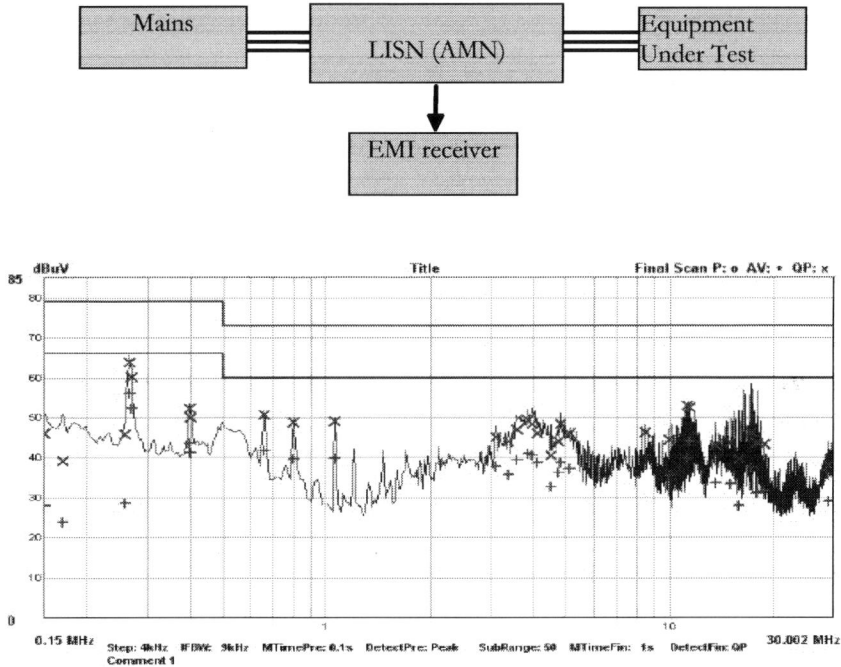

Mains — LISN (AMN) — Equipment Under Test

EMI receiver

dBuV   Title   Final Scan P: o AV: • QP: x

0.15 MHz   Step: 4kHz IFBW: 9kHz MTimePre: 0.1s DetectPre: Peak SubRange: 50 MTimeFin: 1s DetectFin: QP   30.002 MHz
Comment 1

Fig. 5.7: Measurement Instrumentation And Graph

5.20.1 LISN / A.M.N

The LISN performs three functions. Firstly, it acts as a transducer/buffer and isolates the EMI receiver from the mains. Secondly, it functions as a filter and ensures that the EUT is fed with a pure mains supply. Thirdly, the LISN stabilizes the impedance of mains to 50 Ω to match with the 50 Ω input impedance of the EMI receiver.

5.20.2 EMI Receiver

The Fig. 5.8 shows a block diagram of typical EMI receiver as specified in EN55016–1–1. It is nothing but a super–heterodyne radio receiver. The signal at the input first passes through an input attenuator and then via a pre-selector to the RF amplifier. The pre-selector is a bandpass filter which allows only a small band around the tuned frequency to pass through. This is essential to reduce measurement errors since the receiver usually measures pulsed EMI which has a large number of frequency components. Exposing the front end to such a large band will cause overloading the RF amplifier generating spurious responses leading to measurement errors.

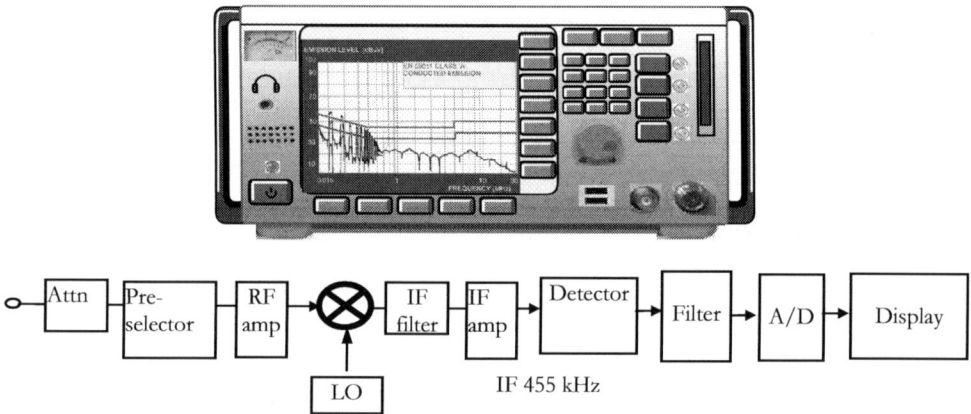

Fig. 5.8: EMI Receiver

The mixer stage which follows the RF amplifier, mixes or heterodynes the incoming signal with signal generated by the local oscillator (LO) to generate an intermediate frequency (IF). The bandwidth of the IF filter, which follows the mixer is called resolution bandwidth. The final IF is fed to the detector stage the output of which is displayed on a screen or printed by a printer/plotter. Three types of detectors are prescribed by EN standards viz. peak, quasi-peak (QP) and average.

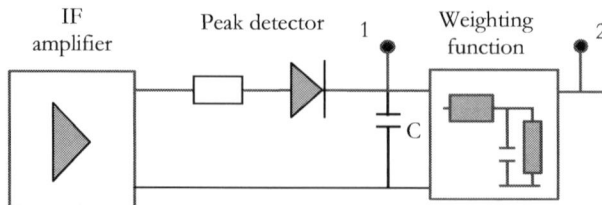

Fig. 5.9: Basic Detector

The Fig. 5.9 shows a basic detector circuit used by most EMI receivers. Average and QP detectors are realized by a peak detector followed by appropriate *weighting function*. In case of un–modulated sine wave voltages at receiver input, all the detectors give equal output readings. The scenario is different for pulsed input and in such a case, the output of the QP detector depends on the pulse repetition frequency (PRF) of the input.

5.20.2.1 Peak detector

A peak detector is characterized by very short time constant for charging and an extremely long time constant for discharging. As shown in the Fig. 5.9, the output at point '1' (before the weighting function) is the output of a peak detector. In the absence of the weighting function (essentially a pulse shaping network) the capacitor 'C' has no path to discharge. The peak detector therefore latches on to maximum value and holds it on for a long time. As a result, the peak detector gives peak maximum value of signal at any moment. Fig. 5.10 shows the response of a peak detector to a train of pulses with changing PRF. As can be seen the output of the peak detector is maximum and does not depend on signal PRF. Normally, measurements with a peak detector can be very fast and will result in highest output.

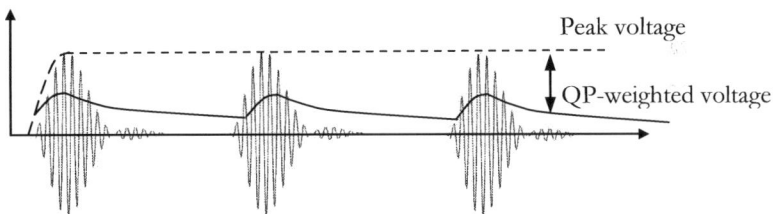

Fig. 5.10: Detector Response Low PRF

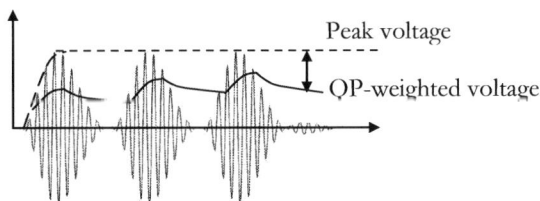

Fig. 5.11: Detector Response High PRF

5.20.2.2 Quasi–peak (QP) detector

This is a peak detector weighed by appropriate charge and discharge time constants such that the QP detector *weighs* signals on basis of *annoyance factor* so that a signal with a higher PRF produces a higher output in the QP detector. As shown in the Fig. 5.10, at low PRF, the output of the QP detector is low. As the PRF increases (see Fig 5.11), the output also increases.

5.20.2.3 Average Detector

EMI receivers also employ average detectors achieved by employing a low pass filter (called video filter) after the peak detector whereby broadband components are averaged. Averaging occurs when the bandwidth of the video filter (called video bandwidth) is narrower (less) than resolution bandwidth. The Fig. 5.12 shows the response of an average detector. The input to the detector is shown to consist of two components —one is broadband noise component which is superimposed on a narrowband component (shown dotted). Since the detector employs a low pass filter, only the narrowband component is allowed to pass through while the broadband component is blocked so that the output of the detector   is   only   the   narrowband   component.

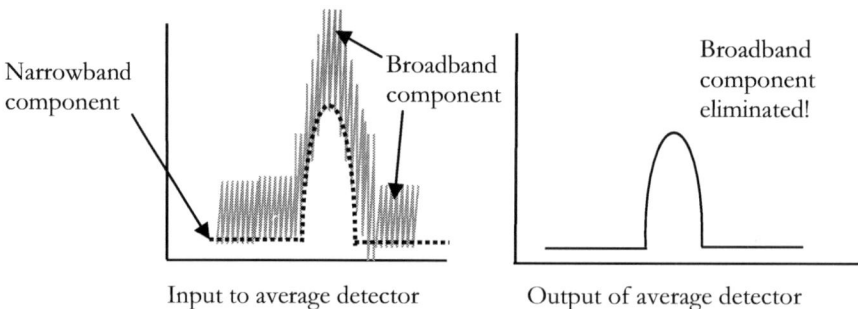

Input to average detector   Output of average detector

Fig.5.12: Output Of Average Detector

5.21   CONDUCTED EMISSION LIMITS

All EMC standards dictate that the emission of an equipment shall be below certain limits. EN standards identify two types of limits depending upon the environment in which the equipment is going to be installed and made operational.

Class 'B' equipment is equipment suitable for use in domestic establishments and in establishments directly connected to a low voltage power supply network which supplies buildings used for domestic purposes. Class 'B' equipment should meet Class 'B' limits. Class 'A' equipment is equipment suitable for use in all establishments other than domestic (i.e. industrial use equipment). Class 'A' equipment should meet (i.e. emission should be below) Class 'A' limits.

For conducted emission, EN standards specify QP limits and average limits while using a QP and average detector respectively. EN55011 identifies higher limits for equipment with power rating greater than 20 kVA. It goes even further and divides equipment into following groups:

> **Group 2 ISM equipment:** Contains all ISM equipment in which radio–frequency energy in the frequency range 9 kHz to 400 GHz is intentionally generated and/or used in the form of electromagnetic radiation for the treatment of material, inspection and analysis or for transfer of electromagnetic energy. Examples are industrial induction heating equipment, microwave ovens, HF surgical equipment etc.
>
> **Group 1 ISM equipment:** Includes all ISM equipment which is not classified as group 2.

The tables 5.3 and 5.4 specify CE limits for class 'A' equipment and class 'B' equipment respectively as specified by EN 55011 (for group 1) and EN 55032. Same limits are prescribed for CE measurement by the EN 60601 series of standards under the medical devices directive and EN 301489 series of standards under the radio equipment directive. The measurement unit is dBμV, which simply put is μV converted into dB as per the formula below:

$$\text{Voltage in dB}\mu V = 20\log (\text{Voltage in } \mu V)$$

For example,
$$1 \ \mu V = 0 \ \text{dB}\mu V$$
$$1 \ \text{mV} \ (1000 \ \text{uV}) = 60 \ \text{dB}\mu V$$
$$1 \ V = 120 \ \text{dB}\mu V$$

While doing an actual CE measurement, the limits are superimposed on the graph. The Fig. 5.12 shows class 'B' limits superimposed on the graph.

Table 5.3: Class 'A' limits

Frequency (MHz)	Limits (dBµV)	
	Quasi– Peak	Average
0.15 to 0.5	79	66
0.5 to 30	73	60

Table 5.4: Class 'B' limits

Frequency (MHz)	Limits (dBµV)	
	Quasi– Peak	Average
0.15 to 0.50	66 to 56	56 to 46
0.5 to 5	56	46
0.5 to 30	60	50

Deciding compliance can be somewhat tricky as EN / CISPR standard measurements are a little complex due to the fact that there are two type of limits viz QP and average and corresponding detectors.

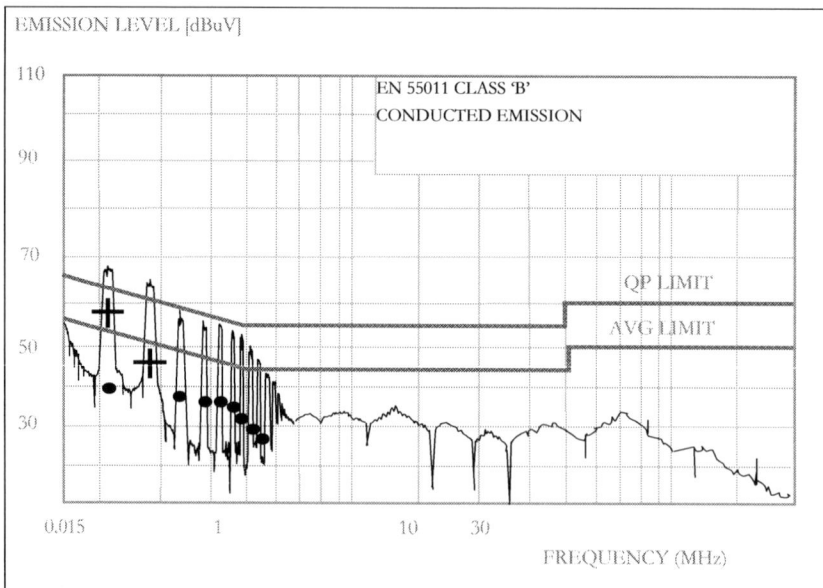

Fig. 5.12: CE Graph

The Fig. 5.12 gives a typical conducted emission graph. Here the frequency (in MHz) is plotted on the X–axis and amplitude (in dBμV) is plotted on the Y–axis. The QP and average limits are shown as bold lines parallel to the X–axis. Although EN standards prescribe measurement with a quasi–peak and average detector, direct measurement with these detectors can consume a lot of time. Taking advantage of the fact the measurement with a peak detector is very fast and gives the highest reading, EN standards allow an initial measurement with a peak detector.

The requirement is that, irrespective of the detector used, emission must be below the average limit so as to declare the EUT as compliant. The typical procedure followed is that an initial scan (called a *pre–scan*) is carried out over the measurement range with a peak detector. If all the peaks are below the average limit, further measurement with QP and average detector is not required (as the emission will invariably be either equal to or below that carried out with a peak detector) and the equipment can be declared as compliant. If, after initial measurement with a peak detector, some of the peaks are above QP limit (first two peaks in graph 5.12), measurement is repeated over such peaks with a QP detector (shown by '+' sign in graph 5.12). After QP measurement, if the peaks are below average limit, the equipment can be declared as compliant but if they are above the QP limit, the equipment is non-complaint. However, even after QP measurement, if some peaks remain below the QP limit but are above the average limit, the measurement is repeated over such peaks with an average detector (shown by a dot in graph 5.12). Now if the peaks are below average limit, the equipment is compliant, otherwise it is a case of non–compliance.

5.22   RADIATED EMISSION (RE) MEASUREMENT

The basic test set–up for radiated emission (RE) measurement is as shown in the Fig. 5.13. The radiated emission of equipment under test (EUT) is intercepted by an antenna kept at a certain distance (called *test distance*) usually 10m from the EUT. The antenna is the transducer, which converts the radiated *field* generated by the EUT into an equivalent *voltage*. This voltage is then measured by an EMI receiver over a certain frequency range and is compared with the limit to decide compliance.

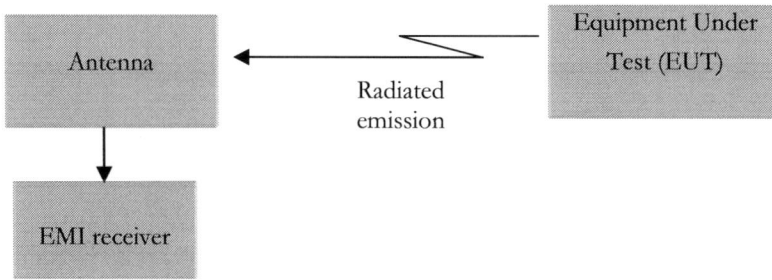

Fig. 5.13: Basic Test Set–Up

5.23   MEASUREMENT INSTRUMENTATION

5.23.1 EMI Receiver

EMI receiver measures the voltage appearing across antenna terminals. Most receivers cover frequency range specified by EMC standards for conducted and radiated emission and therefore EMI receivers used for radiated emissions measurement are the same as used for conducted emissions measurement as described in previous sections.

5.23.2 Antennas

Antennas are the basic transducers used for radiated emissions and susceptibility testing. The choice of antenna depends on the standard being referred for testing, the frequency range of measurement and the type of field being measured. EN55016–1–4 prescribes two types of antennas namely bi–conical and log–periodic.

5.23.2.1 Bi–conical antenna

A typical bi–conical antenna preferred by EN55016–1–4 for measuring the electric component of electromagnetic field in the frequency range of 30 to 230 MHz, is as shown in the Fig. 5.14. It is a modified dipole antenna wherein the dipoles have been *flared* into a conical shape which gives the antenna its name.

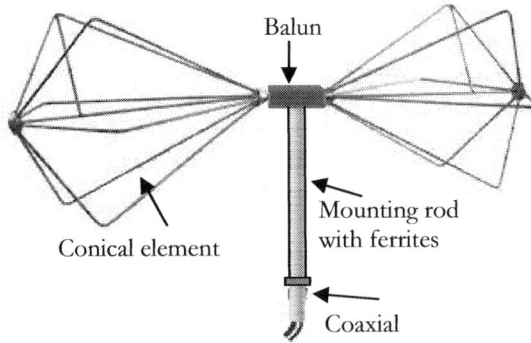

Fig. 5.14: Biconical Antenna

5.23.2.2 Log–periodic antenna

A typical log–periodic antenna (also called as log periodic dipole array or LPDA) is as shown in the Fig. 5.15. This is the preferred antenna type specified by EN55016–1–4 for electric component of electromagnetic field measurement in the frequency range of 230 to 1000 MHz.

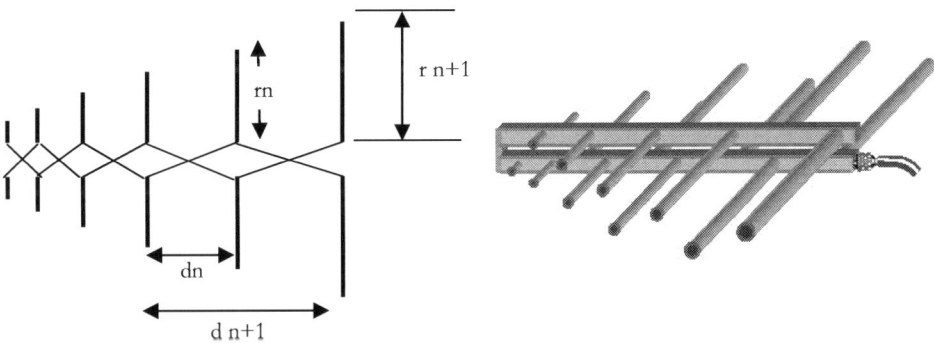

Fig. 5.15: Log–periodic Antenna

5.23.2.3 Hybrid antenna

It is clear from sections 5.23.2.1 and 5.23.2.2 that if one has to carry out measurement from 30 MHz to 1GHz one would require two antennas and that the measurement will have to be interrupted for changing the antennas at 230MHz. To overcome this problem, antenna manufacturers have come up with a unique solution that of combining bi–conical and log–periodic antennas into a single hybrid antenna.

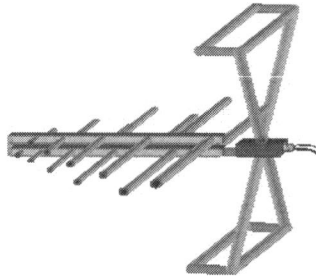

Fig. 5.16: Hybrid Antenna

It is commonly known as *Bi–log* or *Bi–coni–log* or *Ultra–log* or by other brand names. The general construction is shown in Fig. 5.16. It has a log–periodic array in front and a structure resembling a bow–tie (or other complex shape) that has the characteristics of a bi–conical antenna at the back with both the antennas being mounted on the same beam.

5.24 FREQUENCY RANGE OF MEASUREMENT

Radiated emission testing is carried out over a frequency range starting from 9 kHz and extends up to 18 GHz. The range of 9kHz to 30 MHz is where the magnetic field is measured, while from 30 MHz to 18 GHz is the range where electric field is measured.

5.25 LIMITS

As we have seen, EN standards identify two types of limits depending upon the environment in which the equipment is installed i.e. Class 'A' and Class 'B'. The units are dBμV/m which is the standard unit of field strength. As an example, the table 5.5 gives QP limits for group 1, class 'A' equipment (rated power ≤ 20kVA) as per EN55011 and EN55032 from 30MHz to 1000MHz at 10m test distance. For equipment with rated power greater than 20kVA, the limits are 10dB higher. The class 'B' limits are 10dB lower than the corresponding class 'A' limits. EN55032 calls for radiated emission measurement for a further frequency range of 1GHz to 6 GHz for certain category of equipment that use an internal RF energy source or tunes to frequencies above 108MHz. For all group 2 equipment, EN55011 allows unrestricted emissions over fundamental ISM frequencies. For frequencies other than those falling in the ISM bands, it gives limits for magnetic fields in the frequency range of 150kHz to 30 MHz and electromagnetic field limits from 30 MHz to 1 GHz.

Table 5.5: EN55011(group 1) & 55032 Limits

Frequency (MHz)	Quasi–peak Limits measured on a test site
	Group 1, class A 10m test distance dB(μV/m)
30–230	40
230–1000	47

5.26   MEASUREMENT SITE

Radiated emission measurement is carried out in an open area test site (OATS) which is the cheapest and most widely accepted method so far as test set–up and equipment cost is concerned. Alternate test sites include anechoic chambers, fully anechoic rooms (FAR) and transverse electromagnetic (TEM) cells. Let us take a look at these sites in.

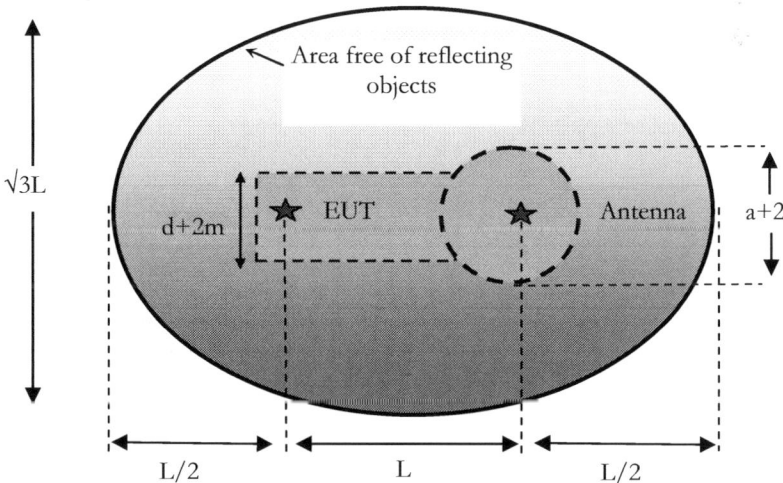

Fig. 5.17: OATS top view

5.26.1 Open Area Test Site (OATS)

The Fig. 5.17 shows the top view of an OATS specified by EN55016–1–4. The site should be flat, clear of overhead wires, and nearby reflecting surfaces. The distance between the EUT and antenna (shown as 'L') is 10m or 3m. The minimum test area (shown dotted) is in shape of a keyhole, the diameter of the circle is a+2m, where 'a' is antenna dimension while the width of the rectangle is d+2 m where 'd' is the maximum EUT dimension. This test area is covered by a ground reference plane made up of metallic sheet or a metallic mesh. The equipment under test (EUT) is kept on a non-metallic turn table. The antenna is mounted on a mast, which can vary the antenna height from 1m to 4m above ground plane.

5.26.2. Anechoic Chambers

Measurement in an OATS has a drawback (as will be clear after we see the test procedure) in the sense that the antenna tends to capture ambient EMI also and that the measurement becomes difficult or sometimes impossible, in presence of high ambient. To overcome this drawback, RE measurement can be carried out in shielded test enclosure like anechoic chamber (also called as absorber lined shielded enclosure (ALSE)). The walls of the chamber are made of a special alloy material which give them high conductivity as well as robustness. These walls act as a shield and help maintain a very low ambient within the enclosure.

5.27 MEASUREMENT PROCEDURE

During testing the EUT is kept on a turntable at the recommended test distance (normally 10m) from the antenna. In case of high ambient the testing can be carried out at 3-meter test distance but the limits are scaled (increased) to account for closer test distance either by a value specified by the standard or by using the expression 5.1

$$L1* d1 \ = L2 * d2 \qquad \ldots 5.1$$

L2: New limit, L1: Old limit, d2: New distance, d1: Old distance.

The Fig. 5.18 shows a typical radiated emission plot.

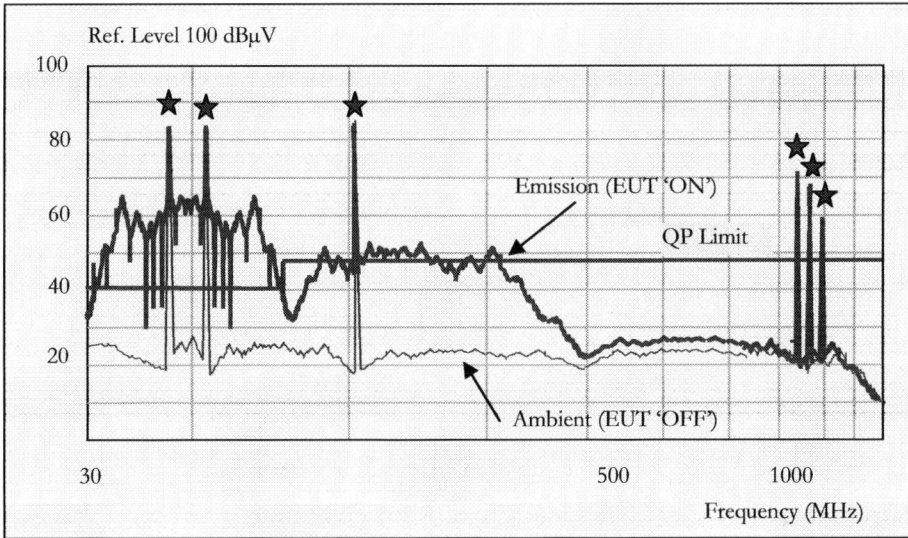

Fig. 5.18: Typical radiated emission plot

In case of testing in OATS, EUT is kept OFF initially and ambient is noted. Ambient should be ideally 6dB or if possible 20dB below limit. EUT is then switched ON. The EUT is rotated in azimuth by rotating the turntable and height of antenna (elevation) varied from 1 m–4 m so as to get maximum emission. The antenna is connected to the EMI receiver and emission peaks are measured in QP mode. When the EUT is OFF, the receiver records ambient EMI (lower graph of Fig.5.18). Most of the ambient is well below limit barring a few peaks (marked by the star). While the top graph shows the emissions when the EUT is switched ON. One can see the difference between measured emission levels and the ambient. Some peaks are common to both emission and ambient (marked by the star). These are from nearby broadcast transmitters either radio FM or TV as well as from cellular base stations. Since these were present when the EUT was OFF, it is clear that these were not generated by the EUT and hence can be discarded. A table of peaks is then drawn up. Appropriate transducer factor called antenna factor (AF) are added that convert measured voltage 'V' into equivalent electric field 'E' according to the expression 5.2. The final emissions (after adding AF) should be below QP limit.

$$E \ (dB\mu V/m) = V \ (dB\mu V) + AF \qquad \dots 5.2$$

5.28   CONDUCTED IMMUNITY TESTING

To establish electromagnetic compatibility an equipment must also be subjected to immunity/susceptibility tests in order to ensure that its operation is not disrupted in the presence of continuous radio–frequency noise and transient pulses which may be present in its operating environment. The source of transient pulses may be natural phenomena like electrostatic discharge (ESD) and lightning strikes or it may be artificial like conducted repetitive transients and ringing transients generated by fault currents. The source of continuous RF fields can be radiated noise generated by RF transmitters and power lines. Conducted immunity tests involve the application of a single transient or a series of transients or continuous wave EMI on power and communication cables (superimposed over mains/control voltage) accompanied by monitoring the function of the equipment under test so as to ascertain whether its operation has disrupted, the extent of disruption and whether the disruption is acceptable or not. The Fig. 5.19 shows the basic immunity tests as per EN standards. In this section we are going to concentrate on the conducted immunity tests. They can actually be divided into two broad categories viz. transients (like fast transients, surges and oscillatory waves) and continuous wave (like conducted RF immunity).

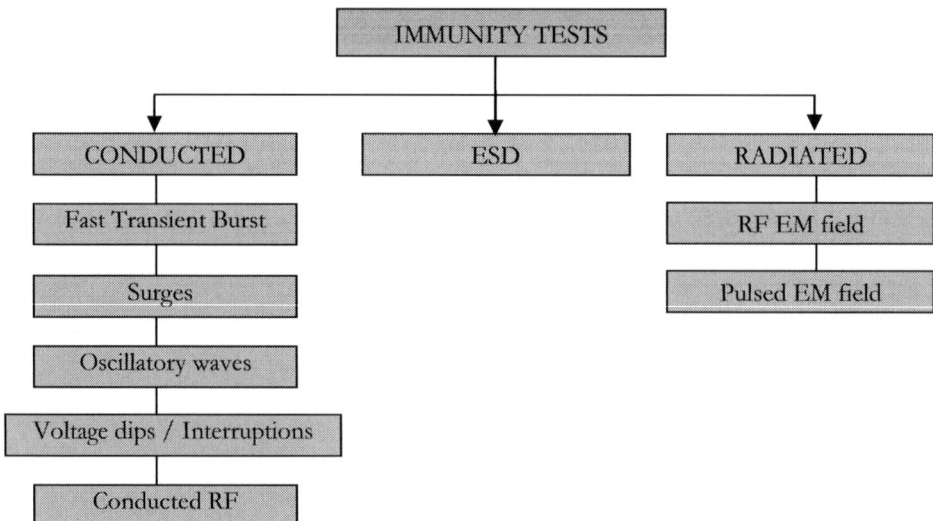

Fig. 5.19: Immunity Tests

5.29 GENERAL TEST SETUP

A test setup for conducted immunity testing is as shown in Fig. 5.20. It consists of either a transient or a RF generator that generates the specified simulations which are then coupled onto mains and/or I/O lines of the EUT using either a coupling–decoupling network (CDN) or a coupling clamp.

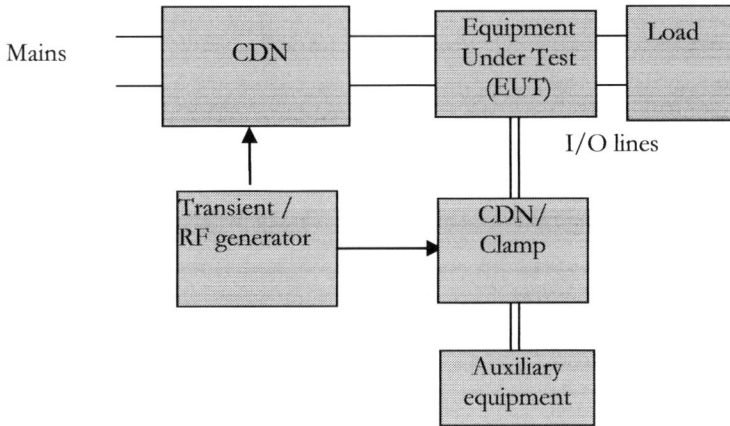

Fig. 5.20: General Test Set-up

5.30 ELECTRICAL FAST TRANSIENTS / BURST (EFT/B)

Switching of inductive loads, relay contact bouncing or switching of H/V switchgear can generate repetitive transients. Such transients are referred to as electrical fast transients or bursts (EFT/B). The amplitude can range from a few hundreds to a few thousands of volts. The transient can also cause an arc across switch contacts, leading to radiated EMI. The word *burst* is used because such transients normally occur in sudden bursts or packets with a lull in between, each packet corresponding to one opening of a contact. The effect of such transients can be simulated by the EFT/B test as given in the basic EMC publication EN 61000–4–4. The test uses specialized generators which generate the EFT/B in a controlled manner. Coupling–decoupling networks (CDNs) or capacitive clamps are then used to couple these on to power supply lines and control /signal lines respectively.

5.30.1 EFT/B Pulse Shape

The wave–shape as specified by EN 61000–4–4 is as shown in the Fig. 5.21. The Fig. 5.21a shows repetitive bursts, each burst is a *packet* of pulses of 15 ms burst width (or duration) and burst separation (or period) is 300 ms. The individual burst is shown in Fig. 5.21b, the frequency at which the pulses occur (the PRF) is 5 kHz. The individual pulse, shown in the Fig. 5.21c, has a rise time of 5ηs and pulse width of 50 ηs. EN 61000–4–4 also specifies an optional 100 kHz PRF in which case, the burst duration is 0.75 ms.

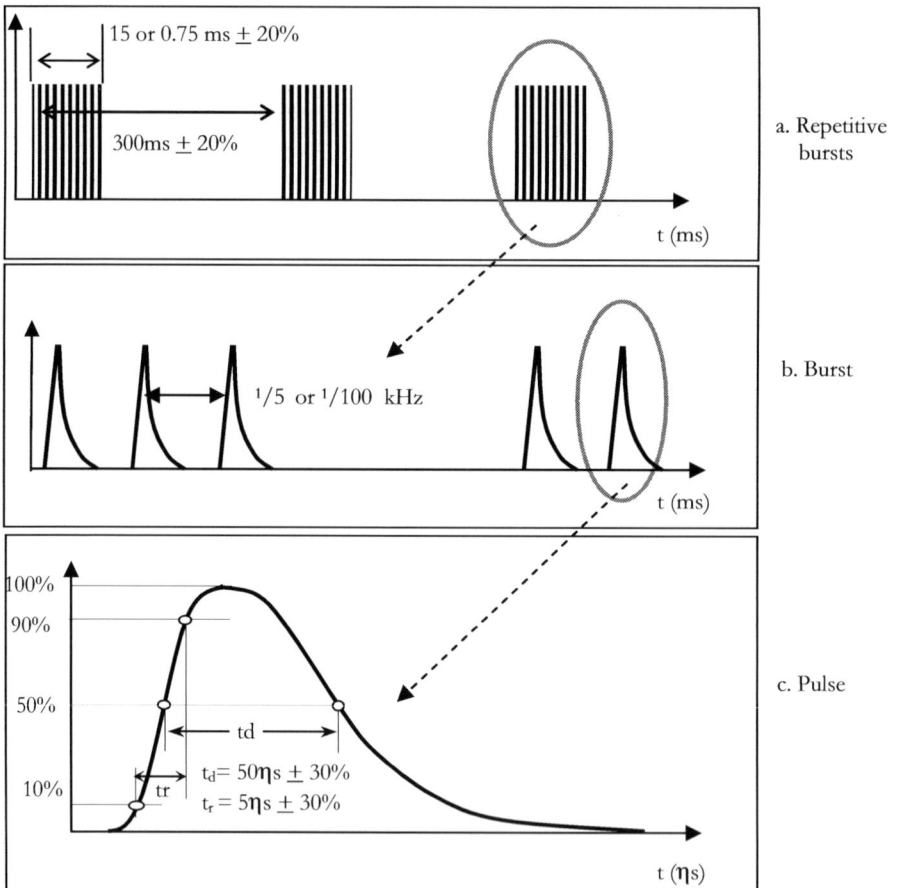

Fig. 5.21: EFT/B Pulse And Burst Characteristics

This typical burst is only a representative of real-life situation and is the best that can be achieved by a test generator. The source impedance of the generator is 50 Ω and the output is unbalanced. The generator generates specified amplitude calibrated to 50 Ω load but the actual voltage impressed on the EUT depends upon the EUT source impedance. The transients generated are coupled in non–symmetrical mode on to the power supply port of EUT using coupling/decoupling network (CDN) through a 33 ηF capacitor. For coupling EFT/B to input/output (I/O), control or communications cables of the EUT, a non-contact method using a capacitive coupling clamp is specified.

5.30.2 Laboratory Test Setup

The Fig. 5.22 shows how an equipment is tested for EFT/B in a laboratory. The equipment under test (be it table–top or floor standing) is placed above a ground reference plane (GRP) (a flat metal sheet whose potential is used as a common reference) and insulated from it by 0.1m thickness insulating support.

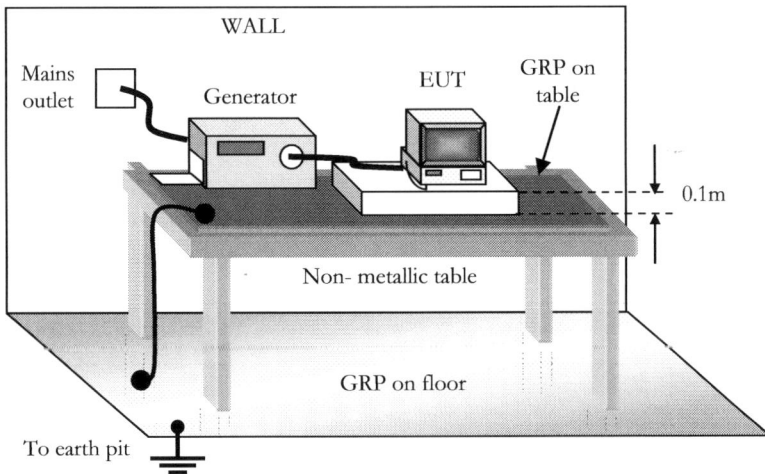

Fig. 5.22: EFT/B lab set–up for table–top equipment

The EFT/B generator is placed directly on the GRP and firmly bonded to it by a low inductance bond (i.e. a metal strap). The standard gives a test set–up drawing (see Fig. 5.22 where a table–top equipment is shown kept on a non–metallic table. The table itself is shown placed on a GRP on the floor, while there is another GRP kept on the table, connected to the GRP on the floor, by a cable. The EUT is shown to be kept at a distance of 0.1m over the *table* GRP. The height of the table is not specified, but can be assumed 0.8m as is with standard laboratory tables. For floor standing EUT, the table (and the table GRP) is not required.

5.30.3 Conducting An EFT/B Test

The test set–up is arranged as in Fig. 5.22. The test severity level is then decided. EN 61000–4–4 proposes four severity levels viz 0.5 kV (level–1), 1 kV, 2 kV and 4 kV (level–4) for EFT/B. The EUT is to be tested at any one level as dictated by the product standard being referred to. The transients are then applied to the EUT power supply, I/O or communication ports and earth. In case of testing on power mains, the transients are coupled via a coupling/de–coupling network to power supply and protective earth ports. Single phase power mains is tested in combinations of L, N, PE, L+N, L+PE, N+PE and the usual L+N+PE. While testing on I/O and communication port, the transients are applied using a capacitive coupling clamp. The performance of the EUT is monitored during the entire duration of the test and after the test. The section 5.43 gives information on the evaluation of test results.

5.31   SURGE TESTING

A surge is an over–voltage from switching & lightning transients which is high energy in nature. The major mechanisms by which the surge voltages are produced can be manmade like those caused by capacitor bank switching, load changes in power distribution system, resonating circuits associated with switching devices, various system faults such as short circuit and arcing faults. The natural mechanism includes lightning transients. An indirect lightning strike (like a strike between or within clouds) can induce voltages/current on the conductor of circuits outside/inside a building. The effect of surges can be simulated by the surge test as given in the basic EMC publications EN 61000–4–5, EN 61000–4–12 and EN 61000–4–18. Surge testing involves the use of specialized simulators or generators which generate the surge waveforms as defined by the standards. These are then superimposed on mains and communication lines of the equipment using coupling–decoupling networks (CDNs).

5.32   TYPES OF SURGES

5.32.1 Combination Wave Or Lightning Surge

This type of surge is normally generated due to lightning (direct/indirect) strikes, operation of lightning protecting devices, lightning current flowing through ground, capacitive bank switching, short circuit and arcing faults. It consists of a voltage and a current pulse with one peak each. The basic standard EN 61000–4–5 deals with testing of equipment for this type of surge. The reason it is called combination wave is because when the pulse is simulated using a generator, two types of pulses are generated depending upon whether the generator is open–circuited or short–circuited. When the generator is open–circuited, a single voltage pulse of 1.2 µs front time (similar to rise time) and 50 µs pulse duration (similar to pulse width) is generated, while under short–circuit conditions a single current pulse of 8 µs front time and 20 µs pulse duration is generated.

The typical combination wave voltage and current wave–forms as specified by EN 61000–4–5 are given in Fig. 5.23a and 5.23b respectively. The front time "Tf" of the surge voltage is a virtual parameter defined as 1.67 times the interval "T" (see Fig. 5.23a) between the instants when the impulse is 30 % and 90% of the peak value. The front time as per the standard is 1.2us ± 30%. The standard defines the pulse duration "Td" to be equal to the pulse width "Tw" i.e. "Td" equals 50µs+20%.

5.32.2 Combination Wave Telecom Surge

Defined by the standard EN 61000–4–5, such a surge is generated due to lightning transients on telephone lines and is also a combination wave. When the generator is open circuited, the output voltage pulse has a front time of 10 µs and a pulse duration of 700 µs. While under short circuit conditions, output current has a front time of 5 μs and a pulse duration of 320 μs. These pulses are referred to as *10/700 telecom* surges.

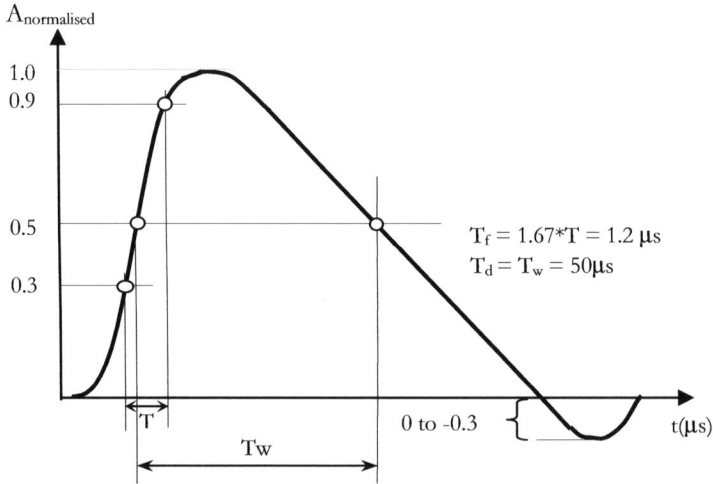

Fig. 5.23(a): Combination surge voltage waveform

$T_f = 1.67*T = 1.2\ \mu s$
$T_d = T_w = 50\mu s$

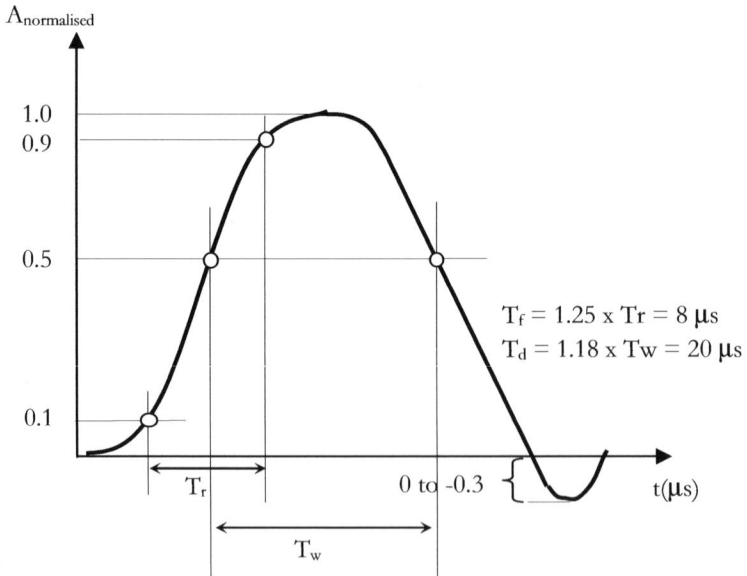

Fig. 5.23(b): Combination Wave Surge Current Waveform

$T_f = 1.25\ x\ Tr = 8\ \mu s$
$T_d = 1.18\ x\ Tw = 20\ \mu s$

5.32.3 Ring Wave Surge

This is a *ringing* transient (meaning there are more than one oscillation in a single pulse) and is typically induced in low–voltage cables due to switching of electrical networks and reactive loads, faults and insulation breakdown of power supply circuits. Lightning is another source of ring wave transient. The ring wave is actually a *short* oscillatory transient of maximum two–and–a–half cycles (i.e. five peaks) and amplitude of each peak is 60% of previous peak.

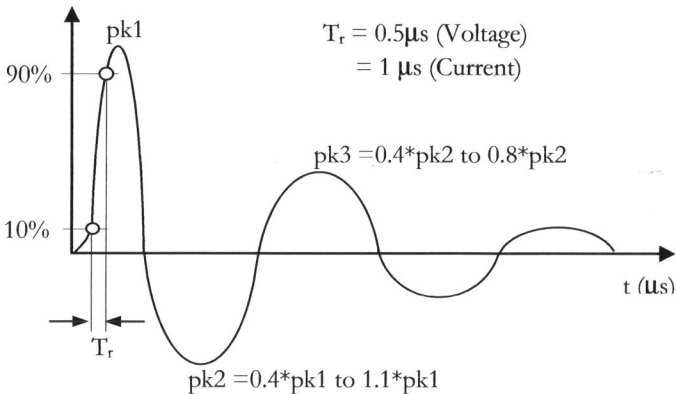

Fig. 5.24: Ring Wave Surge

During simulation of this pulse in a laboratory, when the generator is open circuited, the first peak of the voltage pulse has a rise time of 0.5 μs (see Fig. 5.24) while the time period of the wave is 10 μs (i.e. a frequency of 100 kHz). When the generator is short circuited, the first peak of the current pulse has a rise time of 1μs while the time period is the same as the voltage pulse. The basic standards EN 61000–4–12 deals with testing for this type of surge.

5.32.4 Damped Oscillatory Surge

These are oscillatory transients generated due to switching of isolators in high voltage and medium voltage (HV/MV) open–air stations and particularly due to the switching of HV busbars. The opening and closing of HV isolators give rise to a transient which has a sharp front time (rise time) of some tens of nanoseconds which is followed by oscillations with fundamental frequency ranging from about 100kHz to few tens of MHz. The standard EN 61000–4–18 defines a waveform for commercial applications which consists of a number of cycles decaying to 50 % of the peak value between the fifth and tenth periods as shown in Fig. 5.23.

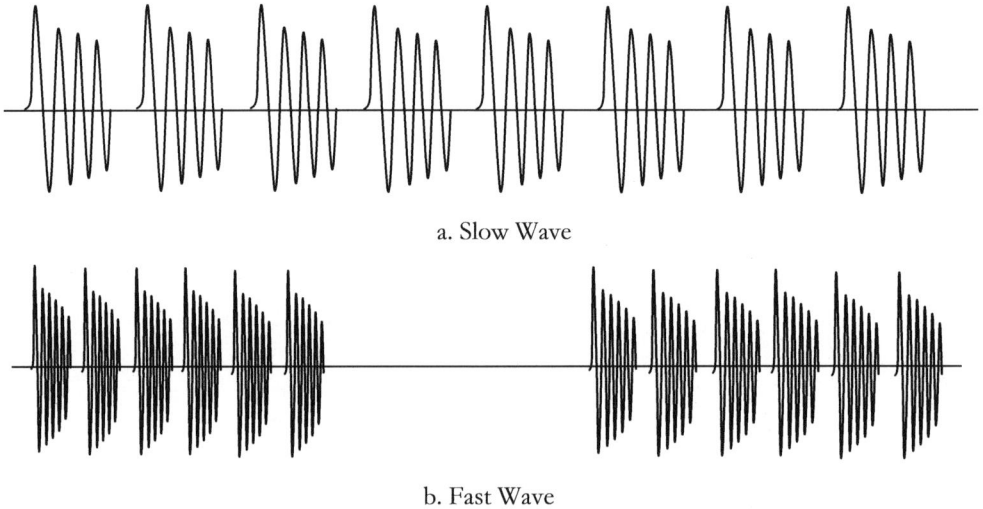

a. Slow Wave

b. Fast Wave

Fig. 5.25: Damped Oscillatory waves

The standard identifies a slow damped oscillatory wave which has an oscillation frequency of 100 kHz and 1 MHz. The former is observed in large sub–stations while the latter is considered representative of typical sub–stations. Slow waves are repeated at rates ranging from a few hertz and a few kilohertz (see Fig. 5.25). The repetition rate depends upon the distance between switching contacts. Related standards call for repetition rates of 40 per second for 100 kHz oscillation frequency and 400 per second for 1 MHz oscillation frequency.

The standard also defines fast damped oscillatory waves with frequencies of 3 MHz, 10 MHz and 30 MHz which are generated by switchgear and control gear of substations and also in installations exposed to high–altitude electromagnetic pulse (HEMP). The nature of the fast waves is such that it consists of a *burst* or packet of individual waves (rather like EFT/burst). In each *burst* the waves are repeated at a rate of 5 kHz. The burst duration is 50ms, 15ms and 5 ms for 3 MHz, 10 MHz and 30 MHz frequencies respectively while the burst period is 300ms.

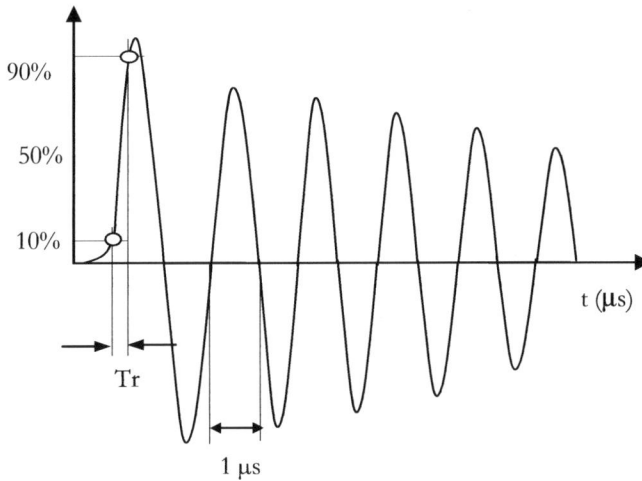

Fig. 5.26: Damped Oscillatory Surge

5.33   CONDUCTING A SURGE TEST

The laboratory test set up for surges is similar to the EFT/B except for the GRP on the ground. The first step for conducting a surge test is to arrange test setup as specified by the standard. The test severity level is then decided. EN 61000–4–5 specifies four severity levels viz. 0.5kV, 1kV, 2kV and 4kV, and referred to as level –1, 2, 3 and 4 respectively. The product standard will specify one of these levels and EUT is subjected to surge of that level.

For combination wave surge, the basic standard calls for five positive and five negative surges to be applied at an interval of 1 minute. For AC/DC mains and unshielded unsymmetrical I/O lines, the standard calls for application of the surge both in common mode (CM) and differential mode (DM). In CM, the generator O/P impedance is 2 Ω while for CM it is 12 Ω. For unshielded symmetrical and shielded I/O lines, the standard calls for testing only in common mode and CDNs as specified by the standard have to be used. In case of shielded high-speed communication lines, the surge can be directly applied on the shield. The length of the cable under test is specified as 10m. The severity level for differential mode is always one level below that in common mode, since testing in differential mode is twice as severe as testing in common mode. For AC mains there is a further requirement of syncronising the surge transient at zero crossings, positive peak &

negative peak of the AC cycle. The test level is increased from lowest level (i.e. 0.5kV) the specified level in such a way that all lower levels are satisfied. For example, for testing at 2kV, the test has to be carried out at 0.5kV, 1kV and 2kV. This is to take into consideration the non–linear current–voltage characteristics of the EUT.

For ring wave surge, the severity levels are the same as with combination wave surge. The basic standard EN 61000–4–12 also calls for five positive and five negative transients to be applied. The time interval between two successive pulses depends upon the output impedance of the generator. The standard specifies three output impedances for the generator viz 200 Ω, 30 Ω and 12 Ω and the time interval between successive pulses is 1s, 6s and 10s respectively. All available ports like power supply, I/O and communication ports have to be tested. For power supply ports the standard calls for a generator output impedance of 12 Ω for ports connected to major feeders and 30 Ω for ports connected to outlets. The standard mentions that phase syncronisation may be required for EUTs that incorporate semiconductor-based surge arrestors in the input power circuitry. The generator output impedance in case of I/O lines has to be kept at 200 Ω.

For damped oscillatory wave test, the basic standard EN 61000–4–18 specifies three severity levels viz 0.5 kV, 1 kV and 2 kV. The test is to be carried out at 100 kHz and 1 MHz oscillation frequencies with a pulse repetition rate of 40 per second and 400 per second respectively. The test voltage is applied to power supply, input/output and communication ports in both polarities. The standard specifies test duration not to be "less than" 2s and that it is left to the product standards to specify the exact duration. There is no requirement of phase syncronization of the transients with the power supply. The performance of the EUT is monitored during the entire duration of the test and also after the test. The section 5.43 gives information on the evaluation of test results.

5.34   Selection Of Severity Levels For EFT/B And Surge Tests

Following guideline is given in the basic standards for selection of a particular test level. However, if product standard exists for a particular product, the level dictated by that standard takes precedence.

Level 1: Well protected environment

This level is specified when the EUT is installed in such an environment where transient suppression devices are used in switched control circuits, there is proper isolation between control and measurement circuits and power lines or high severity lines, the power supply is filtered and power cables are shielded with shield grounded at both ends. Example of such an environment is a computer lab.

Level 2: Protected environment

This level can be used when the EUT's environment is such that switched control circuits are partially using transient suppression, there is proper separation of all circuits from other circuits associated with environment of higher severity levels, the power supply is not filtered but there is physical separation of unshielded power supply and control cables from signal and communication cables. A typical example could be control room of industrial and electrical plants.

Level 3: Typical industrial environment

Such environment is characterized by absence of suppression components in control circuits. Poor separation of industrial circuits from other circuits associated with high severity levels. Dedicated cables for power supply, control, signal and communication lines. Poor separation between power supply, control, signal and communication cables. Availability of earthing system represented by conductive pipes, ground conductors in cable trays and by ground mesh. e.g. open-air H. V. sub–station.

Level 4: Severe industrial environment

Such environment is characterized by no suppression components in the control and power circuits. No separation of industrial circuits from other circuits associated with environment of higher severity levels. No separation of unshielded power supply and control cables from signal and communication cables. Use of multicore cables, common for control and signal lines. Example is a power generating station.

5.35   CONDUCTED SUSCEPTIBILITY–CW

The conducted RF susceptibility test is also known as injected currents (especially EN 55015) or RF common mode voltage. The objective is to test immunity of the equipment to conducted disturbance induced as current in mains and /or I/O cables by continuous wave (CW) electromagnetic fields of nearby RF transmitters. The standard EN 61000–4–6 calls for injection of continuous RF current from 150kHz to 230 MHz on mains and I/O cable of the EUT and studying its response to such simulations. Certain standards like EN 61000–4–16 calls for injection of continuous wave (CW) current from 15Hz to 150kHz onto power mains and I/O lines of the EUT and it typically simulates the harmonics and inter–harmonics generated in power distribution system.

5.35.1 Basic Test Set-Up

The Fig. 5.27 shows basic test set–up for conducted RF susceptibility test as depicted in the basic standard EN 61000–4–6. The radio frequency to be injected is sourced by a test generator which normally consists of a signal generator followed by an amplifier. The signal generator output is amplified by the amplifier. The standard EN 61000–4–6 requires that the signal generator, at the minimum, must have an output impedance of 50 Ω.

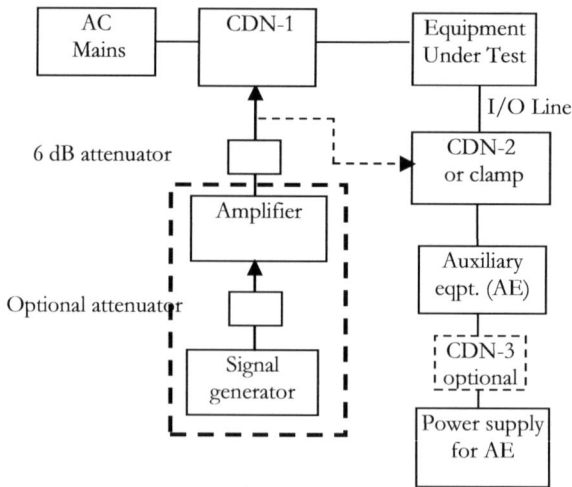

Fig. 5.27: General Set–Up For Conducted RF Susceptibility

The amplifier, on the other hand, must be capable of producing sufficient powers so that the simulated RF voltage levels are generated as specified by the standard. The amplifier output is given to a CDN or EM clamp which couples it to the mains or I/O lines respectively. The standard calls for testing of one line at a time. The CDN for the line which is not being tested (CDN–2 in the Fig. 5.27) should have its RF input terminated by a 50Ω termination. If a clamp is being used instead of CDN–2, then the auxiliary equipment must receive its mains through a CDN (CDN–3). If there are more than one auxiliary equipment, then all lines that are not being tested, should be decoupled.

5.35.2 Coupling Decoupling Networks (CDNs)

The Fig. 5.28 shows a type 'M–3' CDN for coupling RF on to power mains (the letter 'M' stands for mains).

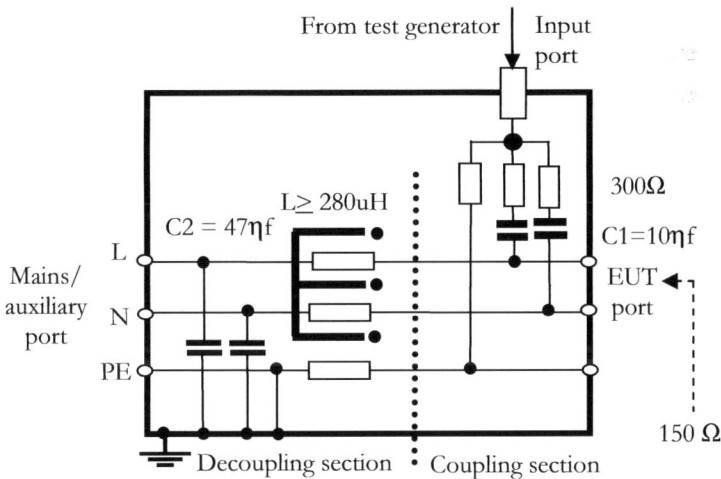

Fig 5.28: Type M3 CDN

The number following the letter M represents the number of lines i.e. M–3 for single phase (three lines: live, neutral and PE) and M–5 for three-phase. As shown in the Fig. 5.28, the RF generated by the RF generator is coupled onto the lines via the capacitor C1 in series with a 300Ω resistor. Since there are three lines, the effective resistance per line is 100Ω. Considering 50Ω output impedance of the test generator, the CDN presents 150Ω output impedance at EUT side. The inductor 'L' and capacitor 'C2' on decoupling side do not allow the RF to go back to the mains. In addition to M–type, the standard EN 61000–4–6 identifies further

CDN types to couple RF on various other type of lines. These are given in table 5.6 wherein the first column identifies the cable type, the second column gives the CDN type while the third column lists the differences with the M–type. The standard EN 61000–4–16 calls only for the type 'T' CDN. While testing, the CDN must be firmly bonded to the ground reference plane.

Table 5.6: EN 61000–4–6 CDNs

Cable type	CDN	Differences with M type
Unscreened supply mains	M–x	——
Unscreened non–balanced I/O lines	AF–x	None
Unscreened balanced I/O lines	T–x	Includes 6mH differential mode chokes in coupling circuit and between decoupling cap C2 and chassis
Screened I/O lines	S–x	Coupling cap C1 absent (Coupling on shield only)
x : Number of lines		

5.35.3 Laboratory Set–Up For Conducted RF Susceptibility

The Fig. 5.29 shows basic laboratory test set–up for carrying out a conducted RF susceptibility test. The EUT, the CDN, the coupling device etc must be kept on a ground reference plane (GRP) made up of copper/aluminium sheet. For table top equipment, the GRP can be located on a table while for floor standing equipment the GRP is on the floor. The GRP is essential so as to establish a common ground reference, since the output of the signal generator and the amplifier is referenced to earth. The CDN/clamp is firmly bonded to the GRP by a very short non–inductive strap, while the EUT is kept 10 cm above the GRP with the help of an insulating support.

All cables that exit the EUT are supported at a height of at least 30 mm above the GRP. The CDN or other coupling devices should be at a distance greater than 0.1m and less than 0.3m from the EUT. The GRP must be 0.2 m larger than the boundary of EUT and all coupling devices and ancillary equipment. It should be bonded to local facility earth.

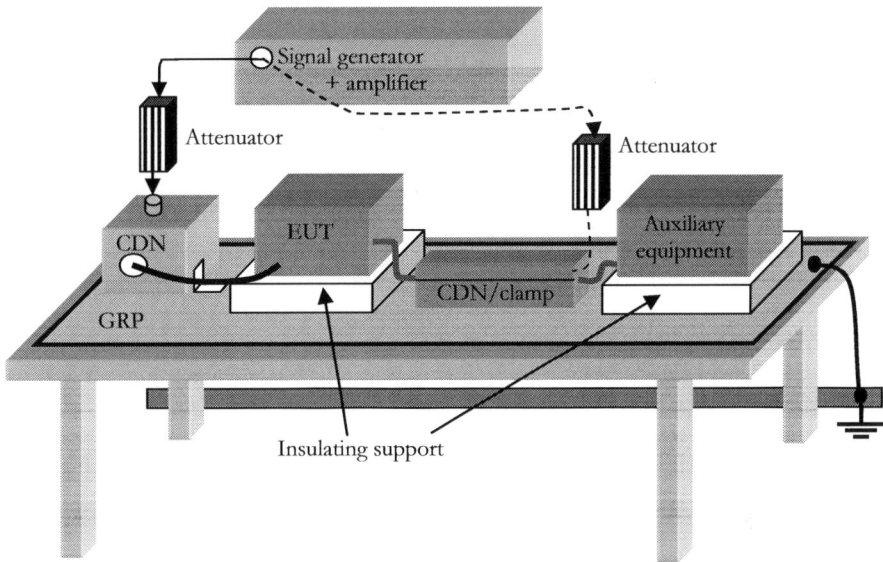

Fig. 5.29: Conducted RF Immunity Test Setup

5.35.4 Test Severity Levels

The standard EN 61000–4–6 identifies three severity levels viz. 1V, 3V and 10V and their corresponding logarithmic values i.e. 120dBμV, 130 dBμV and 140 dBμV respectively. These are unmodulated RMS carrier levels and the EUT is tested to one of these levels depending upon its operational environment while referring to the basic standard. A guide for the selection of test levels is given in EN 61000–4–6, according to which equipment operating in low level electromagnetic environment characterized by absence of radio transmitters in the vicinity and located more than 1km away from high power radio and TV transmitters, have to be tested to level–1 i.e. to 1V.

An equipment operating in a typical commercial environment (a moderate electromagnetic environment) characterized by the presence of low power portable radio transceivers (less than 1W rating) in the vicinity (but not in close proximity) has to be tested up to level–2 i.e. 3V. Equipment operating in a typical industrial environment where portable trans-receivers of rating 2W or more are operating nearby or high power transmitters are operating in the vicinity or process control equipment generating RF energy in the ISM band are operating close by are to be tested up to level–3 i.e. 10V. If a product standard is being referred to, the EUT is tested to the level as given by the product standard.

5.35.5 Performing A Conducted RF Test

For performing a conducted RF test, the set–up is arranged as shown in the Fig. 5.30. It shows testing on an EUT which has one mains input and two I/O lines to be tested. All the lines are routed through respective CDNs (or clamp). The lines which are not under test are either disconnected or connected to un–terminated CDNs. The test is carried out, one line at a time, with RF inputs of CDNs on other lines (line 2 and 3 in the Fig. 5.30) are to be terminated by 50 Ω impedance. If a clamp is being used on a particular line (line 2 in the Fig. 5.30), the mains supply of that auxiliary equipment (AE) must be given through CDN.

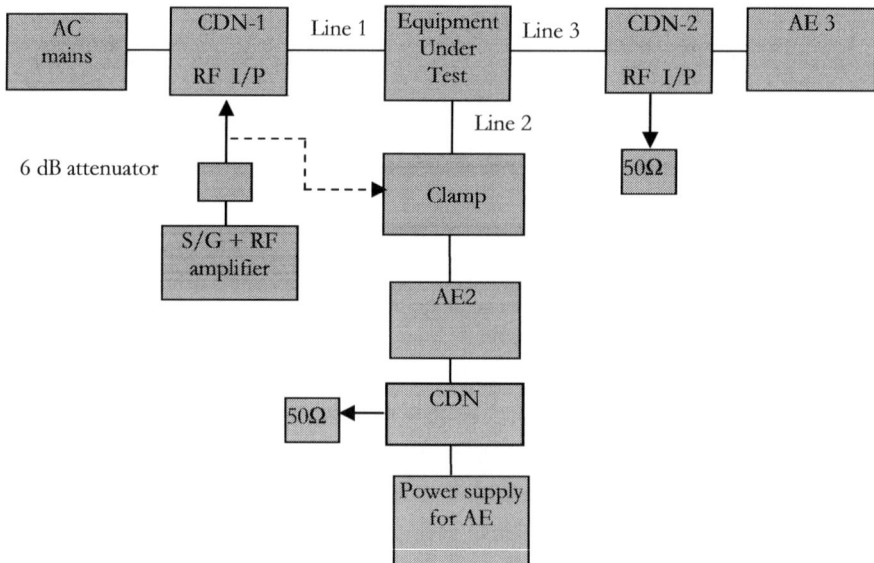

Fig. 5.30: Example Of Conducted RF Test Set-up

The standard EN 61000–4–6 calls for testing in simulation frequency range of 150kHz through 230 MHz. The test is then started wherein the signal generator sweeps the frequency from simulation frequency range of 150 kHz to 230 MHz and the signal generator output level is maintained as per chosen severity level. The frequency is normally swept incrementally and the maximum allowable step size is 1% of the preceding frequency value. The dwell time at each frequency is 0.5s minimum. Normally it is required to monitor the EUT operation during the entire duration of the test and after the test as well. See section 5.43 for evaluation of test results.

5.36   ELECTROSTATIC DISCHARGE TEST

Static electricity discharge, commonly referred to as electrostatic discharge, or ESD, is a transfer of charge between two materials at different electro–static potentials. This phenomenon is fairly ubiquitous and almost of a daily occurrence (especially during winter). When a person walks on a vinyl carpet or sits on a plastic chair, he may get charged to static voltage due to transfer of electrons from his body to the carpet/chair or vice versa. He may get charged to tens of kilovolts without noticing. But the moment he touches any electronic equipment, a discharge current flows through the person's body and he gets a shock. If the discharge path happens to be *direct* i.e. through the electronic equipment (which is not designed and tested for immunity to such discharges) there is a potential danger of damage. On the other hand, *indirect discharge* occurs when a person (or charged object) discharges on any nearby object/equipment and the resultant arc radiates a strong local electromagnetic field which couples to the victim equipment.

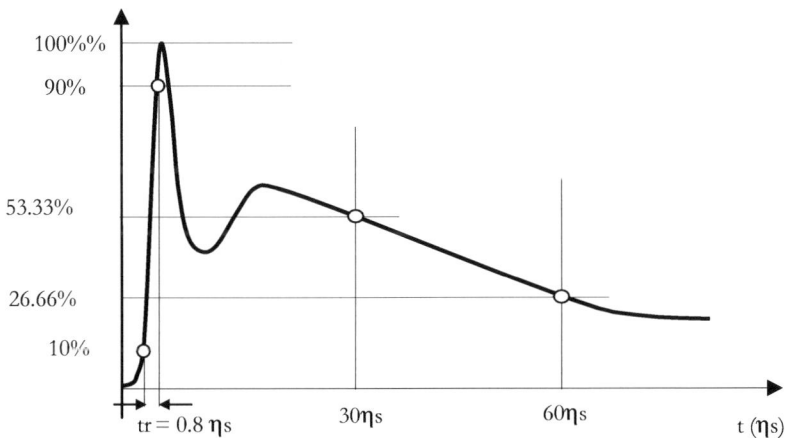

Fig. 5.31: ESD Current Wave–Shape (Not to scale)

5.36.1. ESD Simulator

Since the ESD test is based on charged human being as a source, the basic standard EN 61000–4–2 prescribes a static charge source modelled on the human body and represented by a series combination of a 150 pF capacitor and a 330 Ω resistor. The standard prescribes a discharge current wave–shape (as shown in Fig. 5.31) that is observed in most practical cases and which is also easy to simulate in a laboratory. It has a very sharp rise time of 0.8ηs. The ESD simulator is essentially a *gun* which has the circuit (as shown in Fig. 5.32) incorporated into it.

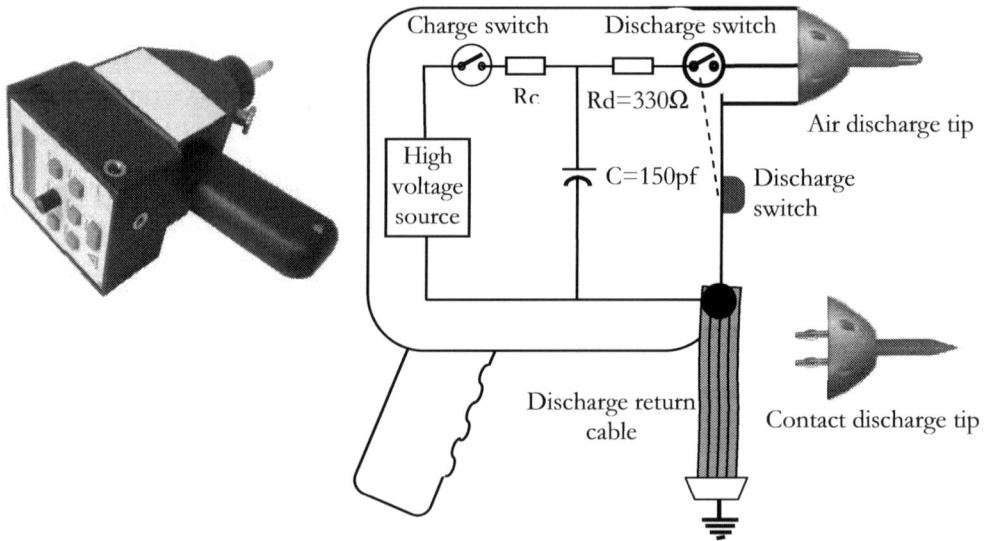

Fig. 5.32: ESD Simulator/ Gun

A HV supply (capable of generating up to 16.5 kV) is used to charge an energy storage capacitor 'C' of value 150 pf through a charging resistor 'Rc' in the range of 50–100 MΩ, the charging being controlled by a charging switch. This capacitor is discharged through a resistor of 330 Ω via a *discharge tip*. When the discharge switch is pressed (i.e closed), the charge appears on the tip. ESD standards normally specify two type of discharge tips. The one for air discharge has a rounded shape more like the human finger, while the other, for contact discharge, is pointed. The other end of the storage capacitor (and the HV source) is connected to a discharge return terminal which in turn is connected to GRP via a *return* cable. The gun must have two modes of operation the *contact* and *air* discharge modes while it must also be capable of giving a single discharge at repetition rate of 1 pulse/sec and also continuous discharge at 20 pulses/sec for exploratory purpose.

5.36.2 Severity Levels

EN 61000–4–2 identifies four severity levels as given in table 5.7 and if one is referring to this basic standard then the equipment has to be tested to the severity level appropriate to the equipment installation environment. This level can be selected as per guidelines given in the standard. The environment is classified according to the relative humidity (RH) and use of antistatic material. In case of environment where antistatic material is used, then the standard specifies severity level of 2 kV where the minimum RH is 35% and a level of 4kV where the RH is 10%. In case of environment where synthetic material is used, then the standard specifies severity level of 8 kV where the minimum RH is 50% and a level of 15kV where the RH is 10%. However, if one is referring to a product standard then the equipment has to be tested to the level given in that standard.

Table 5.7: ESD Severity Levels

Level	Test voltage	
	Contact	Air
1	2kV	2kV
2	4kV	4kV
3	6kV	8kV
4	8kV	15kV
x	Special	Special

5.36.3 Laboratory Test Setup

The laboratory set–up specified by EN 61000–4–2 for ESD testing of floor standing EUT is as shown in Fig. 5.33. The EUT is placed over a ground reference plane (GRP) made up of metallic sheet but isolated from it by an insulating support 0.1m thick. EUT cables that are routed over the GRP are also isolated by insulating material 0.5mm thick. The standard does not specify the material used for the insulating support. The GRP projects 0.5m beyond the EUT. The ground strap of the ESD gun is connected to the same plane and is routed 20cm away from the EUT.

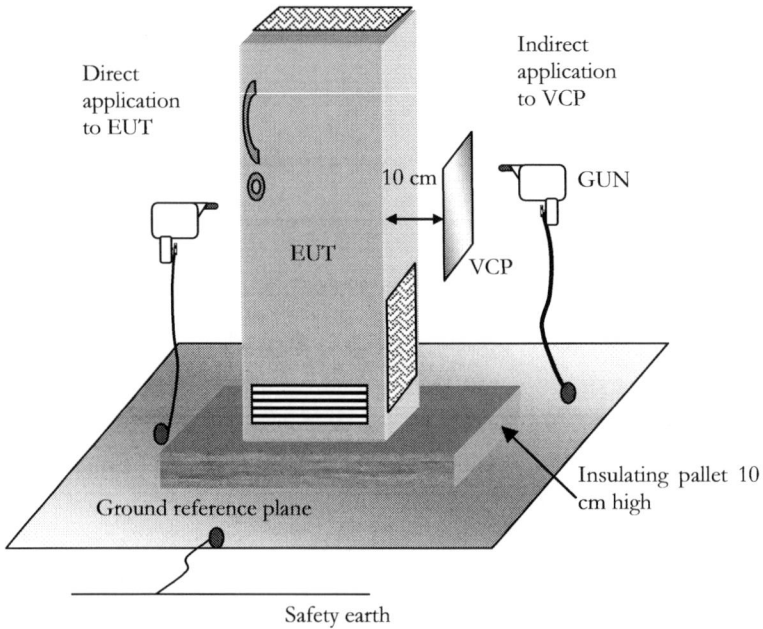

Fig. 5.33: ESD Test Set–up (Floor standing EUT)

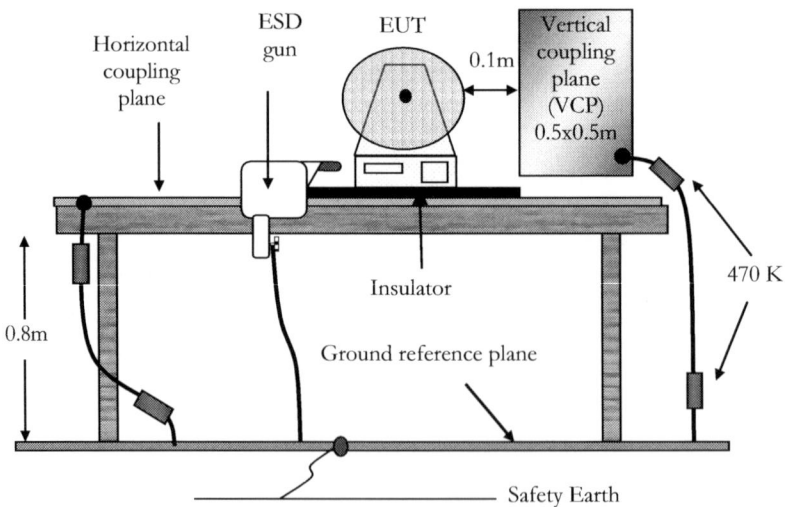

Fig. 5.34: ESD Test Set–up (Table-top EUT)

For table–top testing (Fig. 5.34), a horizontal coupling plane (HCP) of dimensions 1.6m x 0.8m, made from Aluminium sheet, is kept directly underneath the EUT. The EUT is isolated from the HCP by an insulator of 0.5mm thickness and is kept in such a way that the EUT is not less than 0.1m away from the edge of the HCP. The EUT cables must also be isolated from the HCP over their entire lengths. A vertical coupling plane (VCP) of dimensions 0.5mx0.5m is also specified to be kept 0.1m from EUT during testing. The HCP is connected to the ground plane via a cable which has two bleeder resistors of 470k at each end but not more than 20mm away from the end. The purpose of this cable is to isolate the HCP during the actual discharge but allow charge to bleed off slowly afterwards (called as soft discharge). The resistors are located at end of the lead so that they present a distributed resistance and the lead's stray inductance and capacitance are isolated from both the HCP and the ground plane.

5.36.4 ESD Test Considerations

There are two types of ESD tests as per EN 61000–4–2, indirect discharge and direct discharge. The indirect discharge is carried out in contact discharge mode where in the gun is brought in contact with the HCP and then the discharge switch is pressed. During the discharge, the gun is held edge–on to the plane, at the centre of the face of the EUT. The EUT has to be rotated to test all four sides. It is to be ensured that each appropriate face of the EUT is 10cm from the edge of the HCP. The other type of discharge i.e. the direct discharge is carried out in two modes namely contact and air discharge. Air discharge method is prescribed for insulating surfaces including conductive surfaces coated with insulating paint. During air discharge, the discharge switch is pressed first and the charged tip is approached as fast as possible to touch the EUT, with the discharge taking place by ionisation of the air gap as the tip nears the EUT. Contact discharge method is prescribed for conducting surfaces and is the same as that followed during indirect discharge.

5.36.5 Carrying Out An ESD Test

During ESD test in a laboratory, the EUT is initially subjected to indirect discharges on VCP and HCP. Direct discharges are then applied to points and surfaces of EUT which are accessible to the operator. Air discharge is typically done on non–conductive points like switches, key–pads, LED indicators, displays etc while contact discharge is applied on metallic switches, screws etc. As per EN 61000–4–2, ten single discharges are applied with a time interval of at least 1 sec on pre–selected points. The ESD gun must be held perpendicular to the surface being tested.

5.37   RADIATED SUSCEPTIBILITY (RS) TEST

An electronic equipment of today is constantly exposed to radiated EMI present in the environment, the more recent culprit being the ubiquitous mobile phone. Other *disturbers* include intentional radio frequency (RF) sources like walkie-talkies, radio/TV transmitters and RF sources used for purposes other than communication (like induction heating or medical diathermy). The equipment is also exposed to unintentional RF generated from SMPS, drives, arcing etc. It has therefore become imperative to test the equipment to radiated EMI in order to establish electromagnetic compatibility. Radiated susceptibility/immunity test, also called as RF electromagnetic field immunity test, involves exposing the equipment to simulated magnetic or RF electromagnetic fields. In the former case it is called as magnetic field immunity and in the latter case it is called as RF electromagnetic field immunity. The amplitude and frequency range of the simulations are dictated by the standard being followed. The operation of the EUT is then monitored for degradation. The extent of allowable degradation (referred to as *performance criterion*) is again dictated by the standard in question and allowable degradation depends upon operational reliability requirements.

5.38   GENERAL TEST SET-UP FOR RS TEST

A typical test setup for carrying out radiated susceptibility/immunity testing is as shown in the Fig. 5.35, where the required simulation frequency is generated by a signal generator and amplified to the desired severity level by an RF amplifier. It is then fed to an antenna or a radiating system in order to generate the radiated field to which the EUT is exposed to. The most important requirement for radiated susceptibility/immunity test is a shielded enclosure, which contains the simulated field within a small area and does not allow it to escape. As shown in the Fig. 6.1, absorber lined shielded enclosures (ALSE) commonly called as anechoic chambers are normally recommended for radiated susceptibility / immunity testing. Alternative low-cost test enclosures like TEM/ GTEM (Giga-hertz transverse electromagnetic) cells and shielded strip-lines can also be used provided the standard recognises it as a legitimate method. Because of certain compromises involved, most standards relegate these low-cost enclosures to pre-compliance testing or testing of small EUTs. RF electromagnetic field immunity testing requires a fully-anechoic chamber where a set of RF absorbers is placed on the floor between the antenna and the EUT. As far as antennas are concerned, EMC standards normally recommend the same antennas for radiated susceptibility/immunity testing as those used for radiated emission measurement since they are reciprocal networks.

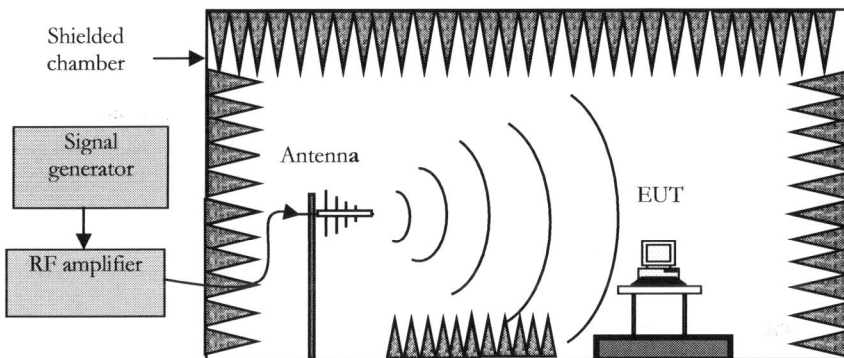

Fig. 5.35: Radiated susceptibility test set up

5.39 SEVERITY LEVELS AND FREQUENCY RANGES

EMC standards provide severity levels and the frequency ranges over which radiated susceptibility test is to be carried out. EN 61000-4-3 calls for testing from 80 MHz to 1000 MHz. For equipment that operate in vicinity of mobile phones and other RF transmitting devices, the standard calls for testing at a higher frequency of 1.4 GHz to 6 GHz. Although the standard asks for the test to be performed without gaps from 80 MHz to 1 GHz, testing in the frequency range of 1.4 GHz to 6 GHz has to be carried out only over those ranges where the RF transmitting devices actually operate. This is particularly applicable to mobile phones where countries/regions use different systems or protocol. EN 61000-4-3 identifies four severity levels viz. 1V/m, 3V/m, 10V/m and 30 V/m. These are unmodulated carrier levels. The frequency is generally amplitude modulated at 80%. The testing is to be carried out at one of these levels as dictated by product standards. The standard EN 61000-4-3 does give a guideline for identifying a particular severity level. If the EUT is located more than 1 km away from radio/television stations and has transmitters/receivers of low power, then it is enough to test up to level 1 i.e. 1V/m. In case the environment is a typical commercial environment, which has low power but portable transceivers (typically less than 1 W rating) but with restrictions on use in close proximity to the equipment then testing may be carried out up to level 2 i.e. 3V/m. In case of severe electromagnetic radiation environment like a typical industrial location which has portable transceivers (2 W rating or more) in close proximity to the equipment but not less than 1 m and high-power broadcast transmitters/ISM equipment in close proximity, the equipment may be tested up to level 3 i.e. 10 V/m. If there are portable transceivers are within less than 1 m of the equipment and other sources of significant interference may be within 1 m of the equipment, the equipment has to be tested to level 4 i.e. 30V/m. In addition, the standard also identifies a fifth level 'x' which is an open level that might be negotiated between the manufacturer and the user.

5.40   TEST PROCEDURE

The procedure for carrying out RF electromagnetic field immunity test can be split into two. Firstly, the set-up has to be calibrated in order to decide the signal generator level and amplifier output power required to generate simulations of desired severity level. After calibration, the actual test is carried out and the EUT performance is evaluated at the targeted severity level over the designated frequency band.   For testing as per EN 61000-4-3, the EUT is placed on a table 0.8 m high (for table top equipment). Floor-standing equipment are mounted on a non-conductive support (to prevent accidental earthing) 0.05 m to 0.15 m above the supporting chamber floor. The test is carried out starting from the lowest frequency and incrementing the frequency to maximum in steps as specified by the standards. In addition, the dwell time i.e. the time for which the signal generator should stay at one frequency is also specified. EN 61000-4-3 specifies a maximum step size of 1%

(of preceding frequency) and a minimum dwell time of not less than 0.5s at each frequency. While sweeping the frequency range, forward power is kept the same as that observed during calibration. The test is repeated after changing the antenna polarisation. The EUT is then rotated and test is repeated so as to expose all faces (i.e. remaining three azimuth faces plus top and bottom faces) to the field. EN 61000-4-3 further states that the EUT shall be fully exercised i.e. tested in all possible operating modes. In case of a large number of possible modes, the modes that are more critical should be evaluated.

Most standards recommend testing in anechoic chambers. These chambers however are very costly and many manufacturing units may not afford them. The best alternative for manufacturers is to go for low cost *pre-compliance* chambers like TEM / GTEM cells / strip-line chamber in order to take necessary precautions during the design phase. When an adequate confidence is gained, they can approach a test laboratory for final compliance testing

5.41 MAGNETIC FIELD IMMUNITY TEST

This test verifies the immunity of the EUT against magnetic fields. EN 61000-4-8 calls for power frequency magnetic field immunity which tests immunity of equipment to magnetic fields generated by power frequency currents flowing in nearby conductors under normal (long term) and fault condition (short term). The set-up (Fig. 5.36) consists of a square loop of 1-meter diameter inside which the EUT is kept. The loop is fed by the required current, generated by a motorised variac (that takes supply from the AC mains) followed by current transformer, to produce a magnetic field at power frequency of 50Hz/60Hz.

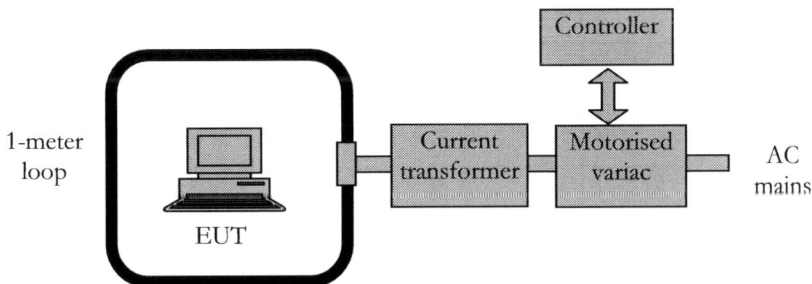

Fig. 5.36: Power frequency magnetic field

A standalone test system is available wherein the variac, the transformer and the controller are enclosed in a single box which has terminals for connections to the loop antenna. The standard specifies four levels viz. 3A/m, 10 A/m, 30 A/m and 100 A/m for long duration test of 1 minute and two levels 300A/m and 1000A/m for short duration tests of 3 seconds. The EUT should be tested in all three orthogonal directions either by rotating the loop or the EUT. The EUT performance is monitored during the entire period of the test for degradation. The standard calls for calibrating the loop antenna with a loop sensor at regular intervals not exceeding one year.

5.42   OTHER EMC TESTS

5.42.1 Harmonic Current Emission

Harmonic emission test embodies the measurement of harmonics of 50 Hz mains current and comparing them with relevant limits prescribed by the standard. The sources of harmonics include devices that use phase angle control such as light dimmers, fan regulators etc. Then there are also devices that employ switching and chopping like rectifiers, SMPS, electronic ballasts etc. Such devices alter the voltage/current wave–shape causing harmonics to be generated. The standard EN 61000–3–2 ("Limits for harmonic current emissions") has been developed precisely to do so. Harmonic emission procedure involves measurement of the amplitudes of the harmonics up to 40th harmonic of the mains frequency (i.e. up to 2 kHz for 50 Hz) and comparing with relevant limits. The standard is applicable to equipment with current less 16 A per phase and mains voltage greater than 220 V.

5.42.2 Flicker Emission

In typical distribution networks, many equipment are connected in cascade. They can draw different amounts of currents at different times which causes the power line voltage to drop or rise (the mains voltage is scarcely regulated in most of the cases). This means that if device used for lighting purpose, a bulb for example, is connected across such a line, its light output is likely to fluctuate. The essence of the word flicker is derived from the fact that this fluctuating voltage causes bulbs to flicker leading to *unsteadiness of the visual sensation* of the human eye which causes stress. Thus, flicker is just a qualitative term, what exactly measured is the voltage fluctuation imposed on a line by the EUT which is compared with limits to ascertain compliance. The limits and method of measurement for flicker is given in the standard EN 61000–3–3 applicable to equipment with current < 16 A per phase and operating voltage greater than 220V.

5.42.3 Voltage Dips And Interruptions

This test is used to simulate various conditions that can appear on a power line due to faults like voltage dips (sudden reduction (non-zero value) in mains voltage) for a few cyles and voltage interruptions (sudden reduction in mains voltage to zero) for a few cycles to a few thousands of cycles. The basic standard is EN 61000-4-11. The test involves simulations of these conditions by dedicated simulators and monitoring operation of EUT.

5.43   EVALUATION OF TEST RESULTS

The test results for all immunity tests have to be classified in terms of the loss of function or degradation of performance of the equipment under test. The following performance criteria have been identified by basic standards:

      Criterion 'A': Normal performance within limits.

      Criterion 'B': Temporary loss of function or degradation of performance which ceases after the disturbance ceases, and from which the equipment under test recovers its normal performance, without operator intervention.

      Criterion 'C': Temporary loss of function or degradation of performance, the correction of which requires operator intervention.

      Criterion 'D': Loss of function or degradation of performance which is not recoverable, owing to damage to hardware or software, or loss of data.

      Performance criterion 'A' is normally applicable to continuous nature of tests like power frequency magnetic fields, RF EM field (RS) and RF common mode. Performance criterion 'B' is normally applicable to tests which are transient in nature like surge, EFT and ESD. Performance criterion 'C' is applicable to voltage interruptions test. On reading the above criteria, one realizes that they are fairly general in nature. To evaluate the performance of an equipment during and after immunity testing, the manufacturer has to specify one or more critical parameters of the product, which give a fair idea about its health. It is not required to test each and every aspect of the product. These parameters are monitored during the test and compliance is decided referring to the above performance criterion.

This can be understood by taking an example. Let us suppose that one is testing a 230VAC –to–25V DC SMPS. The main function of such a product is to generate a DC voltage of 25V $\pm$ 1V from AC and to deliver a certain amount of current (say 500mA max) to a load. This SMPS can therefore be tested with a 50 Ω resistive load and the output voltage monitored. Now consider that a product standard calls for EFT/B test at 2kV and says that the performance criterion is 'B'. If say in the first instance, during the entire period of testing, the output voltage does not vary at all (or varies only by an amount less than 1V) then the equipment complies with the test and meets criterion 'A'. Consider in the second instance, that the output varies by an amount more than 1 V temporarily during testing but recovers automatically to 25V after the test is completed. In this case the product still complies and meets criterion 'B' (as asked by the standard). If, however, the output drops to 15V, stays there even after testing and recovers to 25V only after operator intervention (say in the form of power reset), or does not recover at all, then the equipment does not comply with the test.

The choice of performance criterion depends on whether one is referring to a basic standard or a product standard. When basic standards are used, selection of performance criteria is as per agreement between the manufacturer and the purchaser/user. Otherwise the criteria laid down by product standard or generic standard takes precedence.

5.44   TESTS PRESCRIBED BY PRODUCT STANDARDS – SOME EXAMPLES

As an example, the table 5.8 gives tests, severity levels and performance criteria prescribed by standards that were considered in the previous sections. The conducted tests severities are for AC mains and in both polarities. Tests are also prescribed for DC and I/O lines.

Table 5.8: Severity Levels

Standards	55024 (IT)	61000-6-2 Generic	301489 (Radio Device)	60601-1-2 (Medical)		Perform-ance Criterion
				Professional healthcare	Home healthcare	
Power Freq. Magnetic Fields	1A/m, 50Hz	30A/m, 50Hz	N/A	30A/m, 50Hz		A
RF EM field	3 V/m 80MHz – 1 GHz 80% AM	10V/m 80MHz – 1 GHz 80% AM	3V/m 80MHz – 6 GHz 80% AM	3V/m 80MHz – 2.7 GHz 80% AM	10V/m 80MHz – 2.7 GHz 80% AM	A
ESD	8kV Air 4kV Contact	8kV Air 4kV Contact	8kV Air 4kV Contact	8kV Contact 2 kV, 4kV,8kV, 15 kV Air		B
Conducted RF Susceptibility	3V 150kHz – 80 MHz 80% AM	10V 150kHz – 80 MHz 80% AM	3V 150kHz – 80 MHz 80% AM	3V 150kHz – 80 MHz 80% AM		A
Surge	2 kV CM 1kV DM	2 kV CM 1kV DM	1 kV CM 0.5kV DM	2 kV CM 1kV DM		B
EFT	1 kV Repetition Frequency 5kHz	2 kV Repetition Frequency 5kHz	1 kV Repetition Frequency 5kHz	2 kV Repetition Frequency 100 kHz		B

5.45   TEST REPORT

The results of EMC tests are normally given in the form of a test report. The format and contents of the report are clearly specified by the basic standards. The report, at the minimum, should clearly identify the EUT and associated equipment e.g. brand name, product type, serial number etc. Although not mandatory, many test laboratories also generally include a short description of the EUT regarding its function and intended use. The report further, should identify the EMC test equipment, e.g. brand name, product type, serial number and its calibration validity. The operating condition and the modes in which the EUT was operated during testing must be clearly documented along–with any special environmental conditions in which the test was performed (say a shielded enclosure). The report should also identify the test severity level, EUT performance level, performance criterion and the parameters to be observed for deciding compliance with any pass/fail criterion or tolerances as identified by the manufacturer, requestor or purchaser or that specified in the generic, product or product–family standard. For conducted and radiated RF susceptibility test, the report should also identify the simulation frequency range, the sweep and dwell time and the frequency steps. The test set up must also be described clearly as regards to EUT cabling, the type and position of CDNs used, cable/CDN terminations, any specific type and length of cables other than specified by the standard if used, any special grounding/shielding if used etc. It is therefore worthwhile for test labs to include a photograph of the test set–up for the purpose of clarity. All the observations/effects on the EUT during or after the application of the test disturbance, and the duration for which these effects persist should be clearly recorded.

CONCLUSION

In this chapter we have discussed fundamentals of EMC, EMC standards and EMC test methodologies. In a way, we have seen how the essential requirements of the EMC directive are translated into EMC tests, how these tests are performed in a lab and the method of deciding compliance. Now, it is required that the equipment incorporates certain design features and if required, incorporate certain components that help reduce emission and increase immunity so as to enable it to comply with the tests prescribed by EMC standards thus giving it a presumption of conformity with the essential requirements of the directive. These design methodologies are discussed in the next chapter.

References

Fundamentals of Electromagnetic Compatibility Vol 1 – Don White.

EN 61000–1–1: Application and interpretation of fundamental definition and terms.

EN 60050–161: International Electrotechnical Vocabulary, chapter 161: EMC.

EN 55011: Limits and methods of measurement of radio disturbance characteristics of industrial, scientific and medical (ISM) radio frequency equipment.

EN 55024: Limits and methods of measurement of radio disturbance characteristics of ITE

EN 55016–1–1: Radio disturbance and immunity measuring apparatus.

EN 55016–1–4: Ancillary equipment – Radiated disturbances.

EN 55016–1–3: Ancillary equipment – Disturbance power.

EN 55016–2–2: Measurement of disturbance power.

EN 55016–2–3: Radiated disturbance measurements

EN 61000–4–2: Testing and measurement techniques, Electrostatic Discharge immunity
EN 61000-4-3: Radiated radio frequency electromagnetic field test.

EN 61000–4–4: Testing and measurement techniques, Electrical fast transient/ burst.

EN 61000–4–5: Testing and measurement techniques, Surge immunity test.

EN 61000–4–6: Testing and measurement techniques, Immunity to conducted disturbances induced by RF fields.

• • •

6

EMC DESIGN METHODOLOGIES

6.1   INTRODUCTION

In this chapter, we are going to discuss various EMC design methodologies like filtering, shielding, grounding/bonding and cable routing which are required to make a product compliant to EMC norms. It is important to understand that these measures should ideally be incorporated at the design stage. It can be shown that the cost of product modification at the end of product development cycle can be many times more than the cost incurred at design stage. In addition, the time required for product modification at fag end of product development cycle can also be very high. Considering the scorching pace at which technology is evolving, product lifetimes have become very short. It could therefore, be possible that by the time the product is launched, there may be no market for the product. Hence, design stage EMC measures are indispensable so that product cost is low and it is introduced in time.

6.2   FILTERING

Filtering is a method of attenuating conducted EMI that is entering or leaving an equipment through the power, signal and control lines by introducing line filters and other suppression components so that the equipment can comply with conducted emission and immunity tests as prescribed by EMC standards.

6.2.1   Power Line Filters

Power line filters typically have a configuration as shown in Fig. 6.1. Since they have to deal with both common mode and differential mode EMI currents present on a power line, they need to employ certain components that suppress common mode and others that suppress the differential mode current. Capacitors and inductors both can be used to attenuate common and differential mode currents, it is just a question of how they are connected.

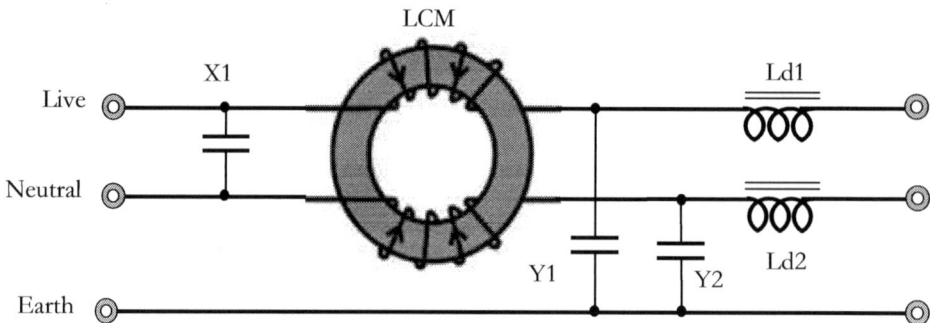

Fig. 6.1: Power Line Filter

In order to attenuate differential mode EMI currents, capacitors need to be connected across the live and neutral wires and hence these are referred to a 'X' capacitors (like the capacitor X1 in the Fig. 6.1) while inductors (that attenuate differential mode) need to be in series with the live and neutral wires ('Ld1' and 'Ld2' in Fig. 6.1). On the other hand, capacitors that attenuate the common mode currents need to be connected across the live & earth and neutral & earth. These are called as 'Y' capacitors (like Y1 and Y2 in the Fig. 6.1). Inductors that attenuate the common mode EMI have a characteristic construction where two coils, one in series with the live and the other in series with the neutral are wound on a single core which is usually in the shape of a toroid. This is called the *common mode choke* shown as 'LCM' in the Fig. 6.1.

The Fig. 6.2 shows one such filter for SMPS applications which is a single stage filter and one can clearly see the common mode choke and the X capacitor. The Fig.6.3 shows a high attenuation line filter which consists of two common–mode chokes (with toroidal cores) apart from X and Y capacitors.

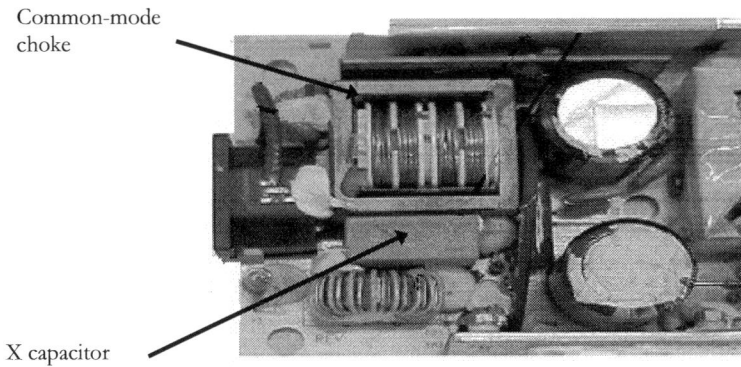

Common-mode choke

X capacitor

Fig. 6.2: Line Filter For SMPS Applications

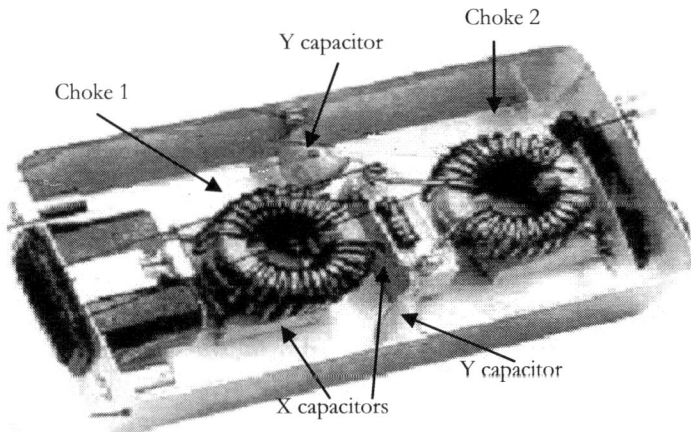

Choke 1

Y capacitor

Choke 2

X capacitors

Y capacitor

Fig. 6.3: High Attenuation Filter

6.2.2   Basic Elements Of Filters

In this section we will take a closer look at typical line filter components namely the inductor, X and Y capacitors. We will also take a look at other filter components incorporated especially for transient suppression like GDT, varistors and silicon avalanche suppressors (SAS).

6.2.2.1 Common mode (CM) choke

The Fig. 6.4 shows a toroidal common mode choke as used in a typical line filter. It consists of a toroidal core over which two windings, one for line and another for neutral, are tightly wound in the same direction. The windings are exactly identical in terms of guage, type of material etc. Since common mode (CM) currents flow in same direction in the line and neutral, fluxes due to such currents (shown by the block arrows) reinforce in the core, giving a very high value of inductance. Differential mode (DM) currents flow in opposite direction in line and neutral so their fluxes mutually cancel each other and therefore the inductor provides very little attenuation for differential mode currents. This is a major advantage considering the fact that heavy fluxes generated by power line currents (which flow in differential mode) also oppose each other in the core which reduces the possibility of core saturation and associated reduction of inductance. This leads to realization of high common mode inductance in surprisingly small inductor size.

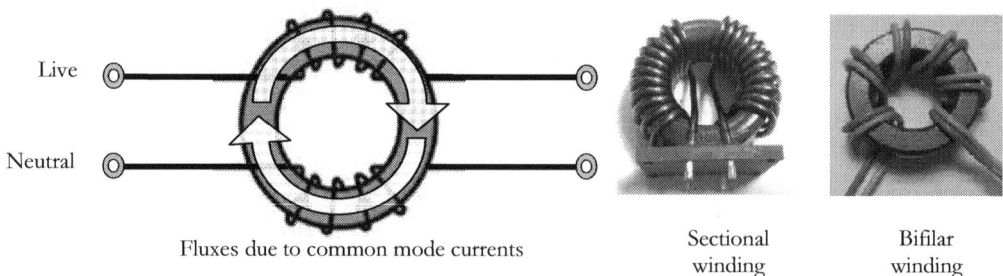

Live

Neutral

Fluxes due to common mode currents

Sectional
winding

Bifilar
winding

Fig. 6.4: Common Mode Choke

Now, for the choke to provide maximum attenuation for common mode currents, it is desirable that all the flux is confined in the core. In practical situation this seldom happens, since there is always a small amount of flux that leaks out. The inductance due to this leakage flux (called leakage inductance) provides small differential mode attenuation. As far as the cores of toroidal inductors are concerned, they are made of ferrites. Ferrites have a cubic crystalline structure with the chemical formula $X\ Fe_2O_3$ where Fe_2O_3 is iron oxide (hematite) and X refers to a combination of two or more divalent metal oxides (e.g. ZnO, NiO, MnO and CuO). Accordingly, ferrite materials come in two types viz–manganese–zinc (MnZn) where 'X' is made up of MnO and ZnO and nickel–zinc (NiZn) where 'X' is made up of NiO and ZnO. MnZn ferrites have a frequency range up to 1 MHz while NiZn ferrites have a frequency range of 30MHz to 1GHz.

6.2.2.2 X capacitors

X capacitors are special type of capacitors connected in a filter across the live and the neutral and attenuate the differential mode interference. They are typically metallised polyester / poly-propelene / polystyrene type. Theoretically X capacitors can be of any value. But high value X capacitors should be provided with suitable discharge resistors across it in order to reduce residual voltage to safe value.

6.2.2.3 Y capacitors

Y capacitors are special type of capacitors connected in a filter across the live and the earth (and neutral and earth) and as such attenuate the common mode interference. Since these capacitors are connected between live and earth, a small amount of current called leakage current flows into the earth which may lead to shock hazard. The value of Y capacitor in a filter is therefore limited by the maximum allowable earth leakage current. The table below gives leakage current values of some standard value Y capacitors.

Table 6.1: Y–Cap Value Vs Leakage Current

Value of Y–Cap	1 ηF	2.2 ηF	3.3ηF	4.4ηF
Leakage current with 1 Y Cap. (mA)	0.072	0.159	0.238	0.340
Leakage current with 2 Y Caps. (mA)	0.144	0.318	0.478	0.679

6.3   OTHER FILTER COMPONENTS

In addition to the components mentioned till now, additional components are often incorporated in a filter or otherwise, especially for transient suppression. Transient *clippers* like gas discharge tubes (GDT) are characterized by their high energy absorbing capability and slow response times. When they conduct, the voltage across them reduces to almost zero and they act as short circuit. Transient *clampers* like metal oxide varistors (MOVs), avalanche diodes, transient voltage suppressor or TVS diodes, silicon avalanche suppressor (SAS) etc are characterized by their low energy absorption capability but faster response times (in µs for MOVs and ηs for SAS). When these devices conduct, the voltage across them does not reduce to zero, but is maintained at a constant value called as *clamping voltage*. Transient *blockers* are special filters which tend to attenuate the transient rather than bypassing it. They can be a combination of snubbers and chokes.

6.4   MULTISTAGE POWER LINE FILTERS

As we have seen in previous chapters, in addition to conducted emission tests in the frequency range of 9kHz to 30 MHz, EN standards also require tests such as disturbance power (for household appliances) which calls for measurement of radiated emission from mains chord that is a result of currents flowing in the frequency range of 30 MHz to 300MHz . EMC standards also call for immunity tests such as conducted RF susceptibility wherein conducted RF noise in the frequency range of 150kHz to 230 MHz is injected in the power mains (in common mode) and the equipment's immunity to such phenomenon is evaluated. In order that the equipment complies with such tests, a power line filter should provide a high attenuation both in CM and DM over a large frequency band of 9kHz to 300 MHz. If all these requirements are to be met using a single filter, one is left with no option but to go for multistage filter which can have a combination of multiple stages of differential mode inductors, common mode inductors, X and Y capacitors etc.

The Fig. 6.5 shows a multistage filter for SMPS wherein the chokes L1 and L2 are rod chokes (values ranging from 2 to 10µH) with NiZn cores which provide differential mode attenuation up to 300MHz in addition to the CM choke L3 (10 to 30 mH) which provides CM attenuation. CY1 through CY4 are ceramic Y capacitors whose values depend on leakage current requirements and range from 0.22 ηF – 1ηF. CX 1 is X capacitor with values ranging from 2.2 to 4.7μF (with suitable discharge resistor across it). For complying with conducted emission, disturbance power and conducted RF susceptibility tests, a multistage filter such as shown in Fig. 6.6 can also be used. The inductor L1 is a current compensated choke with both CM and DM windings on the same core made of MnZn (such chokes offer a 2.5mH to 140mH CM inductance while providing DM inductance from 0.3mH to 19mH) that covers a frequency range up to 30MHz, while L2 is a similar choke wound on a NiZn core, which takes care of frequency range further up to 300 MHz. The rest of the component values are the same as the previous filter.

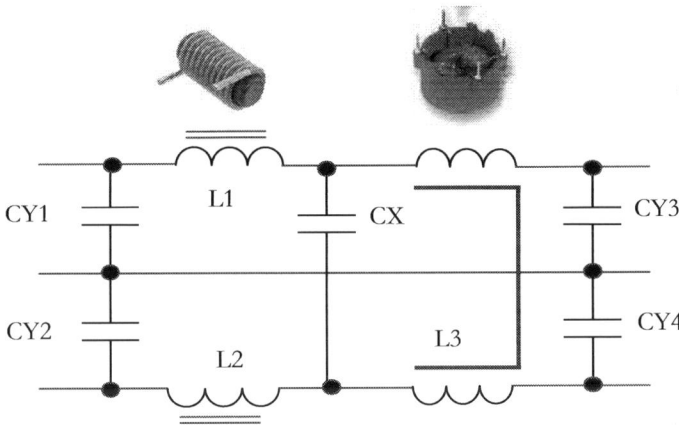

Fig. 6.5: Multi–stage Filter

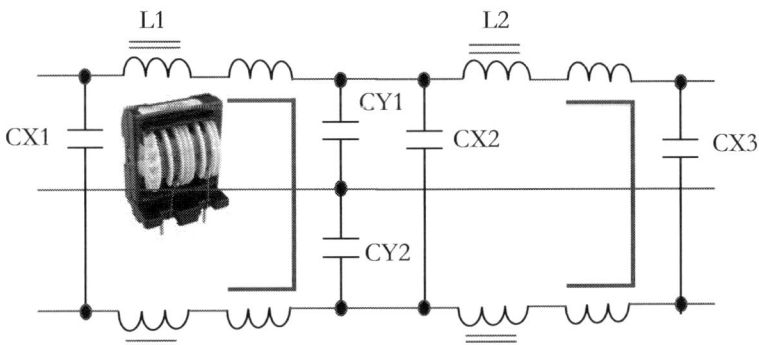

Fig. 6.6: Broadband Multistage Filter

6.5   FILTER INSTALLATION

In order to get the most out of a filter, installation and mounting is as crucial as the design of the filter itself. Incorrect installation methods particularly those related to grounding, bonding and cable routing, can ruin the performance of the best of filters. A filter as a rule (at least those available as standalone modules) must always be enclosed in a shielded metallic box to reduce stray coupling effects and to allow maximum area of contact between the filter and the mounting surface.

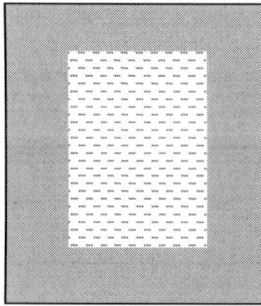

Paint on mating surface removed   Filter placed   Filter bolted for firm bonding

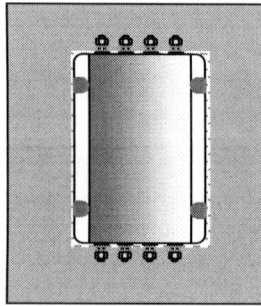

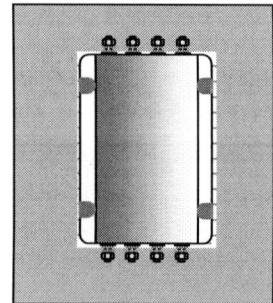

Fig. 6.7: Filter Installation On A Mounting Plate

Almost all of the commercially available filters come in a metallic box which is invariably not painted. During installation on a mounting plate in a typical control panel, the mating surfaces must be stripped off any nonconductive material like paint etc or an area slightly larger than the mating area should be masked during painting or powder coating process as shown in Fig. 6.7. The filter should then be firmly bonded using mounting screws, the bond should be tight allowing maximum surface contact between the filter base and the mounting plate.

The location of filter inside an electronics panel or cabinet is crucial. Filters should be mounted precisely at that point where power mains enter the cabinet as shown in Fig. 6.8. This is to reduce exposed lengths of unfiltered mains which can radiate EMI and pollute the cabinet or radiations from modules within the cabinet may couple on to it. Similarly filters for individual modules should be mounted as close to the module as possible (Fig. 6.9).

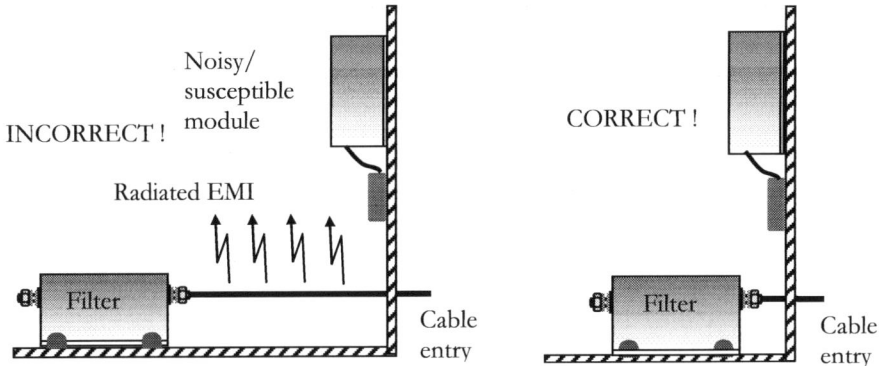

Fig. 6.8: Filter Position From Mains Entry

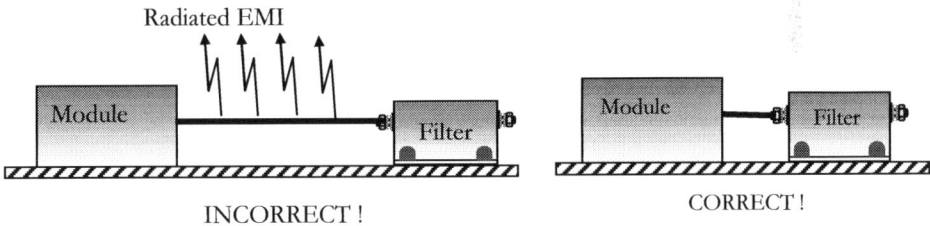

Fig. 6.9: Filter Position From Module

The cable between filter and the module shall be routed close to the chassis / mounting plate as shown in Fig 6.10. This reduces the loop area between the cable and ground and the associated common mode coupling.

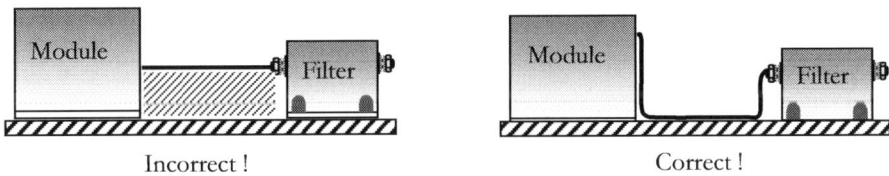

Fig. 6.10: Cable Between Filter And Module

The separation between cables leading to the filter input and those coming from the filter output should be maximized. They should never be routed through the same conduit, they should not cross under any circumstances or even come near (Fig. 6.11). This is to avoid direct coupling of EMI from the input cable to the output cable via stray capacitance that is formed if the wires are near to each other or via the mutual inductance between the wires. This coupling bypasses the EMI rendering the filter ineffective.

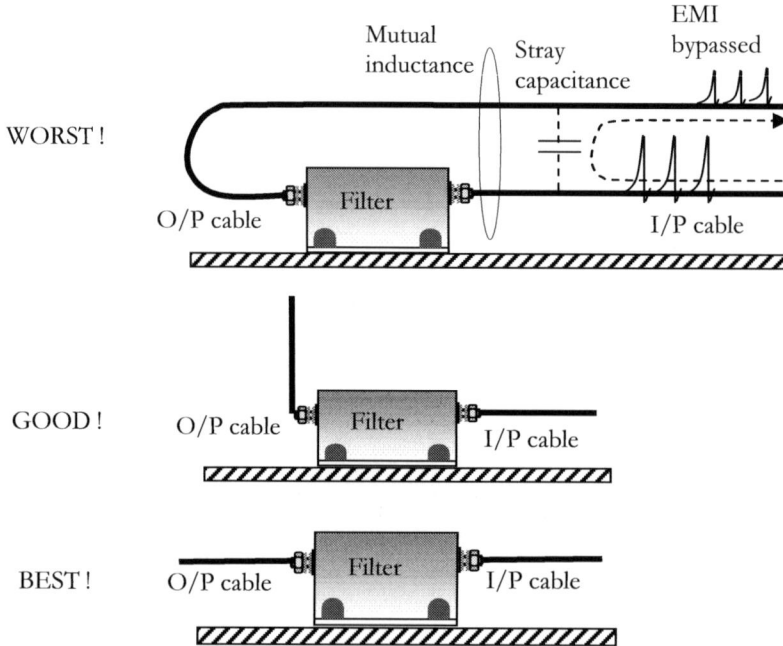

Fig. 6.11: Filter I/P And O/P Cables

6.6   SHIELDING

Shielding is a method of reducing radiated EMI entering or leaving a component, equipment or system. A typical shield works either way i.e. it reduces radiated emissions from an equipment and at the same time increases immunity by attenuating radiated EMI entering the equipment. Shielding   works   partly   by reflection, which is the result of mismatch between impedance of the wave and the impedance of the shielding material and partly by absorption suffered by the wave as it travels through the shield material. Metals are the best option for shielding material not only because they provide the sudden impedance discontinuity for a wave travelling through space or air. Although metals are good for providing shielding, one cannot just use any metal just for the sake of it. The choice must be judiciously made depending on the type of wave –viz electric, magnetic or electro-magnetic, the amount of attenuation required and last, but not the least, the cost of the shield material. Once a proper choice is made, one has to face another hurdle –that of ensuring shielding over the various seams and discontinuities, openings for cable entries, ventilation, displays, controls and keypads that have to be provided in a typical equipment enclosure. This may, at times, prove to be much bigger a challenge than selecting the shield material. The performance of a good shield is in danger of being ruined if proper care is not taken to maintain the shielding integrity over these openings and discontinuities.

6.7   SHIELDING AND EQUIPMENT ENCLOSURES

The shield for a typical equipment is never uniform. There are various penetrations and apertures i.e. openings that have to be provided for cable entries, ventilation, display windows and for keypads and switches. Then, there are also various seams i.e. discontinuities in the shield structure like those at access doors, panel openings, joints etc. Since the overall shielding effectiveness is often compromised by the presence of these openings and discontinuities, they are aptly referred to as shielding compromises. The shielding compromises are depicted in the Fig. 6.12.

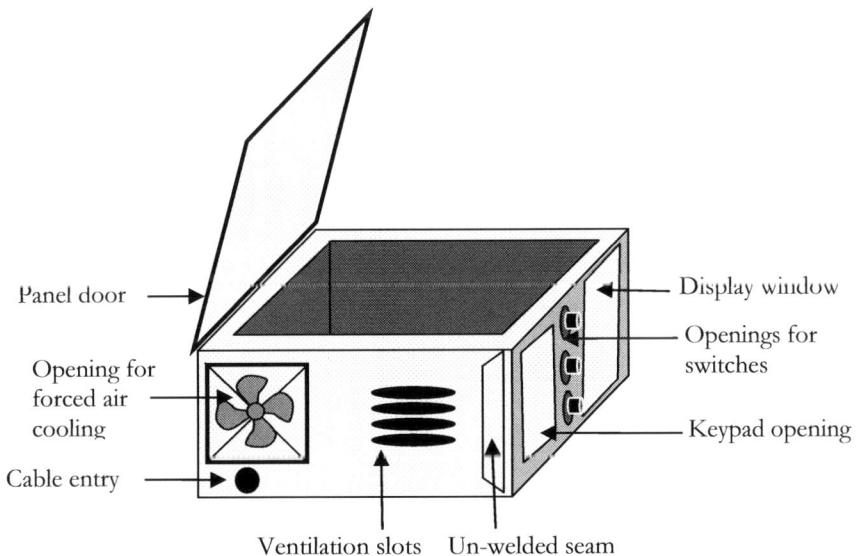

Fig. 6.12: Shielding compromises

The sections that follow discuss various components and mounting methods employed to maintain shielding integrity over these openings.

6.7.1   Ventilation Louvers And Holes

Ventilation louvers have to be provided in an equipment for heat dissipation. If uncovered, they leave a large opening in the shield structure allowing waves to pass through and reducing shielding effectiveness. For ventilation louvers, instead of a large opening, a number of narrow slots can be used but these can allow waves that are polarised along the length of the slots. A better option is to use small holes (dimensions small as compared to the operating wavelength). Small cylinders, which act as waveguide beyond cut-off, can also be used instead of the hole. This can allow the design engineer to have bigger holes but at an increased cost.

If still higher values of shielding effectiveness are required, a fine mesh copper screen can be used that consists of a mesh of wires welded at each crossover point. The screen mesh suffers from a drawback in the sense that it reduces ventilation efficiency and forced air cooling becomes a must. Also, such screens tend to get clogged by dust particles and therefore require frequent cleaning by oil impregnation. A better alternative is to use a honey-comb mesh which is a screen of hexagonal cells made of metal and having a certain width, which are stacked like the cells in a bee hive. The individual cells act as wave-guides beyond cut-off. This allows large openings, consequently increasing ventilation efficiency and reducing maintenance costs. Shielding efficiency of the order of 100-120 dB is achievable by the use of honey-combs.

6.7.2   Keypad And Displays

This section deals with the methods of ensuring shielding over openings which need to be provided for keypads and displays. In case of metal cabinets, it is always better to use individual openings for keys instead of a large opening. In case of non-metallic membrane type keypads, a metallic foil can be introduced between inner and outer membranes which not only shields against RF but also can bypass currents due to ESD.

6.7.3   Openings For Large Displays

For large display windows like CRTs or LCDs (where individual openings are not possible) one can use a shielded glass screen which has a conductive mesh sandwiched between two optically clear glass or polycarbonate sheets or a mesh cast within a plastic sheet.

6.7.4   Leakages At Seams

Every shielded enclosure is bound to have seams at panel doors and joints where two metal sheets are joined together. As a standard rule, enclosures should be bonded at every joint by welding brazing or soldering. Where welding is not possible, seams should be bonded with closely spaced nuts or bolts and as much overlap as possible should be provided. For covers/ panel doors that require frequent opening/closing, conductive EMI gaskets should be used at the periphery (after removing paint on mating surface) to have RF tight seam. Nowadays, conductive finger stocks are available which give better shielding than gaskets. We shall now take a look at various gaskets available for installation.

6.7.5   EMI Gaskets

Gaskets are used to reduce leakages at seams of enclosure doors and removable panels. In addition, they can also be used for RF tight bonding for connectors and EMI filters. Since gaskets work by pressure created by compression, they should also be elastic and should spring back in shape when compression is removed.

6.7.5.1 Knitted wire mesh gasket

The knitted wire mesh type of gasket is available in two types. One type consists of a metal knitted wire mesh throughout and the other type has a wire mesh fabricated over an elastomer core.

6.7.5.2 Foil/Fabric over foam gaskets

As the name suggests, this type of gasket consists of a core made up neoprene/ PVC/ polyurethane/ silicone foam covered by either a reinforced foil made up of special alloy or a conductive abrasion resistant fabric (nickel over copper). These gaskets are provided with conductive self-adhesive strips for easy installation.

6.7.5.3 Immersed wire / stranded wires in silicone sponge

This composite gasketing material is designed for use in suppressing EMI/RFI while at the same time providing an effective, full pressure, environmental seal between mating surfaces. They comprise of a highly conductive composite shielding material consisting of sponge or solid silicone elastomer in which aluminium wires are chemically vulcanized.

6.7.5.4 Conductive elastomer

Also called as volume conductive elastomer gaskets, these are made up of silicone elastomer containing a uniform dispersion of microscopic conductive particles such as carbon, nickel, nickel-plated graphite, silver-plated aluminum, silver-plated copper, silver-plated nickel, silver-plated glass, or pure silver. Many manufacturers also use other elastomers such as fluoro-silicone, fluorocarbon, and ethylene-propylene terpolymer.

6.7.5.5 Metallic finger-stock

Finger stock are metallic gaskets that consists of strips of flexible metal, usually beryllium-copper alloy. This alloy is used since its mechanical spring characteristics far surpasses all other materials and yields the best electrical spring contact available. The finger stock gasket is far more superior than the wire mesh and elastomer gaskets and it offers the highest EMI shielding effectiveness.

6.8 ENSURING SHIELDING EFFECTIVENESS OVER OPENINGS –A SUMMARY

The Fig 6.13 summarises shielding precautions to ensure shielding effectiveness over openings. This can be compared with Fig 6.12. The display is covered with a screen mesh. There are individual openings for keys of the keypad (although a shielded keypad can also be used). The ventilating holes are small while the fan opening is covered by a wire mesh screen. Cable entry is via a circular bulkhead connector mounted on the chassis ensuring 360° contact. EMI gasket has been used for seams on the panel door.

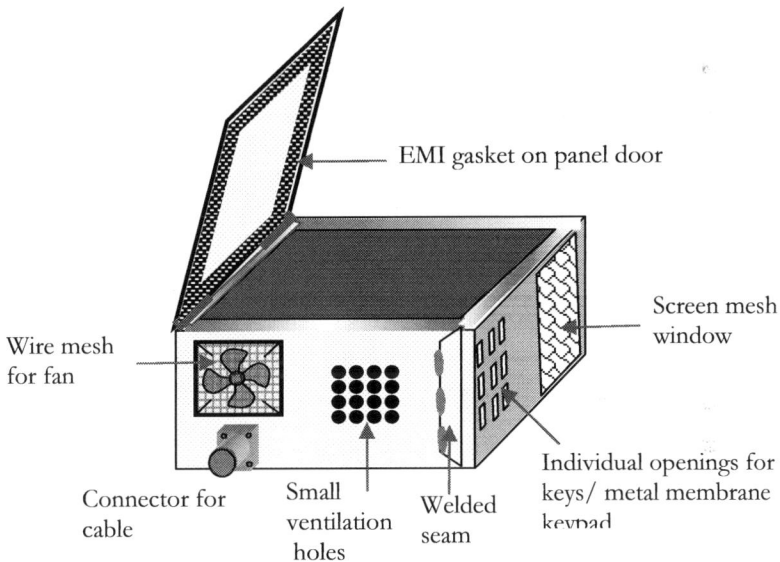

Fig. 6.13: Shielding effectiveness over openings

6.9 GROUNDING

Most of the EMI suppression methods, in one way or the other, use two basic techniques –filtering for conducted EMI and shielding against radiated EMI. Also, ESD and its associated transient noise is reduced by bonding and earthing equipment frames. A low resistance connection to earth is required if any of the interference suppression methods are to work properly. Grounding can be defined as *a connection, whether intentional or accidental, between an electrical circuit or equipment and the earth or to some conducting body that serves in place of the earth.*

On the other hand, bonding is defined as *a permanent joining of metallic parts to form an electrically conductive path that not only ensures electrical continuity but also has the capacity to conduct safely any current imposed on the joint.* Thus, bonding is that means which serves to give a *low resistance connection to earth* which is an essential part of a good ground.

6.10 TYPES OF GROUNDING

6.10.1 Single Point Grounding

Here each subsystem/module has its own ground. These individual grounds are finally connected by shortest route back to a single system ground point by simple wires (Fig. 6.14, left). Such system is advantageous for low frequency and analogue circuits because no common impedances (that cause common mode coupling) exist. At high frequencies (above 1MHz) however, the grounding wires start to exhibit high inductance and consequently start offering high impedance to ground currents. At the same time the capacitive reactance of stray capacitance between the modules starts reducing. The ground currents no longer follow the high impedance path offered by the ground wires to ground, but are rather invited to follow the low impedance offered by the parasitic capacitance to other modules. This causes common mode coupling and can be reduced by reducing the inductive reactance of the ground wires. This is achieved by multipoint grounding.

6.10.2 Multi-point Grounding

In multi-point grounding (Fig. 6.14 right), each subsystem/module is bonded as directly as possible to a common low impedance equipotential ground plane (essentially a continuous sheet of metal). Now a metal sheet offers a far less impedance at high frequencies (above 1 MHz) compared to wires. This is because sheets have more surface area which reduces resistance to high frequency currents that tend to flow on the surface due to skin effect. The sheet also provides multiple parallel paths for ground currents, which reduces the inductance. In such a scenario (when modules are mounted on a ground plane) the ground currents are invited to follow the low impedance of the ground plane to ground rather than going to another module.

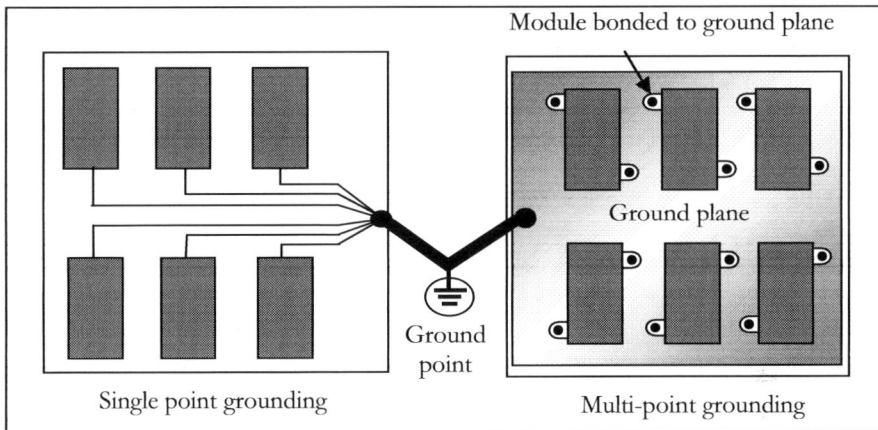

Fig. 6.14: Hybrid Grounding

6.10.3 Hybrid Grounding

Hybrid grounding is used in situations where systems involve both high frequency (digital) circuits and low frequency (analog) circuits as shown in Fig. 6.14.

6.11   REDUCING GROUND IMPEDANCE COUPLING

As we have seen earlier, due to improper grounding and bonding or due to improper grounding practices, the ground impedance tends to increase. This has been shown as a lumped impedance in the Fig. 6.15'a'. EMI currents flowing through this impedance cause a voltage drop 'Vcm' across it, which forces common mode EMI currents I1 and I2 through the circuit. The first way is to eliminate this common impedance completely by grounding the modules at a single point as shown in Fig. 6.15 'b'. The second option is, opening the ground loop by grounding only one of the modules (Fig. 6.15 'c'). But these hold good only at low frequencies up to a few tens of kilohertz, where this impedance behaves only as a resistance. At higher frequencies, stray capacitance 'C' (shown dotted) begins to appear between grounding cables or between the module and the ground, causing common mode currents to circulate again.

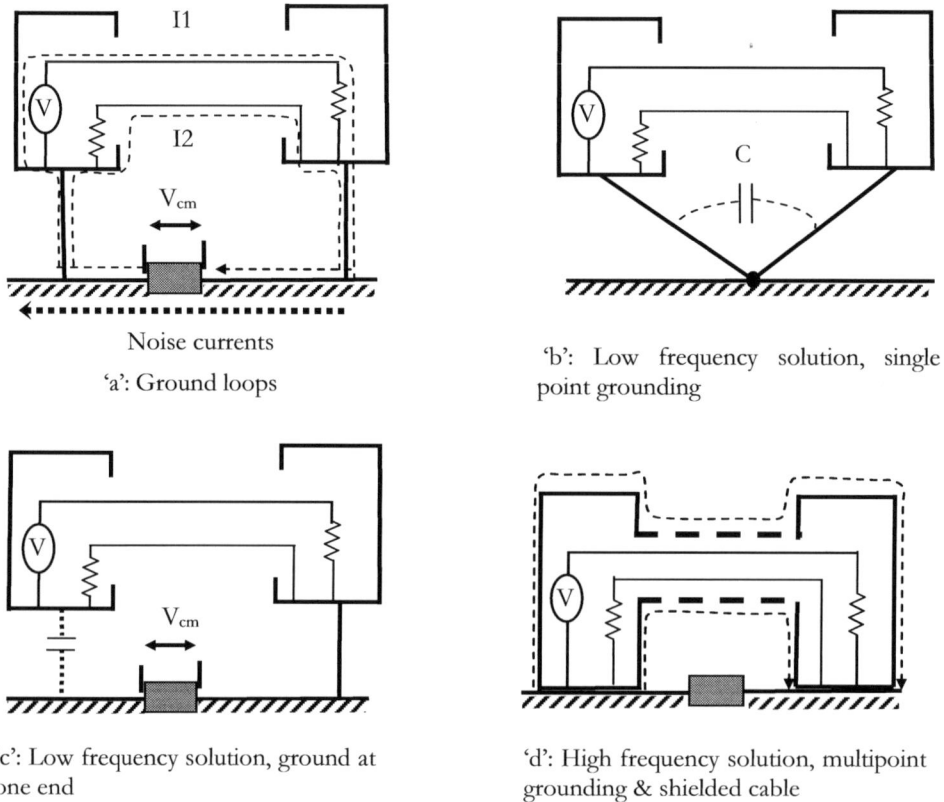

'a': Ground loops

'b': Low frequency solution, single point grounding

'c': Low frequency solution, ground at one end

'd': High frequency solution, multipoint grounding & shielded cable

Fig. 6.15: Reducing ground impedance coupling

Also, at higher frequencies the ground wires themselves start offering high impedance since their inductive reactance and resistance due to skin effect start to increase, causing potential gradients along the wire. At high frequencies therefore, efforts should be directed towards reducing ground impedance which in turn can be reduced first by reducing the inductance i.e. replacing the wires by a metal sheet (a mounting plate) and then to reduce the value of ground resistance by proper grounding and bonding practices. Also interconnecting cables should be shielded so that the common mode EMI currents now flow (see Fig. 10.5d) on the outside of the shield reducing common mode coupling.

6.12   CABLE SELECTION AND ROUTING

It is frequently observed that the overall system often fails to pass EMC tests (particularly emissions) because one crucial aspect of EMC design that of cable routing and selection has not been given proper attention. To complete the EMC design process therefore, one has to select the correct cable type i.e. shielded or unshielded, armoured or un-protected, balance or unbalanced, single core or multi-core, twisted or un-twisted etc depending upon the frequency of operation, voltage and current levels, type of loads and length of cable runs. After identifying the cable type, attention has to be given to the grouping or segregation of cables to avoid interference between various type of circuits. Thirdly, various cables and cable groups have to be properly routed so as to minimize cable-to-cable and cable-to-component coupling.

6.13   CABLE CLASSES

In order to properly segregate cables, they can be split into six classes as per EN 61000-5-2 depending upon their application, voltage or current levels, the frequency of signals they carry, the type of signal they carry (like analogue or digital), the components they connect to etc. These classes are as follows:

6.13.1   Class 1

These are cables that carry highly sensitive signals like low level analog signal (in mV) from transducers, instrument lines, radio receiver antenna cables and cables carrying high frequency digital signals such as ethernet. Such cables are highly susceptible to noise and interference and can be further classified as class 1A that carry analogue signals and class1B that carry digital signals. Class 1A and 1B cannot be bundled or twisted together, although they can run parallel to each other.

6.13.2   Class 2

These carry slightly sensitive signals which are somewhat susceptible such as ordinary analog signals under 1MHz and of the order of 1-10 V and 4-20 mA. They also include low frequency digital signals like RS232 or RS485 and digital I/Os like those from limit switches, encoders and control signals.

6.13.3 Class 3

These are cables carrying noise source that is slightly interfering such as low voltage AC distribution less than 1kV or DC power, linear power cables, control cables for inductive loads such as relay coils, motor brakes, contactors where proper transient suppression techniques like RC snubbers have been employed to control transients at the load. They also include motor cables from output of inverter drives but fitted with output filters as per manufacturer's specifications.

6.13.4 Class 4

These are cables carrying signals from strong noise sources that are highly interfering and includes output cables from frequency converters, inputs to and outputs from adjustable speed motor drives, inputs and outputs from SMPS or DC-DC converters, cables associated with electrical welders, RF equipment like induction welders, wood gluers, microwave equipment and diathermy apparatus, cables to DC motors, to RF transmitting antennas and unsuppressed inductive loads.

6.13.5 Class 5 and 6.

Class 5 cables are those that carry medium voltage while class 6 are those that carry high voltage.

As is clear from the above classes, that it is the class1 and class 4 cables which require special attention during the design or manufacturing process of electronic equipment. Another important thing to remember is that cables have been classified according to the type of signals they carry, but no consideration is given to the effect of external environment on the cables. A typical hostile environment, for example on shop floors or switch yards or power generating stations, may include hand held walkie-talkies being used in vicinity, RF heating devices around, radio or TV transmitters nearby, arc welding equipment in proximity, high power intentional

RF sources etc. When such situations exist, every cable entering or leaving the equipment cabinet should be classified as class 1 or class 4.

6.14   TYPE OF CABLES FOR A PARTICULAR CLASS

Since the various class of cables carry different type of signals, they dictate certain type of cables that can be used for that particular application. Since class 1 cables carry highly sensitive signals, they are highly susceptible to noise and interferences and hence such cables should be fully screened throughout their lengths with screens grounded at both ends by 360° shielded connectors which can be D-type with metal cans or circular bulkheads connectors. Class 2 and 3 cables carry slightly sensitive signals and hence there not as susceptible to interference as class 1 cables, the frequency of the signal is also not high and hence shielded cables grounded at one end will suffice. The shields can be terminated by P-clamps or semi-circular *saddle* clamps although pigtails (where in a small portion of the shield is brought out and soldered by a wire to ground) can also do but may be avoided if possible. Class 4 cables carry strongly interfering signals and can cause emission of EMI that can couple on to other cables in the vicinity. All considerations for class1 cables apply to class 4 too. Now class 5 and 6 cables carry medium voltage and high voltage signals respectively and therefore cannot be individually shielded. Armoured cables are recommended which may be used as shields, but as such are difficult to work with and are not easy to bond at every terminal or joint. Moreover, the armour does not, in any case, provide good shielding at high frequencies.

6.15   CABLE SEGREGATION

During their run in a system, cables of different classes have to be properly segregated i.e. separated by certain minimum distance. This separation distance not only depends on class but also the cable length. The Fig. 6.16 shows recommended minimum segregation distance for different classes of cables based upon a run of 30 metres and assuming that the cables run in close proximity to the protective earth conductor that can be a wire, a metal tray or the mounting plate that forms a local RF reference. Accordingly, class 1 and class 2 cables must be separated by a minimum distance of 150mm (ditto for class 3 and class 4). While class1-2 and class

3-4 should be separated by at least 450mm and class1 and class 4 by at least 600mm. The separation between class 4 and 5 and class 5 and 6 should be 150mm. Now these distances are for a cable run of less than 30 metres. For longer cables, the separation distance is multiplied by the length and divided by 30.

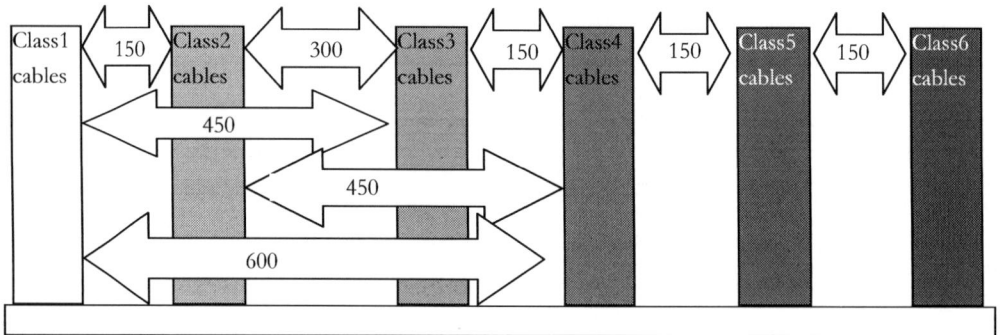

Fig. 6.16: Cable classes (dimensions in mm)

When cables run within a product, they must be physically segregated depending upon their class at all times, long parallel runs should be avoided and if it is not possible they should be more than 150mm apart, cable classes as far as possible must not cross or if at all they have to, they should do so at right angles and the cables should run as close as possible to their local RF reference or to cables carrying return signals. Cables of the same class when routed through the same tray or duct should not be twisted around one another.

6.16   CABLE ROUTING IN ELECTRONICS CONTROL PANEL

The Fig. 6.17 shows cable trunking for a control panel used for controlling a typical machine. Such a panel normally has, among other things, PLCs for controlling various operations, drives for controlling motors, SMPS for supplying DC power to PLC / other devices, contactors and relays for starting and stopping functions. All the components should be mounted on an unpainted mounting plate for proper grounding connections for the modules as well as their cables. In a typical control panel, all cables are routed through conduits which have not been shown for simplicity. The three-phase mains comes from a very hostile environment that is a characteristic of shop floors and carry such signals that can be strongly interfering. Such cables are therefore classified as class 4. It is essential that these are filtered immediately after entering the cabinet which is normally achieved by a suitable line filter. The filtered cable is classified as class 3, which then goes to the SMPS and the drive. The output of the SMPS is again classified as class 4 since it can carry strong interference. The filter at PLC I/P converts this into a class 3 cable which then goes to the PLC.

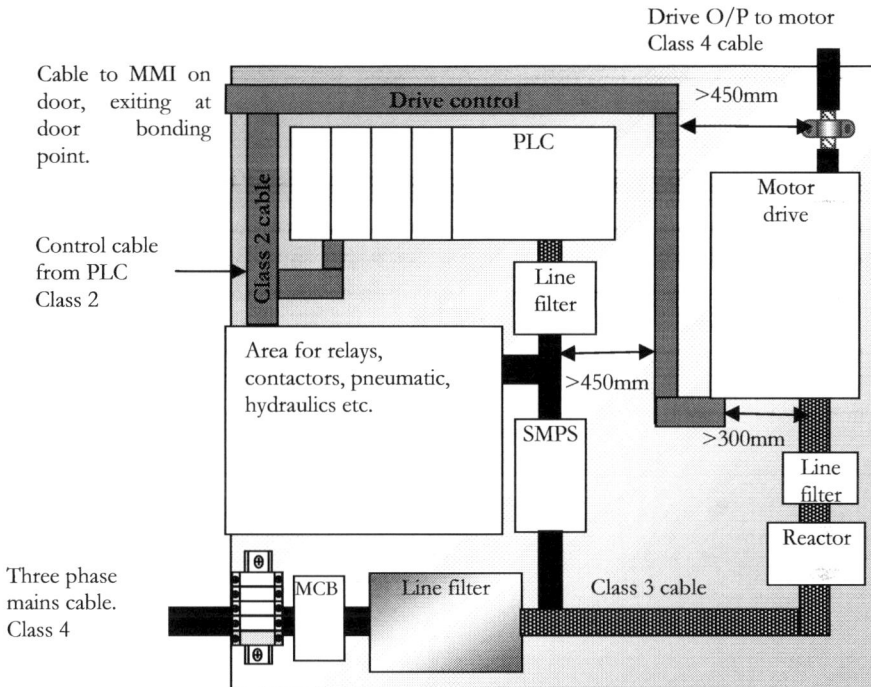

Fig. 6.17: Cable Routing In Electronics Cabinet

The I/O lines of PLC carry control signals that can be slightly sensitive and hence are classified as class 2. Some class 2 lines from PLC go as a control cables to the drive. As a rule, shielded cables must be used for such lines with the shield exposed and bonded to the mounting plate through 360° saddle clamps. Moreover, proper distance has to be maintained between these and nearby class 3 and 4 cables. In the Fig. 6.16, this distance is shown as 300m and 450mm respectively, although it may be less considering small cable lengths (remember this 300mm distance is for lines in excess of 30m). Some I/O cables of the PLC go to the relay and contactor section while some can also go to MMI usually mounted on panel doors. These also must be preferably shielded, with shield bonded to the mounting plate at a point nearest to the PLC. For cables going to the MMI, pigtails can be used for grounding although saddle clamps are better. Such cables should be very close to RF reference (i.e. the mounting plate) as they cross over to front panel door. This can be achieved only if the cross over is along the bonding point of the panel to the door i.e. along the bond straps that connect the panel to the door. If possible, bond straps shall be mounted along these cables or additional bond straps shall be provided along these cables. Many-a-times it is observed that these cables cross over along the door hinges, which is not a good practice, since hinges cannot be relied upon to provide a good bond.

Motor drives are the most notorious components so far as the generation of conducted and radiated EMI is concerned and are the main reasons for non-compliances observed in control panels. Most drive manufacturers give a fairly detailed information regarding the type of cables to be used for input mains supply, the drive output cable going to the motor, frequency input cable and I/O and control cables. All these cables must be invariably shielded. The drive manuals also discuss the cable routing to be followed, how the cables are installed and how saddle clamps and pigtails are to be used to ground the shields. Some even go to the extent of suggesting how much tightening torque has to be used for the screws for proper bonding. Machine manufacturers who casually ignore these guidelines eventually face a daunting task while complying with EMC tests.

Now, the output cables of motor drives can carry strongly interfering signals and therefore classified as class 4. It is imperative that the output and control cables of a drive are shielded with the shield bonded to the mounting plate by 360° connector or saddle clamp at a point nearest to the drive as shown in Fig. 6.16. The shield must also be grounded at the motor end preferably through 360° bulkhead connector at the motor terminal box. The drive control cable, which is class 2, must be separated from the O/P cable (class 4) by a distance more than 450mm and from the mains cable (class 3) by more than 300mm. Even while maintaining these distances, long parallel runs should be avoided. Motor O/P and control cable shall never cross. Crossing of control and I/P mains cable should also be avoided, but if they have to cross, they should do so at 90°. The mains cable to the drive should preferably be a shielded cable with the shield bonded to the mounting plate by a saddle clamp and the power earth (PE) wire of the mains cable should also be fastened under this clamp. For analog control cables, a double shielded twisted pair cable is normally recommended which consists of forward and return pairs individually shielded and the pairs bunched together with the entire bunch enclosed in outer shield jacket. A similar type of cable is recommended for digital control but single shielded multi-core cables can also suffice. Analog and digital signals must have their own separate cables, there cannot be a bigger blunder than running them as wires through the same cable although they are of the same class (class 2)! Relay-controlled signals, provided their voltage does not exceed 48 V, may be run as twisted pairs in the same cable as digital input signals.

CONCLUSION

In this chapter we have seen the various design approaches for a power line filter and also how transient suppression components are incorporated in a filter. Ready-made filters are also available which can be directly incorporated in circuits. These are however general purpose and have to be carefully selected to suit a particular application. The best way is to select a filter depending upon the measured conducted emission values without any filter. Every readymade filter available in the market, comes with insertion loss values plotted as a function of frequency which can be used to select a filter to attenuate emission in particular range. For example, if the emission is dominant in the frequency range of 150kHz to 500kHz, a filter with more insertion loss in this range can be selected. As regards to filters for I/O lines, we have discussed various filter configurations.

We have also seen shielding of equipment against radiated EMI. Good shielding effectiveness not only depends upon choice of shield material but also largely depends on various devices and fixes (meshes, gaskets, coatings etc) used for maintaining shielding effectiveness over various openings that need to be invariably provided. The entire shielded enclosure needs to be properly grounded so that the EMI currents induced in them find a low resistance path to ground. Also, the shielding fixes for openings have to be mounted properly in order to ensure a low resistance bond with the shield structure so as to enable them to work up of their capabilities. Cable selection and routing is generally a subject of system level EMC design where a large number of components are assembled to make a bigger system. It is generally observed that assembling all EMC compliant components does not automatically result in a compliant system. This is because components that make up the system have been tested as a finished product and that its manufacturer does not exactly know in what way his product will be assembled in a bigger system. It is the final system assembler who has to take care to ensure correct assembly of all the components to make a compliant system. Cable selection and routing plays a vital part in this endeavour.

References

"Trilogy of inductors". Design guide: Wurth Electronics.

"EMC and Inductive solutions". Catalogue. Wurth Electronics.

"Ceramic transient voltage suppressors". Epcos Guide.

"Noise reduction techniques in electronic systems". Henry Ott.

"EMC guide for avionics &related ground support equipment ". NAVAIR

"Application manual for power supply noise suppression". Murata.

"Modern Ferrite technology". Alex Goldman, Spinger Science,New York.

"The protection of USB 2.0 applications". Application Note, Wurth Electronics.

"Filtering DC power supply lines for high speed ICs". Toshiro Tsubouchi. ITEM Update 2003.

"Feedthrough filter problems". Dr Jeff Chambers. ITEM update 2001.

"SMD ferrites & filter connectors–EMI problems solved". Alexander Gerfen. ITEM 2001.

"EMI Shielding Engineering Handbook", Parker-Chomerics

"Shielding Design and Calculation", Ron Brewer, Laird Technologies, Tech Notes #105

"RFI shielding windows" article by David Carter, ITEM – 2004.

"Design Guidelines to EMI shielding windows" P. M. Grant, Tecknit EMI shielding products.

"Conductive painting of plastics". Roy Bjorlin, William Chen, Mel Browen, Becea Obuchowski, ITEM 2004.

"Mechanical design tips for EMI shielding". Fred Broekhvizen. ITEM 2001.

"EMC for servo drives" -application note. Kollmorgen, USA.

"Electrical installation guide 2010". Schneider electric, Germany.

"EMC installation guidelines for motor drives". Guide 690 292B. Schaffner.

"EMC installation guide". Eurotherm controls.

"EMC, telecom and computer encyclopedia handbook". 3rd edition. Don White

• • •

7

OTHER DIRECTIVES

7.1   INTRODUCTION

In this chapter we are going to discuss three more directives generally applicable to electrical/electronic products namely the medical devices directive, the RoHS directive and the radio equipment directive. Only unique requirements will be discussed and common requirements regarding placing in the market, free movement, roles of manufacturers/importers/distributors, presumption of conformity, declaration of conformity, affixing rules for CE marking, market surveillance, action on non- compliances, safeguard procedure, penalties, transposition and transitional provisions will not be discussed to avoid duplication. Clauses not related to electrical/electronic equipment are also not discussed.

7.2   MEDICAL DEVICES DIRECTIVE (MDD) (93/42/EU)

This directive applies to medical devices and their accessories and gives essential requirements that manufacturers and importers must meet to sell their devices in the EU. The objective is to provide, maintain or improve level of protection of medical devices to patients, users or third party. Since the MDD covers many types of devices, the specific requirements depend on the risk classification and intended use of the device. Contrary to the LVD and EMC directives, the use of an EU notified body (NB) is required to assess compliance with this directive in most of the cases. Although MDD covers a wide variety of devices, we are going to concentrate only on electrical devices in this chapter.

7.2.1   Scope

The definition of a medical device falling under the scope of the directive is given in article 1. Medical device falling under the scope of the directive is *any instrument, apparatus, appliance, software, material or other article, whether used alone or in combination, intended by the manufacturer to be used for human beings for the purpose of:*

- *diagnosis, prevention, monitoring, treatment or alleviation of disease,*
- *diagnosis, monitoring, treatment, alleviation of or compensation for an injury or handicap,*
- *investigation, replacement or modification of the anatomy or of a physiological process,*
- *control of conception,*

The medical device also includes software intended by its manufacturer to be used specifically for diagnostic and/or therapeutic purposes and necessary for its proper application.

Accessory means *an article which whilst not being a device is intended specifically by its manufacturer to be used together with the medical device.* This accessory enables the use of the medical device as intended by its manufacturer.

What are not considered as medical devices are products which achieve its principal intended action in or on the human body by pharmacological, immunological or metabolic means. For example, medications and pharmaceutical preparations, which treat medical conditions through chemical action or by being metabolised by the body, are not considered medical devices, and are subject to different regulations and requirements. Hence medical devices should not be confused with such classes of products.

Further, devices that are outside the scope of the directive are:

- – In-vitro devices (covered by In-vitro devices directive 98/79/EU)
- – Medicinal products
- – Cosmetic products
- – Blood/blood products
- – Transplants of tissues

Custom-made devices and those intended exclusively for the purposes of clinical trials are excluded from certain of the directive's administrative provisions (declaration of conformity), although they must still meet the essential safety requirements and may only be applied to patients by clinical specialists.

Examples of products which are considered to be medical devices include:

- – Medicine measuring cups
- – Syringes
- – Dental instruments
- – Stethoscopes
- – Thermometers
- – Prescription spectacles and contact lenses
- – Bandages and splints
- – Dental treatment chairs
- – Wheelchairs
- – First aid kits

Equipment which is not considered to fall within the scope of the MDD includes:

- – Toothbrushes
- – Baby nappies
- – Mouthguards
- – Intense Pulsed Light (IPL) therapy for, for example, hair removal.
- – Sunglasses (which are covered under the Protective Personal Equipment Directive)
- – Breathalysers

- Consumer products aimed at comfort

- Products for sport or leisure

- Cosmetic products, including tooth whitening products

7.2.2   Essential Requirements

Article 3 of the directive deals with essential requirements and are detailed in annexure I of the directive. These are as under:

7.2.2.1 Design, manufacturing and packing

- The devices must be designed and manufactured in such a way that, they will not compromise the clinical condition or the safety of patients or the safety and health of users and other persons associated with their use.

- The solutions adopted by the manufacturer for the design and construction of the devices must conform to safety principles. The manufacturer must eliminate or reduce risks, take adequate protection measures including warnings/ alarms if necessary, for those risks that cannot be eliminated and inform users of the residual risks.

- The devices must be designed, manufactured and packed in such a way that their characteristics and performances during their intended use will not be adversely affected during transport and storage taking account of the instructions and information provided by the manufacturer.

- Devices with a measuring function must be designed and manufactured in such a way so as to provide sufficient accuracy and stability within appropriate limits of accuracy and taking account of the intended purpose of the device. The limits of accuracy must be indicated by the manufacturer.

7.2.2.2 Devices connected to or equipped with an energy source

- Devices incorporating electronic programmable systems must be designed to ensure the repeatability, reliability and performance of these systems according to the intended use. In the event of a single fault condition (in the system), appropriate means should be adopted to eliminate or reduce as far as possible, consequent risks.

– Devices, where the safety of the patients depends on an internal power supply, must be equipped with a means of determining the state of the power supply.

– Devices, where the safety of the patients depends on an external power supply, must include an alarm system to signal any power failure.

– Devices intended to monitor one or more clinical parameters of a patient must be equipped with appropriate alarm systems to alert the user of situations which could lead to death or severe deterioration of the patient's state of health.

7.2.2.3 Electromagnetic compatibility

Devices must be designed and manufactured in such a way as to minimize the risks of creating electromagnetic fields which could impair the operation of other devices or equipment in the usual environment. This requirement can best be met by testing as per harmonised EMC standards identified against the directive.

7.2.2.4 Protection against electrical hazards

Devices must be designed and manufactured in such a way as to avoid, as far as possible, the risk of accidental electric shocks during normal use and in single fault condition, provided the devices are installed correctly. This requirement can best be met by testing as per harmonised electrical safety standards identified against the directive.

7.2.2.5 Other requirements

– Chemical, physical and biological properties
– Infection and microbial contamination
– Construction and environmental properties
– Protection against radiation
– Protection against mechanical and thermal risks

7.2.3 Steps To Conformity.

The Fig. 7.1 shows the steps to conformity as per the directive.

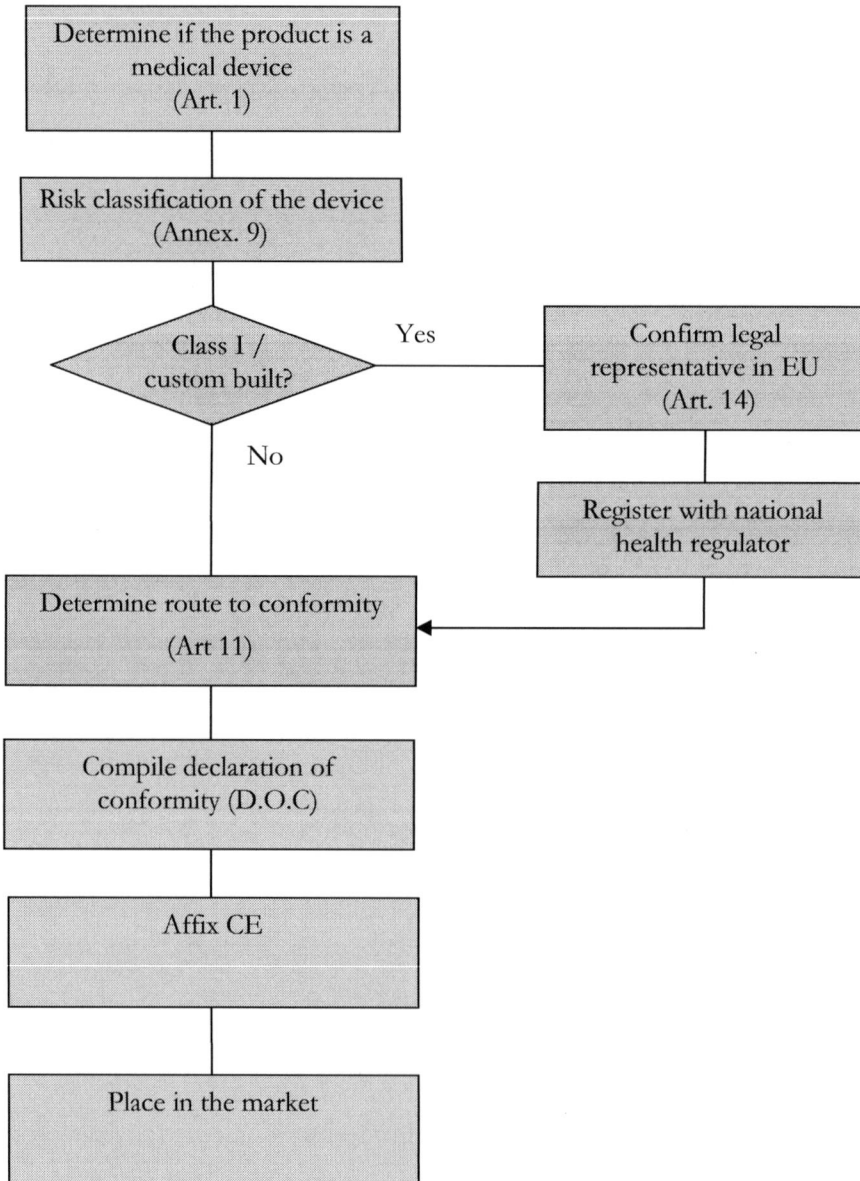

Fig 7.1: MDD Steps To Conformity

The first step is to determine whether the product is a medical device. This is done by referring the scope of the directive as discussed in section 7.2.1. The second step towards conformity is to classify the device according to the risk it poses and is done according to annexure IX of the directive. After that the manufacturer must follow one of the routes to conformity or *conformity assessment modules* prescribed by the directive for his category of equipment. The directive prescribes various modules for conformity assessment depending upon the level of risk the devices poses. Some of the modules require services of notified bodies for certain aspects of conformity assessment. Thereafter the manufacturer compiles the declaration of conformity (DOC) and affixes the CE marking on the product.

It is important to note that if the device is classified as Class I (see section 7.2.3.1) or if it is custom made, the manufacturer must be registered with the competent authority i.e. national health regulator in the member state where they are resident. The competent authority will only register manufacturers who are resident in their own territory so if the manufacturer is not based within the EEA, then he/she must appoint someone who will act as their authorised representative and their name and address must appear on the product instead of the manufacturer.

7.2.3.1 Risk classification

The device risk classification is done as per the rules given in annexure IX of the directive. The rule and corresponding class must be mentioned in the D.O.C. Accordingly the following class of devices are identified:
- Class I
- Class IIa
- Class IIb
- Class III

Before going into the details of classification, it is worthwhile to note some terms that are also the basis of the classification.

Transient use: Devices that are normally intended for continuous use for less than 60 minutes.

Short term use: Devices that are normally intended for continuous use from 60 minutes up to and including 30 days.

Long term use: Devices intended for continuous use for more than 30 days.

Invasive devices: A device which, in whole or in part, penetrates inside the body, either through a body orifice (any natural opening of the body) or through the surface of the body.

Surgically invasive device: An invasive device which penetrates inside the body through the surface of the body (other than through a body orifice) after a surgical operation.

Implantable device: Any device which is intended to be totally introduced into the human body (or to replace an epithelial surface or the surface of the eye) by surgical intervention which is intended to remain in place after the procedure. It also includes any device intended to be partially introduced into the human body through surgical intervention and intended to remain in place after the procedure *for at least 30 days*.

Active medical device: Any medical device operation of which depends on an external source of electrical energy.

Class I devices:

   – Non-Invasive devices *(ECG electrodes, dental patient chairs etc)*
   – Contact with intact skin. *(tourniquet)*
   – Mechanical barrier on injured skin *(dressing, cotton wool, bandaid etc)*
   – Invasive devices thru body orifice for transient use *(gloves, enema devices etc).*
   – Invasive E.N.T devices (up to drum in the ear, up to pharynx in the throat and in nose not effecting mucous membrane) for transient/short-term use.
   – Reusable surgical instruments for transient use (scalpels).
   – Channeling/storing liquids. (Class II A if connected to active devices).
   – Active devices not covered in Class II and III (examination lights, optical surgical microscopes).

Class IIa and Class IIb devices:

The table 7.1 gives examples of Class IIa and IIb devices

Table 7.1: Class IIa and IIb devices

Class IIa	Class IIb
Non-invasive devices for channeling/storing/filtration of body	Non-invasive devices for modifying contents of body fluids (dialysis).

Class IIa	Class IIb
fluids (infusion pumps).	
All active devices intended to administer and/or remove medicines, body liquids or other substances to or from the body (*Nebulizers, suction equipment*)	All active devices intended to administer and/or remove medicines, body liquids or other substances to or from the body in a potentially hazardous way (*Dialysis equipment, ventilators, blood pumps for heart-lung machines*).
Non-invasive devices to manage micro-environment around wound (*poly-film dressing, gauze dressing,*).	Non-invasive devices for wounds that only can be treated by secondary intent (*dressing for severe burns*).
Invasive devices through orifice for short term use (*tracheal tubes*).	Invasive devices through orifice for long term use (*urethral stents*).
All surgically invasive devices intended for short term use (*Temporary filling materials*)	All implantable devices and long-term surgically invasive devices. (*Prosthetic joint replacements*)
Long-term invasive devices for E.N.T. Implantable devices for teeth.	All implantable and long-term invasive devices (other than those in IIA).
Active therapeutic devices to administer energy (*electrical acupuncture, cryosurgery equipment, powered drills*).	Active therapeutic devices to administer energy in hazardous way (baby incubators, high-frequency electrosurgical generators, electrocautery equipment, external pacemakers and defibrillators, surgical lasers, lithotripsy equipment, ttherapeutic X-ray sources etc)
Surgically invasive devices for transient/short term use (*lancets, needles of syringes*)	Surgically invasive transient/short term use devices to administer medicines in hazardous way/ supply energy in form of ionising radiation/has biological effect (*insulin pens, catheters containing radio isotopes*).
Active diagnostic equipment that supply	Diagnostic equipment that supply energy

Class IIa	Class IIb
energy absorbed by human body (diagnostic ultra-sound)	in form of ionizing radiation (Diagnostic X-RAY)
Active diagnostic equipment for monitoring non-life-threatening body parameters and monitoring vital physiological processes (ECG, EEG)	Diagnostic equipment for monitoring such body parameters whose variation can result in immediate danger to human life (ICU monitoring/ alarm devices)

Then there are also some other Class IIb devices like:
- Invasive devices for short term use that undergo chemical change in human body.
- Devices for controlling/monitoring performance of active therapeutic devices.
- Devices for disinfecting/cleansing hydrated contact lenses.
- Contraceptive devices for short/transient use.

Class III:
- Surgically invasive devices (long/transient/short term) in direct contact with heart/circulatory/nervous system to diagnose, monitor or correct (angioplasty balloon catheters, distal protection devices, spinal needles).
- Short / long term surgically invasive devices to be absorbed/having biological impact/undergoing chemical change in body (*absorbable surtures*).
- Long term contraceptive devices (IUDs).
- Devices utilizing animal tissues or derivatives (except those in contact with intact skin) (*biological heart valves*).
- All surgically invasive devices specifically to control, diagnose, monitor or correct a defects of the heart or of the central circulatory system through direct contact with these parts (*cardiovascular catheters*)

7.2.3.2 Conformity assessment modules

These are embodied in article 11 of the directive and indicated in Fig. 7.2. For Class I devices which are neither custom built nor have a measuring function, the route to conformity is EC declaration of conformity i.e. self-declaration with internal production control as discussed in section 1.12. A list of standards identified by the directive is given in annexure-I of this book. The manufacturer has to choose the relevant standard and comply with the requirements. The procedure for choosing a standard has already been discussed in chapter 1.

For medical devices, the manufacturer (or the person placing the device in the market) has an additional obligation of notifying the competent authorities (read EU government) of following incidents:

- Death of serious deterioration of the health of a patient caused by malfunction or deterioration in the characteristics and/or performance of a device (including any inadequacy in the labelling or the instructions for use which might have caused it).

- Any technical or medical reason connected with recall of devices of same type.

For Class I devices with measuring function, in addition to DOC, the manufacturer should get his QS assessed by a notified body (NB) for those aspects of manufacturing concerning conformity to metrological requirements.

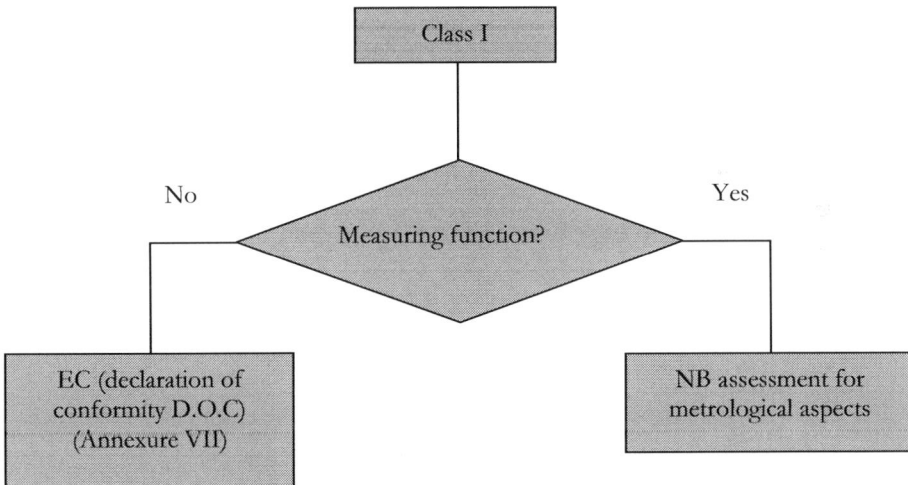

Fig 7.2: Conformity assessment modules Class-I

For Class IIa devices, the manufacturer has to avail services of a NB and the directive gives a choice of modules as depicted in Fig 7.3. Accordingly, the manufacturer can follow the full quality assurance module (see section 1.12) without the design examination part. In lieu of that, he can compile technical documentation as per annexure VII (of the directive) and thereafter follow one of the following (without the EC type examination certificate):

Product QA module (see section 1.12).

OR

Production QA (see section 1.12).

OR

EC verification (see section 1.12).

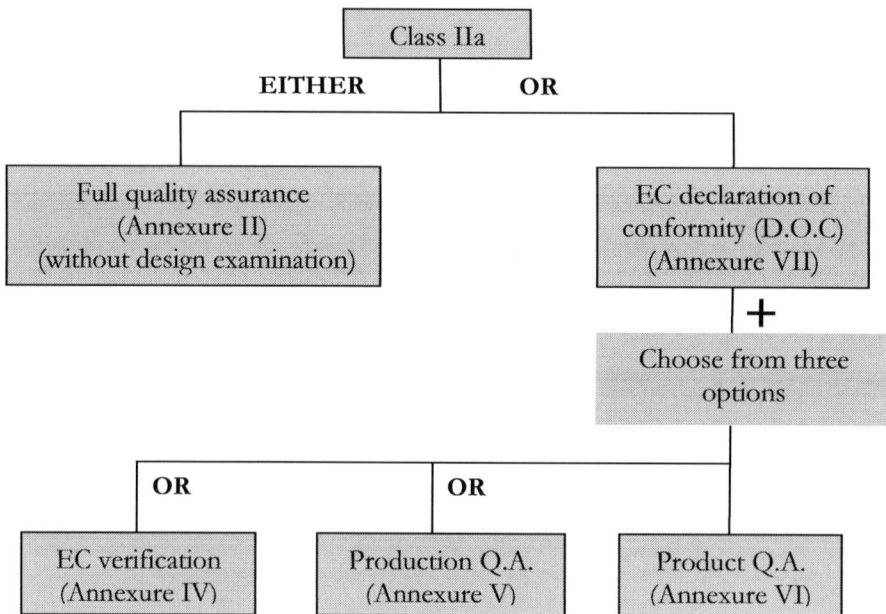

Fig 7.3: Conformity Assessment Modules Class IIa

For Class IIb devices, the manufacturer has to avail services of a NB and the directive gives a choice of modules as depicted in Fig 7.4. Accordingly, the manufacturer can follow the full quality assurance module (see section 1.12) without the design examination part. In lieu of that, he can follow the EC type-examination module (see section 1.12) as per annexure III (of the directive) and thereafter follow one of the following:

Product QA module (see section 1.12).

OR

Production QA (see section 1.12).

OR

EC verification (see section 1.12).

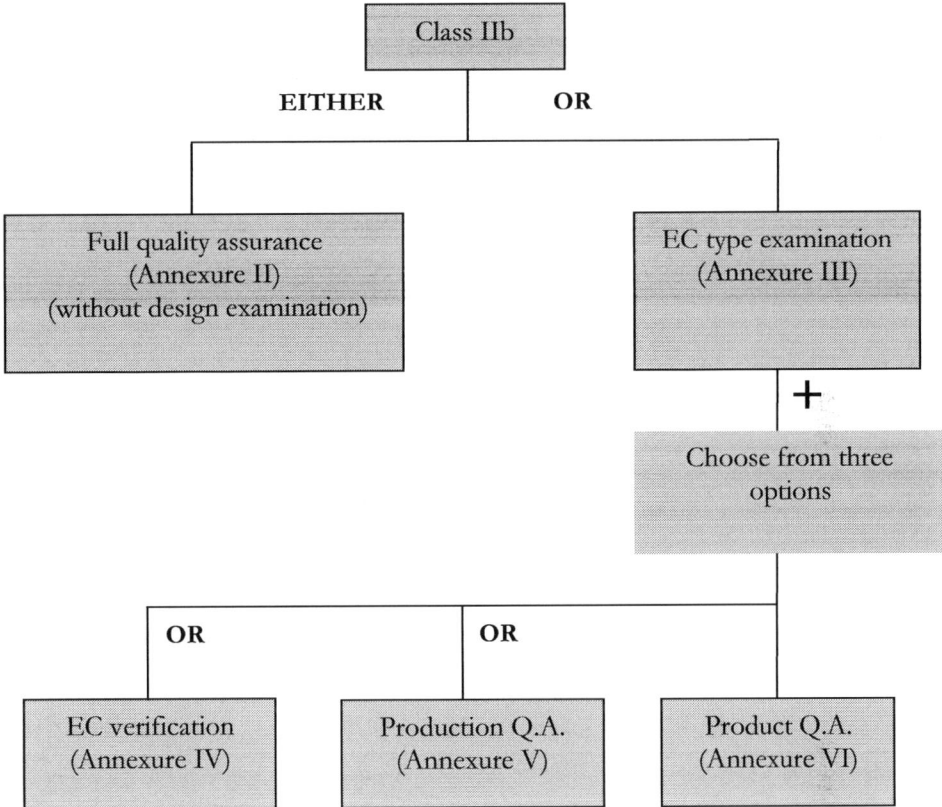

Fig 7.4: Conformity Assessment Modules Class IIb

For Class III devices, the manufacturer has to avail services of a NB and the directive gives a choice of modules as depicted in Fig 7.5. Accordingly, the manufacturer can follow the full quality assurance module (see section 1.12) with the design examination part. In lieu of that, he can follow the EC type-examination module (see section 1.12) as per annexure III (of the directive) and thereafter follow one of the following:

Production QA (see section 1.12).

OR

EC verification (see section 1.12).

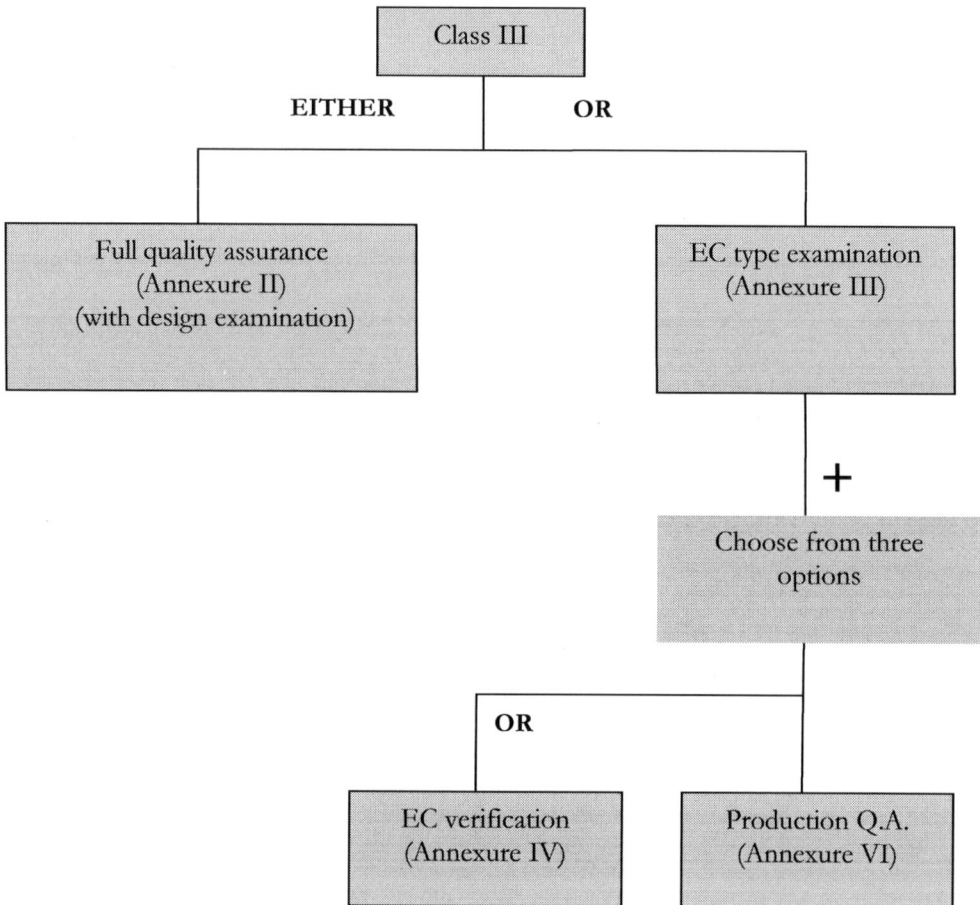

Fig 7.5: Conformity Assessment Modules Class III

7.2.4 Technical Documentation, DOC and CE Marking

After carrying out the relevant conformity assessment procedure, the manufacturer compiles the technical documentation in support of his conformity claim and then signs the DOC. Both the technical documentation and the DOC should be kept with the importer or authorised representative within the EU so that it can be produced, if demanded, by the surveillance authorities for five years after the last sample of the device model/type has been placed in the market. This is followed by affixing the CE marking. The format of the DOC is the same as explained in section 8.2 with following additional requirements:

- Device classification (class and rule)
- Notified body name and ID number (if applicable)
- Route to compliance (example: Annex II, V, VII)

For Class IIa, Class IIb and Class III devices where the services of NB are required, the CE marking has to be accompanied by the identification number of the NB.

7.2.5   Information Supplied By the Manufacturer

Each device must be accompanied by the information needed to use it safely and to identify the manufacturer, taking account of the training and knowledge of the potential users. As far as practicable and appropriate, the information needed to use the device safely must be set out on the device itself and/or on the packaging for each unit or, where appropriate, on the sales packaging. If individual packaging of each unit is not practicable, the information must be set out in the leaflet supplied with one or more devices. Instructions for use must be included in the packaging for every device.

7.2.6   Labelling Requirements

The device label must bear the following particulars:

- Name or trade name and address of the manufacturer. For devices imported into the community market, the label (or the outer packaging or instructions for use) should also contain the name and address of the importer or of the authorized representative.
- The details strictly necessary for the user to identify the device and the contents of the packaging.
- Where appropriate, the batch code, preceded by the word 'LOT', or the serial number.
- Any special storage and/or handling conditions, special operating instructions, warnings and/or precautions.
- Year of manufacture for active devices.

7.3   THE RoHS DIRECTIVE (2015/683/EU)

RoHS stands for Restriction on the use of Hazardous Substances. The original directive 2002/95/EC was published in 2002. The directive laid down rules on the restriction of the use of hazardous substances in electrical and electronic equipment (hereafter referred to as EEE) with a view to contributing to the protection of human health and the environment. It sought to restrict the use of more than the permitted levels of six substances namely lead, cadmium, mercury, hexavalent chromium, polybrominated biphenyl (PBB) and polybrominated diphenyl ether (PBDE) flame retardants. It included eight product categories out of ten of the Waste Electrical and Electronic Equipment (WEEE) directive. At that time, it was not a CE directive. The directive was recast in 2011 as 2011/65/EC as a CE marking directive and included all ten WEEE product categories. In 2015, four additional substances called *phthalates* were added to the list taking the number of banned substances to ten. The directive was re-designated as 2015/683/ EC and as such is identical to 2011/65/EC except for the annexure 2 which adds the four new substances.

7.3.1   Scope

Article 2 gives the scope of the directive. Accordingly, the directive applies to all EEE with a voltage rating not exceeding 1000 Vac and 1500 Vdc. The product categories to which the directive applies in given in annex I of the directive and includes:

– **Large household appliances**

Examples are refrigerators, washing machines, dish washing machines, electric stoves/ hotplates, microwaves, electric fans, air conditioners etc.

– **Small household appliances**

These include vacuum cleaners, irons, toasters, grinders and the like.

– **IT and telecommunications equipment**.

PCs, laptops, printers, fax, telephones, cell phones etc.

– **Consumer equipment**.

TVs, radios, cameras, audio amplifiers, music systems etc.

– **Lighting equipment.**

CFLs, LED lamps, sodium vapour lamps etc.

– **Electrical and electronic tools.**

Sewing machines, drills, electric saws, spray painting guns etc.

– **Toys, leisure and sports equipment.**

Video games, treadmills etc.

– **Medical devices.**

– **Monitoring and control instruments including industrial monitoring and control instruments.**

Thermostats, smoke detectors, fire alarms etc.

– **Automatic dispensers.**

Vending machines, ATMs.

Other EEE not covered by any of the categories above.

7.3.2 Excluded Equipment

The directive does not apply to:

– Military use equipment.

– Equipment designed to be sent into space.

– Large scale stationary industrial tools.

Industrial tools are an assembly of machines, equipment and/or components, functioning together for a specific application that are permanently installed and de-installed by professionals at a given place and are used and maintained by professionals in an industrial manufacturing facility or R&D facility. Examples of industrial tools are CNC lathes, bridge-type milling and drilling machines, metal forming presses, newspaper printing presses etc.

– Large scale fixed installations.

Fixed installations have been defined in the EMC directive (see section 4.5). Examples may include production and processing lines, including robots and machine tools (industrial, food, print media etc.), passenger lifts, conveyor transport systems, automated storage systems, electrical distribution systems such as generators, railway signalling infrastructure, fixed installed cooling, air conditioning and refrigerating plants or heating systems designed exclusively for non-residential use.

Now the term *large* is subjective since the directive does not clearly define what exactly is *large scale*. Hence it becomes the responsibility of the manufacturer, importer, or any other economic operator involved to assess whether his tool or installation benefits from either exclusion. Since the Indian manufacturer is based outside the EU, it largely depends upon the importer as to how he defines large scale since he will be the person responsible for placing the product in the market.

- Means of transport for persons or goods, excluding electric two-wheel vehicles which are not type-approved.
- Non-road mobile machinery made available exclusively for professional use;
- Active implantable medical devices.
- Photovoltaic panels intended to be used in a system that is designed, assembled and installed by professionals for permanent use at a defined location to produce energy from solar light for public, commercial, industrial and residential applications.
- Equipment specifically designed solely for the purposes of research and development only, made available on a business-to-business basis.

7.3.3 Banned Substances

As per article 4 (which can be considered to be essential requirements) of the directive, all EEE (including associated cables, spare parts for repair and reuse, and any parts subsequently used for updating / upgrading functionalities) placed on the market must not contain more than the permitted values of ten banned substances listed in annexure II of the directive which are as follows :-

- Cadmium (Cd): < 0.01%
  Major activity affected: Some cadmium solder alloys.
- Lead (Pb): < 0.1%
  Major activity affected: Solder.
- Mercury (Hg): < 0.1%
  Major activity affected: Switches and relays.
- Hexavalent Chromium: (Cr VI) < 0.1%
  Major activity affected: Zinc Passivation.
- Polybrominated Biphenyls (PBB): < 0.1%

Major activity affected: Flame retardants in plastics.

–   Polybrominated Diphenyl Ethers (PBDE): < 0.1%

Major activity affected: Flame retardants in plastics.

–   Bis(2-Ethylhexyl) phthalate (DEHP): < 0.01%

–   Benzyl butyl phthalate (BBP): < 0.01%

–   Dibutyl phthalate (DBP): < 0.01%

–   Di-isobutyl phthalate (DIBP): < 0.01%

The percentage indicates maximum concentration values by *weight* in homogenous materials. For example, if a homogenous material weighs 100gm, then the allowable amount of lead will be less than 0.1gm or 100mg. The directive defines homogenous material as:

–   one material of uniform composition throughout (examples are individual types of plastics, ceramics, glass, metals, alloys, paper, board, resins and coatings) OR

–   a material, consisting of a combination of materials, that cannot be disjointed or separated into different materials by mechanical actions such as unscrewing, cutting, crushing, grinding and abrasive processes.

Using these interpretations, a plastic cover (for example) would be a homogeneous material if it consisted exclusively of one type of plastic that was not coated with or had attached to it (or inside it) any other kinds of materials. In this case, the maximum concentration values of the directive would apply to the plastic. On the other hand, an electric cable that consists of metal wires surrounded by non-metallic insulation materials would be an example of something that is not homogeneous material because mechanical processes could separate the different materials. In this case the maximum concentration values of the directive would apply to each of the separated materials individually. A semi-conductor package (as a final example) would contain many homogeneous materials, which include the plastic moulding material, the tin electroplating coatings on the lead frame, the lead frame alloy and the gold-bonding wires.

7.3.4   Exemption In Specific Applications

Although the directive seeks to ban prohibited substances, certain exemptions in specific applications are necessary because viable alternatives have not yet been identified or are not available considering the technology or because of operational reliability requirements. Annexure III of the directive lists exemptions (at the time of writing this book) in specific applications which are as follows:

- Mercury in fluorescent lamp (CFL) for general purposes not exceeding 5 mg per lamp for power rating $\geq$50 W and < 150 W and not exceeding 15 mg per lamp for power rating > 150 W. For special purpose lamp not exceeding 5 mg per lamp.

- Mercury in metal halide lamps (MH) and other lamps not mentioned in the annexure.

- Lead in glass of cathode ray tubes, electronic components and glass of fluorescent tubes not exceeding 0.2% by weight.

    Lead, or more specifically lead oxide, is often used in glass for electrical and electronic equipment to obtain specific characteristics, such as radiation protection (CRTs, medical applications), filtering (photography, image processing) and strengthening purposes (e.g. production of fluorescent tubes).

- Lead in high melting temperature type solders (i.e. lead based alloys containing 85% by weight or more lead).

- Lead in solders for servers, storage and storage array systems, network infrastructure equipment for switching, signaling, transmission as well as network management for telecom. This would include all servers, power supplies, display devices and similar electronic units that are incorporated into network infrastructure equipment. It would also include all cables and cable assemblies, and all connectors and connector assemblies used to provide interconnections for network infrastructure equipment but is not intended to include desktop or notebook computers, telephones, fax machines or consumer type like modems, switches etc.

- Lead in electronic ceramic parts excluding ceramic capacitors (e.g. piezo-electronic devices).

- Cadmium and its compounds in electrical contacts on account of the reliability required of the apparatus on which they are installed.

- Lead and cadmium in optical and filter glass. These are used in optical glass and filter glass to obtain specific properties and meet quality standards, for a wide variety of applications including in the photo industry (e.g. camera lenses), in projectors, scanners, printers and copiers.

– Lead in solders to complete a viable electrical connection between semiconductor die and carrier within integrated circuit flip chip packages.

– Lead halide as radiant agent in high intensity discharge (HID) lamps used for professional reprography applications.

– Lead as activator in the fluorescent powder (1 % lead by weight or less) of discharge lamps when used as sun tanning lamps containing phosphors such as BSP.

– Lead in solders for the soldering to machined through hole discoidal and planar array ceramic multilayer capacitors.

– Cadmium alloys as electrical/mechanical solder joints to electrical conductors located directly on the voice coil in transducers used in high-powered loudspeakers with sound pressure levels of 100 dB (A) and

– Lead in soldering materials in mercury free flat fluorescent lamps (which, e.g. are used for liquid crystal displays, design or industrial lighting).

– Lead oxide in seal frit used for making window assemblies for Argon and Krypton laser tubes.

– Lead in cermet-based trimmer potentiometer elements.

– Lead in the plating layer of high voltage diodes on the basis of a zinc borate glass body.

– Lead in solders for the soldering of thin copper wires of 100 μm diameter and less in power transformers.

7.3.4   Conformity Assessment Procedure

For declaring conformity to the essential requirements of the directive, the route to be followed is module 'A' i.e. *Self-declaration with internal production control* (refer section 1.12). To comply with essential requirements the manufacturer can refer a harmonised standard for tests and measurements on materials as per article 16 (which gives presumption of conformity) or can use other means like demonstrating compliance to essential requirement using technical justification as per article 13. A description of the entire assessment process should be included in the technical documentation as evidence of compliance/ due-diligence.

Considering that it not practically possible for a manufacturer (especially of equipment that use large number of components sources from different vendors) to test all the materials for banned substances, the demonstration of conformity can be done by what is called as *due-diligence* followed by sample testing (if required). By due diligence it is meant that a manufacturer, importer or distributor of EEE, should be able to show that he/she took all reasonable precautions, steps and measures to ensure compliance and has exercised all precautions to avoid committing the offence. This will form a part of legal defense in case of problems. This defense can be used if manufacturer proves that he had taken all reasonable steps or precautions and has taken all efforts to avoid committing the offence. To demonstrate due diligence, a manufacturer should be able to show that they have introduced a series of appropriate and effective processes to check production control and material supply, and ensured that those checks were being carried out. It is important to note that, in the present scenario, where Indian manufacturers are producing both RoHS and non-RoHS products, avoidance of contamination/mixing of RoHS and non-RoHS components is as important as selection of RoHS compliant components.

Compliance of components and equipment to RoHS is ensured by following a system of checks at various stages of the production process as given below:

- Selecting vendors that can supply RoHS compliant components.
- Inventory/stores control (that checks mixing compliant and non-compliant materials/components).
- Controlling production process to avoid contamination.
- Final inspection and QA checks.
- Maintaining records of evidence and documentation for components and process control.
- Staff training and experience.

All procedures regarding this due diligence should be documented and the process should be explained in the technical documentation in support of the conformity claim.

7.3.4.1 Selection of compliant vendors

While selecting compliant vendors, two type of evidences may be considered namely, a test report and supplier's declaration of conformity. In order to demonstrate compliance with RoHS, a test report must be up to date and clearly show that the product contains less than the maximum level of hazardous substances, including detailed information about each material or component supplied. Following points need attention:

– The report should be from accredited laboratory.

– The test report should refer only to the actual components in the finished product.

– The report should clearly show that each homogeneous material used in the product contains less than the maximum concentration value (MCV) of each hazardous substance.

– The methodology used in testing for hazardous substances must be clearly listed in the report. The manufacturer should be familiar with testing methods and should question suppliers if they are unsure how the component was tested.

– Compliance information needs to be clearly understood and test report should be in English or translated into English.

– In case of electronic components, the main non-conformity is the solder material. The report should ensure that results of the solder test is contained within the report.

In case report is not available, supplier's declarations may be considered as evidence and supporting information. However, it is the manufacturer's responsibility to ensure that any declaration is valid. Here are some points to be noted:

– The declaration should refer specifically to the goods supplied and should not be generic.

– There should not be any caveats. Wording such as *to the best of my knowledge* and *no intentional use* may devalue a supplier's declaration

– It should be ensured that the declaration is signed by a person who is authorised to do so.

– Any document provided by a supplier in support of his/her declaration should be accurate and they should demonstrate the facts they are intended to prove.

A suggested process for assessing vendor through his/her declaration is shown in the Fig. 7.6

Component/Material

No further
action

No further
action

Yes to
all 3

No

Yes

No

From same vendor?
Regularly checked in
prev. 3 yrs?
Always compliant?

No

Assessed in
past 12
months?

Yes

High risk of
banned
substance?

Yes

No to any
one

Analyse every batch
until confident that
risk is low

Reassess

No

Yes

Marked to
indicate
compliance?

Yes

High risk
material?

Yes

Previous
declarations
accurate?

No

Request materials
declaration from
supplier

Yes

No

Yes

Certificate
obtained ?

Yes to
either

Analyse
random
batches

New part
from new
supplier ?

No

No

Yes

Analyse before
using

Yes

Doubt over
reliability of
declaration ?

No

No further
action

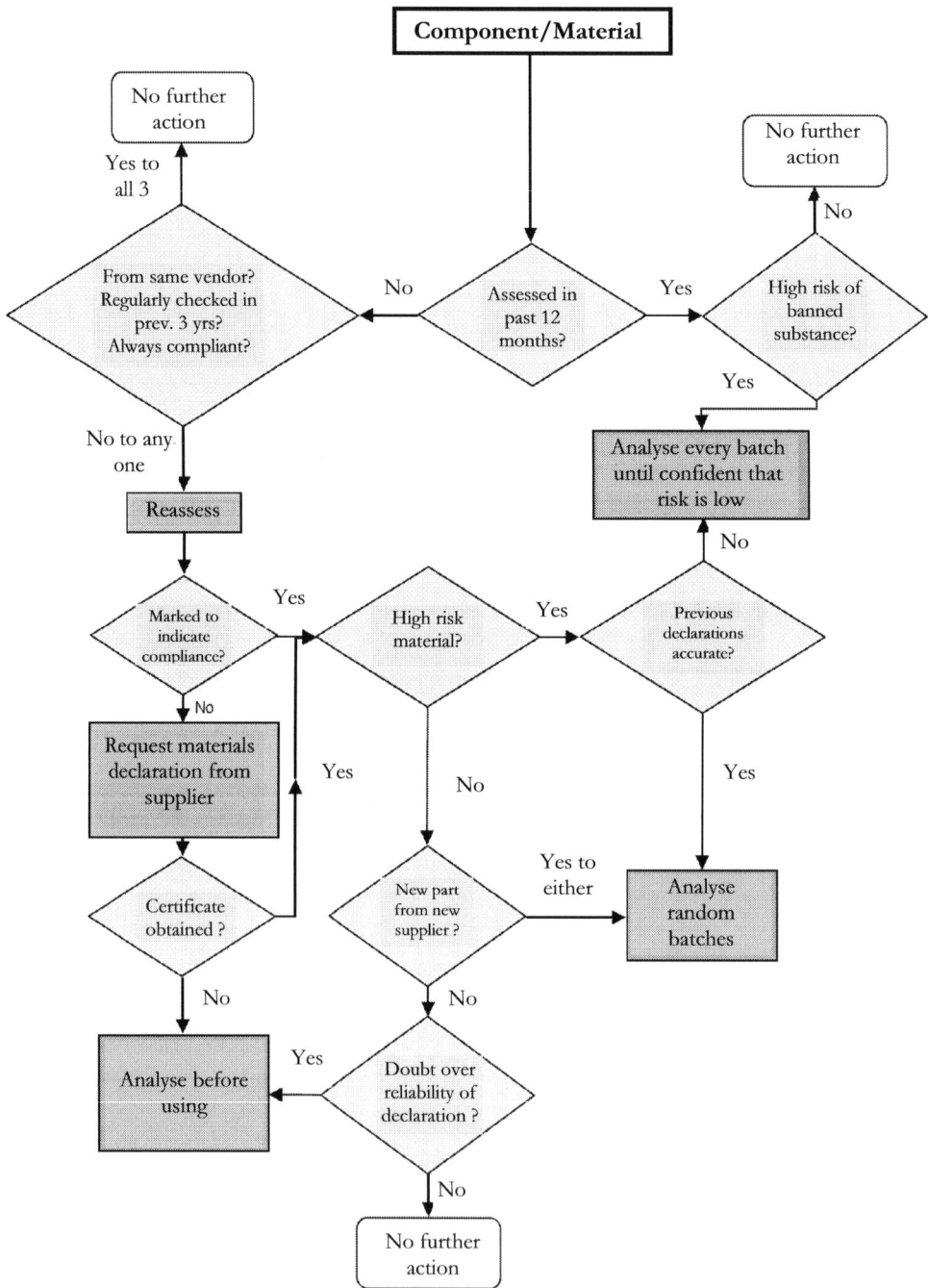

Fig 7.6: Vendor Assessment

7.3.4.2 Inventory/stores control

When inventory has to handle both compliant and non-compliant items, care should be taken to avoid mixing and issuing of wrong component. It is worthwhile to completely segregate compliant and non-compliant items and if possible, they should be in separate rooms with clear partition. Component code should clearly identify compliant and non-compliant items and personnel should be well educated regarding the codes. Appropriate check should be insured while issuing/retrieval of items, so that only the correct item is issued. All the procedures should be well documented in the QS.

7.3.4.3 Controlling assembly/production process

Contamination can also occur during production/ assembly process. For example, if the same wave soldering machine is used for manufacturing compliant and non-compliant product, contamination can occur while replacing the solder bath with compliant/ non-compliant one. The best way to avoid contamination is to have two different production lines for compliant and non-compliant products. If this is not possible, proper checks need to be in place while sample preparing, soldering, assembly and packaging.

7.3.4.4 Final inspection and test

Appropriate procedures should be developed to weed out non-compliances during final inspection and test. In case of critical product like medical devices, each sample coming out of the production line may be inspected. If this is not possible owing to a large volume of production, sample inspection can be carried out. If possible, sample testing can also be considered using an X-ray absorption spectrophotometer permanently installed in the section. It is worthwhile to appoint a designated compliance officer who is responsible for managing compliance issues.

7.3.4.5 Packaging and despatch

Contamination can happen in this last stage of the production process if RoHS compliant products are packed in non-compliant packing material. Hence packing material should also be appropriately marked for easy identification and appropriate checks should be made to ensure compliance.

7.3.4.6 Maintaining records of evidence, documentation and review

The manufacturer should be able to document his/her control systems, operate them effectively and review them regularly. All records regarding vendor reports, declarations, production control, QA checks and evidence of final inspection should be stored at a secure place and are retrievable when the need arises. A manufacturer should also consider auditing the systems at regular intervals to ensure that they remain fit for purpose and any identified failures are resolved at the earliest opportunity. If a programme of checks has been developed, it must work. Having a system in place that nobody follows is as bad as having no system at all. This is something that demands periodic or even constant monitoring. In creating a system, one must consider all aspects of business, from the design stage through to after sales. Any risk to the system of controls have to be identified and all steps should be taken to eliminate the risk. These should be reviewed from time to time.

7.3.4.7 Staff training and experience

A manufacturer should make sure that all employees are aware of all established control systems and their importance in producing compliant products. All employees should be given appropriate training from time to time regarding RoHS so as to maintain awareness during the production process.

7.3.5 Technical Documentation, DOC And CE Marking

All evidence of compliance which includes testing as per harmonised standards or documented due diligence including copies of test reports and vendor declarations should be included in the technical documentation. Thereafter the manufacturer has to sign the DOC. Both the technical documentation and the DOC should be kept with the importer or authorised representative within the EU so that it can be produced if demanded by the surveillance authorities for *ten years* after the last sample of the type has been placed in the community market.

7.4 THE RADIO EQUIPMENT DIRECTIVE (RED) (2014/53/EU)

This directive concerns the placement of radio devices where radio frequency (RF) energy is intentionally generated. It not only ensures electrical safety and electromagnetic compatibility of a radio device within its scope but also ensures an efficient use of radio spectrum to avoid harmful interference (especially due to out of band emissions). Although radio receivers themselves may not cause interference, reception capabilities are also a part of efficient use of the spectrum and as such the directive seeks to ensure that receivers should operate as intended and should be protected from harmful interference.

7.4.1 Scope

Article 1 gives the scope of the directive. Radio equipment under the scope of the directive is defined (as per article 2) as:

An electrical or electronic product, which intentionally emits and/or receives radio waves for the purpose of radio communication and/or radio determination, OR

An electrical or electronic product which must be completed with an accessory, such as antenna, so as to intentionally emit and/or receive radio waves for the purpose of radio communication and/or radio determination

Now the term radio communication is communication using radio waves (i.e. waves below 3000GHz propagating in free space as electromagnetic waves). The term 'radio determination' is the determination of the position, velocity and/or other characteristics of an object, or the obtaining of information relating to those parameters, by means of the propagation properties of radio waves. Rather than giving a list of radio equipment within scope, the annexure I of the directive gives a list of equipment outside the scope. The equipment ***outside the scope*** include:

- Radio equipment *exclusively* used for activities concerning public security, defence, state security and the activities of the state in the area of criminal law.
- Radio equipment used by radio amateurs like radio kits for assembly and use by radio amateurs, radio equipment modified by and for the use of radio amateurs and equipment constructed by individual radio amateurs for experimental and scientific purposes are excluded.

- Marine equipment which has to be carried on ships which are subject to International Maritime Organisation (IMO) Conventions is excluded.
- Airborne equipment are excluded.
- Custom built (built on the basis of a specific request from a specific customer) evaluation kits (circuit with ICs and support components for evaluation and development) destined for professionals to be used solely at research and development facilities for such purposes.
- Products that use RF energy for purposes other than radio communication and/or radio determination like:
    - Inductive warming and heating appliances
    - Wireless power transfer
    - High frequency surgical equipment and systems
    - Inductive heating appliances
    - Cabling and wiring

Certain equipment that fall in scope are:

- Radio equipment used by radio amateurs if commercially available (if supply is regular and there is a business content).
- Unmanned aircraft (remote controlled) with an operating mass of less than 150 kg e.g. drones.
- Marine equipment on recreational crafts.
- Ground aviation radio equipment.
- Active antennas.
- Amplifiers and other electronic equipment (e.g. filters, splitters, convertors, transverters, antenna tuners, switches) intended to be connected to an antenna,
- Radio broadcast receivers.
- RF modules
- Radio equipment installed in vehicles
- RFID tags

7.4.2   Essential Requirements

These are given in article 3 of the directive. Accordingly, radio equipment should be constructed so as:

7.4.2.1 Article 3.1a: Health and safety requirements as per the LVD (with no voltage limit) are ensured.

7.4.2.2 Article 3.1b: Adequate level of electromagnetic compatibility as per EMC directive is ensured.

7.4.2.3 Article 3.2: Equipment uses and supports the efficient use of radio spectrum in order to avoid harmful interference.

There are also separate essential requirements to be introduced in the future for certain category of products depending upon adoption of certain decisions taken by the commission (so called *delegated acts*) to fulfil such requirements. Such products may also require registration via a central registration system. Presently no such delegated acts are in place and consequently no registration is required but may be required in the future

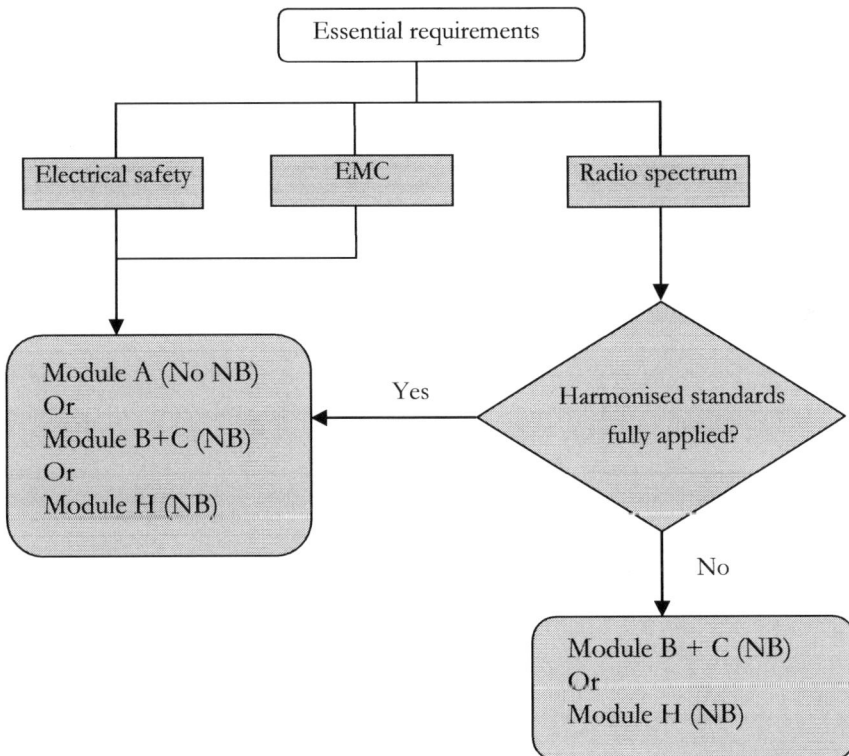

Fig 7.7: Conformity Assessment Procedure for RED

7.4.3   Conformity Assessment Procedures.

The conformity assessment procedure is given in Fig. 7.7. For essential requirements given in 7.4.2.1 and 7.4.2.2 the directive gives a choice of using one of the three modules (i.e. NB involvement is voluntary):

Module A: Self declaration with internal production control (voluntary application of harmonised standards) (see section 1.12)

OR

Module B + C: EC type examination + conformity to type (see section 1.12)

OR

Module H: Full quality assurance (without design examination)

For essential requirements given in 7.4.2.3, the choice of modules remain the same except that conformity to harmonised standards is compulsory for module 'A'. If harmonised standards have not been referred or have been partially referred, then NB involvement is compulsory and module A cannot be used.

7.4.4   Harmonised Standards for RED

Although the essential requirements refer to EMC and LVD, tests to meet these requirements are given in the EN standards developed by ETSI. For presumption of conformity, the radio equipment must comply with the tests and evaluations as per relevant standard. The standard EN 301489–1 gives common technical requirements, while ETSI EN 301489-2 to 301489–32 give specific requirements for a particular type of RF product. The EMC limits and test methods are based on EN55032 (see section 5.21) and 61000-3 series for emissions while EN 61000-4 series for immunity. Also see table 5.8 section 5.44 for EN 301489 EMC requirements. The annexure II gives a list of EN 301 489 series of standards that give presumption of conformity with article 3.1b (related to EMC). Example of standard selection has been discussed in section 1.13.

7.4.5   Technical Documentation, DOC And CE Marking

If module A is being referred to then the manufacturer has to comply with all tests as per ETSI standards (or can use other methods without presumption of conformity). Thereafter he/she has to compile technical documentation (refer chapter 8) and sign the DOC. The RED calls for inclusion of a description of accessories and components including software that allows the radio equipment to operate as intended. The drawing up of DOC is followed by affixing the CE marking.

CONCLUSION

In this chapter, we have seen three important directives namely the MDD, the RED and the RoHS. We have considered them separately since there are only procedural differences in the directives. So far as MDD and RED are concerned, products should meet EMC and safety testing as per relevant harmonised standards and these have already been discussed in the previous chapters. All the procedures carried out for CE marking along with technical reports need to be documented i.e. compiled in the form of technical documentation or technical file which is a crucial part of the CE marking process. The contents of the technical documentation are discussed in detail in the next chapter.

References

93/42/EC: Council Directive of 14 June 1993 concerning medical devices

2014/53/EU: Council Directive of 8 June 2011 on the restriction of the use of certain hazardous substances in electrical and electronic equipment

2015/683/EU): Council Directive of 16 April 2014 relating to radio equipment and repealing Directive 1999/5/EC

Medical devices guidance document –MDDEV 2.4: 2010

Guide to the Radio Equipment Directive 2014/53/EU 19 December 2018

• • •

8

TECHNICAL DOCUMENTATION

8.1   INTRODUCTION

Technical documentation is an important part of the CE marking process as it documents the entire conformity assessment process. As we have seen in chapter 1, the technical documentation may be demanded anytime by surveillance authorities and as such should unambiguously describe the equipment and the conformity process. It is important to note that in the event of any safety incident involving the equipment, the technical documentation will come to the rescue of the manufacturer/importer as it proves due diligence on the part of the manufacturer and can be a part of legal defence (if required). Each directive specifies the contents of the documentation which are more or less similar barring certain additional requirements considering specific application of a particular directive. The documentation can also be in soft copy form. The documentation should include the following:

8.2   DECLARATION OF CONFORMITY

This is usually the first page of the documentation. The Fig. 8.1 shows suggested DOC format for SMPS considered in section 1.13 for which EMC, LVD and RoHS directives are applicable. The DOC format for a high frequency surgical equipment considered in section 1.13 is as shown in Fig 8.2. Since this DOC is for medical device, it should also identify the class and rule according to which the classification was done (see section 7.2.3.1). Again, most of the medical devices refer to conformity assessment module that requires the assessment to be done by a NB. In that case the DOC should also mention the name, identification number of the NB, the EC type examination certificate number and date. The DOC format for a radio device is as given in Fig. 8.3. It should be noted that the above contents of D.O.C are the bare minimum. The manufacturer may include additional information if deemed necessary.

8.3   PROFILE OF MANUFACTURER AND IMPORTER

This section should provide detailed information of the manufacturing company like brief company history, type of products being manufactured, quality policy, etc. It should also include detailed address of the manufacturer and importer (if the manufacturer is non-European) along with contact details.

8.4   EQUIPMENT DESCRIPTION

This section should provide in detail a general description of the electrical equipment so as to enable EU surveillance authorities to clearly understand the equipment and its intended use. This detailed information will protect the manufacturer (or his representative) in event of a hazard as a result from wrong use.

EU DECLARATION OF CONFORMITY

We,

Name and address of the manufacturing company
Name and address of the importer / authorised representative

Hereby declare under our sole responsibility that-

Object of declaration: *Switch Mode Power Supply*
Model and Model variants (if any): *MXCC-A, MXCC-B, MXCC-C*

conforms with the essential requirements of EU legislation 2014/35/EU (EMC), 2014/30/EU (Electrical equipment designed for use within certain voltage limits or LVD) and RoHS (2011/65/EU) based on the following:

I)   For Electromagnetic Compatibility:

Testing in full as per harmonised standard *EN55032*: EMC of multimedia equipment - Emission requirements and EN 55024: I.T. equipment -Immunity characteristics.

II)   For LVD:

Testing in full as per harmonised standard EN 60950: Safety of I.T. equipment.

III) For RoHS:

EN 63000: Technical documentation for the assessment of electrical and electronic products with respect to the restriction of hazardous substances.

This declaration is only valid when:

—   The product is installed and put in to service by our representative.
—   Operated and serviced by qualified personnel.
—   No interlocks are defeated.
—   All EMC and safety measures recommended are taken care of.

Place and Date   : *El Dorado, 10th Jan 2020*

Signed for and behalf of   : *Manufacturer's name*

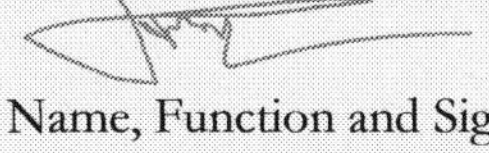

Name, Function and Signature (of signatory)

Note : All test reports, EMC assessment, safety risk assessment and technical justification are part of the technical documentation no. 12345 available at above address of the importer.

Fig. 8.1: D.O.C format for SMPS

EU DECLARATION OF CONFORMITY

No: 430-17   Valid up to: 23/12/2022

Manufacturer   *: Name and address of the company*

Importer   *: Name and address*

This declaration is issued under the sole responsibility of the above manufacturer for

Object of the declaration   *: Electrosurgical generator*

Model   *: EG 3CC*

Classification rule and class   *: Rule-9, Class IIb*

The object of the declaration described above is in conformity to the following legislations:

1.   Medical Devices Directive (MDD) 93/42/EU

2.   RoHS Directive 2011/65/EU

Based on the following:

Conformity assessment for MDD

Module   *: Annex II (Full quality assurance)*

Standards applied   *: EN 60601-2-2: 2009, EN 60601-1-2: 2015*

The notified body XXXX, identification no. YYYY performed the assessment and issued the certificate(s):

Title of Certificate   *: EC Design Examination Certificate*

Certificate No.   *: ZZZZ dt. _____ , valid up to _____*

Certificate No.   *: WWW dt. _____, valid to 2019-02-28 for QS based on EN-13485 : 2012*

Conformity assessment for RoHS directive

Documentation as per *EN 63000,* essential requirements of the directive, due diligence and technical justification

Place and Date   *: El Dorado, 10th Jan 2020*

Signed for and behalf of *: Manufacturer's name*

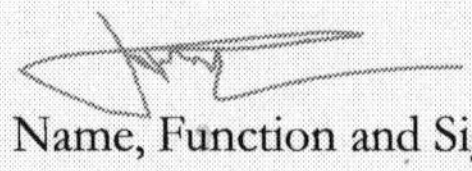

Name, Function and Signature (of signatory) : *Name and Designation*

Fig. 8.2: D.O.C format for Medical Electrical Equipment

EU DECLARATION OF CONFORMITY

No: 1111-11

Object of declaration

Radio equipment   : Handheld *Wireless Set*

Model   : *Double speak xxx*

Batch no.   : *2020:yyxx*

Manufacturer   : *Name and address of the company*

Importer   : *Name and address*

This declaration is issued under the sole responsibility of the manufacturer who declares that the above equipment (object of declaration) is in conformity with the following legislations:

1   Radio Equipment Directive (RED) 2014/53/EU

2   RoHS Directive 2011/65/EU

The conformity assessment procedure followed is as per Annexure III of the RED

Conformity to the essential requirements of the legislation(s) have been demonstrated by using the following standards:

Health and Safety (Art. 3.1(a))	: EN 62368-1: 2018, EN 62311: 2008
EMC (Art. 3.1(b))	: EN 301489-1: 2019, EN 301489-5: 2016
Spectrum (Art. 3.2)	: EN 300296: 2016
RoHS	: EN 63000, EN 62321, due diligence & tech. justification

The notified body XXXX, identification no. YYYY performed the type examination and issued type examination certificate no. *ZZZZ dt.* _____ , *valid up to* _____

Place and Date   : *El Dorado, 10ᵗʰ Jan 2020*

Signed for and behalf of   : *Manufacturer's name*

Name, Function and Signature (of signatory) : *Name and Designation*

Note : The supporting technical documentation is available with the above identified importer.

Fig. 8.3: D.O.C format for Radio Device

The description can include:

— Principal of operation

— Intended function

— Working

— Operating instructions

— Instructions for servicing and maintenance

— Technical specifications (like electrical, mechanical, environmental etc),

— Installation requirements/instructions.

— Limitations for installation (like environmental, altitude etc) and

— Prohibited use

   The manufacturer's explanation may be like "*hazardous location is any location where a potential hazard, either a fire or an explosion, can exist because of the presence of flammable, combustible, or ignitable materials. These materials can consist of gases, vapours, liquids, dust, fibers, etc. The equipment must not be installed in such an environment*".

— Detailed warning and precautions while storing, operating and servicing the product should be included to prove that the user/service personnel have been sufficiently warned.

8.5   EQUIPMENT DESIGN

This section should include conceptual design and manufacturing drawings and schemes of components, sub-assemblies, circuits, etc. It is best to provide all circuit diagrams. If this is not possible on account of proprietary issues, detailed block diagram can be given. In this case the custodian of the circuit diagrams can be identified who can produce them if demanded by surveillance authorities. Descriptions and explanations necessary for understanding of said drawings, diagrams and should also be included.

8.6   PHOTOGRAPHS

All photographs (internal and external) should be given like all views, internal components, photos of critical assemblies and design features to overcome hazards (like those against electric shock, EMI (filters, surge arrestors, shielding), guarding against mechanical hazards etc.) should be properly highlighted in the photographs. This will protect the manufacturer (or his representative) in event of an unauthorized change of component by the user which may cause a safety hazard (like removal or change of mechanical guarding, removal of safety components and EMI suppression components without informing the manufacturer) and which can be readily identified based on the photographs. A photograph of the equipment label allowing its identification should also be included.

8.7   BILL OF MATERIAL (BOM)

A detailed BOM should be included (can be in soft copy form like CD). Critical components like those for electrical safety (MCBs/ELCBs, fuses/thermal cut outs etc) and EMC (like line filters, surge arrestors, EMI suppression capacitors, EMI suppression ferrites etc) should be clearly identified with model and type approvals. This detailed information will protect the manufacturer (or his representative) in event of an unauthorized change of component by user which may cause a safety hazard.

8.8   PRODUCTS VARIANTS

Many-a-times manufacturers apply CE marking on all product variants based on testing of one variant. Normally, that variant which has the maximum components/features is subjected to the conformity assessment process and based on it, a blanket declaration for all variants (without testing/assessment) is given in the DOC. The section should include product variants clearly identifying how the variant differs from that which was subjected to the conformity assessment process (like identification of change parts/components) and a justification as to how this change would not affect the conformity. The justification could be that the changes are just cosmetic, or parts will only be deleted and not added (except safety and EMC critical components) etc. It should be noted that, if deletion or addition of a change part affects conformity, then the product cannot be considered as a variant and should be ideally subjected to separate conformity assessment.

8.9   SPECIFICATIONS OF PACKAGING AND LABELLING

If the product is likely to be moved frequently in its lifespan then instruction for proper packing and un-packing should be included along with precautions while transportation including proper labelling of the package (like *fragile/ top/ lift here* etc). This will save the manufacturer in case any component becomes loose and causes a safety hazard.

8.10   RATIONALE FOR CONFORMITY

As seen in previous chapters, EMC directive requires documented EMC assessment of the product, LVD requires documented risk assessment while the RoHS requires documentation as per EN63000 and due diligence on part of the manufacturer. This section should provide an explanation as to why the product complies with the essential requirements of the applicable directives. Details are discussed in sections 8.11 to 8.15.

8.11   EMC ASSESSMENT AND CONFORMITY

As discussed earlier, the EMC directive requires an EMC assessment to be carried out. However, if a harmonised standard has been fully applied (i.e. all tests as per the standard have been carried out and complied with) then there is no need for separate EMC assessment. Nonetheless, the table 8.1 gives an example of EMC assessment. The manufacturer's EMC assessment may be as follows:

> **Type of assessment**
>
> *A mixed type of EMC assessment has been carried out wherein technical justification has been given for some immunity tests and the harmonized standard BS EN 61326:2010 (Electromagnetic compatibility (EMC) Electrical equipment for measurement, control and laboratory use – EMC requirements Part 1: General requirements) has been applied.*

Nature Of apparatus:

The apparatus is a _____ used in typical industrial environment and on equipment designated for use in industrial environment.

Intended Use

It is used primarily in industrial applications.

Location of use:

It is usually incorporated in equipment located on shop floors.

EMC Environment

The environment in which the equipment is supposed to operate is a typical industrial environment characterized by absence of suppression components in control circuits, poor separation of industrial circuits from other circuits associated with high severity levels., dedicated cables for power supply, control, signal and communication lines, poor separation between power supply, control, signal and communication cables, Availability of earthing system represented by conductive pipes, ground conductors in cable trays and by ground mesh.

Type of disturbances created by the equipment

Due to the type of components used like SMPS, drives, PLCs, the equipment may generate EMI –conducted and radiated both. The emission of this particular equipment is however, below specified limit.

Type of disturbances affecting the equipment:

Sensitive components like PLC which, if not tested for EMC, are likely to be affected by high and low energy transients (like surges, burst and electrostatic discharges) which are generated in a typical industrial environment. The equipment uses PLC which is compliant to EMC standards and bears the CE marking. Furthermore, the equipment has been actually tested for immunity and has complied with all applicable tests.

Performance criterion for immunity

The performance criterion as given in EN 61326:2012 been applied for all applicable immunity tests

EMC phenomenon for assessment and their justification (table 8.1)

Table 8.1: Example of EMC assessment

Phenomenon	Justification for compliance
EMISSION	
Conducted and Radiated Emission	Testing has been carried out as per EN _____ . The equipment complies with specified limit with sufficient margin. (For details see section _____ of report no. _____ dt _____copy of which is attached in Section 5 of this documentation)
Harmonics	All components liable to generate harmonics like SMPS, PLC and drives bear CE marking and complying to EN 61000-3-2. Hence the equipment is deemed to comply with this test.
Voltage Fluctuations and Flicker	The equipment is of continuous current rating and does not draw discontinuous current. It is an insignificant load factor on the power network, and does not create daily peak power demands. Hence this test is not applicable.
IMMUNITY	
Power Frequency Magnetic Field Immunity	The equipment does not incorporate components that are susceptible to magnetic fields. Hence the equipment is deemed complied without testing.
Radiated Susceptibility	The components that are sensitive to such fields like PLC have CE marking and complying with _____ standard. Therefore, the equipment is deemed to comply with this test.
Electrostatic Discharge	Testing has been carried out as per EN _____. at a severity level of 4 kV (contact discharge) and 8 kV (air) on equipment body, screws etc. The equipment worked satisfactorily during and after the test and no deviations from normal operations were observed. (For details see section _____ of report no. _____ dt _____copy of which is attached in Section 5 of this documentation)
Conducted RF Susceptibility	Testing has been carried out as per EN _____ . at a severity level of 10 V. The equipment worked satisfactorily during and after the test and no deviations from normal operations were observed. (For details see section _____ of report no. _____ dt _____copy of which is attached in Section 5 of this documentation)

Electrical Fast Transients	Testing has been carried out as per EN _____. at a severity level of 2 kV. The equipment worked satisfactorily during and after the test and no deviations from normal operations were observed. (For details see section _____ of report no. _____ dt _____copy of which is attached in Section 5 of this documentation)
Surge	Testing has been carried out as per EN _____ at a severity level of 2 kV (CM) and 1kV (DM). The equipment worked satisfactorily during and after the test and no deviations from normal operations were observed. (For details see section _____ of report no._____ dt _____copy of which is attached in Section 5 of this documentation)
Voltage Dips and Interruption	Meets all requirements of EN _____ (For details see section _____ of report no. _____ dt _____copy of which is attached in Section 5 of this documentation).

The above considerations and the results of the relevant test indicate that the product meets all EMC requirements.

8.11.1 Other Considerations

Certain other information although not compulsory, may also be given such as EMC design considerations and/or calculation results, statistical evaluations, theoretical studies or other examinations carried out, information on shielding, cable selection and routing, filters (reference should be made to the EMC critical components in the BOM), general or specific requirements taken to limit emission of disturbances and worst case selection criteria for declaring variants (reference should be made to the product variants section of the documentation) for which conformity assessment procedures have not been applied.

A description of critical components which have a bearing on EMC performance of the equipment like SMPS, drives, PLCs, DC-DC converters etc and critical EMI suppression components like line filters, surge arrestors etc. It may be given in tabular form that includes type, model, manufacturer, certifications, mounting position etc.

8.12   RISK ASSESSMENT AND CONFORMITY TO LVD

The manufacturer, to establish conformity with LVD, should get his product tested to the relevant safety standard and after complying with all evaluations should carry risk assessment that compliments (not replaces) safety evaluation. In case if the harmonised standard has been applied partially, then technical justification in the form of descriptions of the solutions adopted to meet the safety objectives of this directive, for those clauses which have not been applied, should be given. The risk assessment carried out should be documented and it may be given in the following way:

> *The product has been tested as per standard _____ and it complies with all applicable clauses. In addition to the conformity assessment as per above standard, risk assessment has been carried out and given below:*
>
> *(The entire process of risk assessment as discussed in chapter 3 should be given here).*

In addition, the final result of the risk assessment can be documented as shown in the table 8.2. The left column lists the hazards while the 2nd column indicates whether the hazard is relevant to the equipment and the 3rd column documents the solution for reducing the risk associated with the relevant hazard.

Table 8.2: Risk Assessment Documentation For LVD

Risk	Relevant Y/N	Fulfilled by
1. Protection against electrical hazards.		
Electric Shock	Y	**I. Protection by insulation** All accessible pins are at SELV. Voltage is not exceeded under normal conditions and under single fault, voltage rise is below 70 V See pg ___of report no. ___ SELV circuits (secondary of power transformer) are separated from primary by double insulation. Screen inside the transformer is connected to

Risk	Relevant Y/N	Fulfilled by
		earth. Parts with hazardous working voltage separated from the secondary by double insulation Opto-coupler input and output are separated by reinforced insulation (compliance checked by inspection and voltage withstand test (see pg ___ of report ___). Secondary circuit and chassis/frame is connected to protective earth (PE). Heat sink is connected to PE. **Protection By separation** Operator cannot access circuits/wiring carrying hazardous voltage. Compliance was checked by inspection, test finger and test pin. See pg ___of report no. ___ Creepage distances have been maintained between tracks carrying hazardous voltage in PCB, in power transformer and optocoupler considering RMS working voltage > 150V and < 300V, pollution degree 2 and material group III Clearances have been maintained for peak working voltage of 420V, transient overvoltage of 1500V and pollution degree 2 (see page ___ of report no, ___) **Selection of components.** All critical components bridging single insulation (like MOVs and Y capacitors) are certified as per their respective product standard. **Overload and abnormal operation** Following faults were simulated Short-circuiting of Y caps and MOVs (Fuse blown but equipment safe) Short-circuiting of power transformer primary and secondary (Fuse blown but equipment safe). Cooling fan was stopped. Temperature rose to 55 degrees C. Thermal cut-out

Risk	Relevant Y/N	Fulfilled by
		operated and equipment tripped (rendered safe). For details see pg__ of report no __. **Testing** Ground continuity test has been carried out on third pin of power plug at 10A for two minutes. Ground resistance was found to be less than 0.1 OHM (see pg __ of report no, ___) Hi-pot test was carried out at 1500V DC for 1 minute (humidity conditioning done for 48h) between line-neutral shorted and earth. No flashover was observed. Instructions for proper handling of the equipment to prevent shock hazard is given in section _____ of the operating manual.
Leakage current	Y	Leakage current (earth leakage and touch current) test carried out and the leakage current was found to be 0.7 mA in normal and single fault condition (such as neutral open and reversed line and neutral) which is much below the 3.5mA limit for movable and pluggable equipment
Energy storage	Y	Values of X capacitor has been optimized. Discharge resistor has been provided across the X capacitor to slowly bleed off the stored charge and reduce it to tolerable level. Capacitor discharge test has been carried out as per section ___ of report no. ____ .
Stored charges	Y	Transfer of ESD charges from operator to the equipment can take place during normal operation. The equipment has been tested to 8 kV of ESD as per relevant EMC standard (see section of EMC report no_____)
Arcs	N	--
Burns	Y	The equipment uses a high intensity

Risk	Relevant Y/N	Fulfilled by
		mercury lamp that may cause burns if used in close proximity to the human body. Warning sign indicating high temperature and burn hazard has been fixed near the lamp and near ventilation opening. Appropriate warning instructions have been provided in section ___ of the equipment manual
2. Protection against mechanical hazards		
Instability	Y	Improper mounting on ceiling, may result in equipment coming loose and falling. Detailed mounting instruction have been given on pages ___ of the manual. User has been adequately warned about the dangers of wrong mounting.
Breakdown during operation	N	--
Falling or ejected object	N	--
Inadequate surfaces, edges or corners.	Y	Considering that injury may result during operation and cleaning due to corners and edges, these have been smoothened and rounded to mitigate the hazard.
Moving parts, especially where there may be variations in the rotational speed of parts.	N	--
Vibration	N	--
Improper fitting of parts	N	--
3. Protection against other hazards		
Explosion	Y	There is a rare possibility that the mercury lamp may explode. The risk increases when lamp is used beyond its replacement cycle. Appropriate instruction/ warning is

Risk	Relevant Y/N	Fulfilled by
		provided in section __ of the instruction manual of the product and the user is adequately warned.
Hazards arising from electric, magnetic, and electromagnetic fields, other ionizing and non-ionizing radiation	N	--
Electric, magnetic, and electromagnetic disturbances (EMI)	Y	The equipment employs such components as SMPS and digital controller board. The equipment has been tested to EMC standard EN55032 and EN 55024 to take care of the electromagnetic interference. Please refer EMC report no._____ Instructions of not using the equipment in proximity to wireless equipment and cell phones included in instruction manual.
Optical radiation	Y	The equipment uses a high intensity mercury projector lamp of 230W and 4200 lumens. If viewed from close directly, grievous harm may result to the eye. The user is appropriately warned regarding this hazard in pages ___ of the manual.
Fire	Y	There are some components in the circuits which may be a source of fire. However, the enclosure acts like a fire enclosure preventing the spread of fire outside the enclosure. All components inside the fire enclosure are of flammability class V-2, Outside the fire enclosure, materials for components and other parts (including mechanical enclosures, electrical enclosures and decorative parts), are of

Risk	Relevant Y/N	Fulfilled by
		flammability class HB.
Temperature	*Y*	*There can be a high temperature hazard near the lamp and the ventilation exhaust. Appropriate instructions have been provided in the manual not to cover these ports.* *Appropriate warning signs have been placed near these openings to war the user of high temperature hazard.* *Single fault has been simulated in which the ventilation port was blocked. Thermal cut-out was found to be activated and equipment switched off in fail safe mode. See page ___ of report no. ____.*
Acoustic Noise	*N*	*--*
Biological and chemical effects	*N*	*--*
Emissions, production and/or use of hazardous substances (e.g. gases, liquids, dusts, mists, vapour) Unattended operation.	*N*	*The equipment uses a high pressure mercury lamp. There is a rare possibility that the mercury lamp may explode. If the lamp bursts the gas contained in the lamp will be released in the form of smoke.* *The user is appropriately warned regarding this in the instruction manual page ___.*
Connection to and interruption of power supply	*Y*	*Appropriate instruction is available in the instruction manual.*
Combination of equipment	*N*	*--*
Implosion	*N*	*--*
Hygiene conditions	*N*	*--*
Ergonomics	*N*	*--*

8.13 SPECIFIC REQUIREMENTS FOR MEDICAL DEVICES DIRECTIVE

Although the MDD is a separate directive, it requires testing to be carried out as per the standard EN 60601. Basic safety evaluation should be carried out as per *EN 60601-1-1: General requirements of basic safety and essential requirements*. EMC testing is required to be carried out as per the collateral standard *EN 60601-1-2: Electromagnetic disturbances —and tests*. The testing can be carried out by any third-party test lab for Class-I equipment as NB involvement is not required for these. For other classes of equipment, testing will be carried out by NB. Nonetheless, a copy of the test report as evidence of compliance has to be included in the report section of the documentation. The contents of the test report should be as per table 8.3.

Table 8.3: Contents Of Test Report As Per MDD

Item	Information
Name and location of the test facility	Name and location of the test lab should be given here. Accreditation status can also be given.
Names and functions or equivalent identification of the persons authorizing the test report	Authorised signatories of the report should be identified.
Description of the equipment or system	Device name, model number and manufacturer should be given here. Reference of appropriate page of the EMC test report (copy of which is attached in reports section of the technical documentation) can also be given.
Description of the basic safety and essential performance including a description how the basic safety and essential performance were monitored during each test.	Reference may be given to the relevant page of the test report which gives the parameters with tolerances which were monitored to establish performance of the EUT.
Equipment software / firmware version	That software version loaded during the test should be given.
Prototype or production version of the device	Additionally, the relationship of the model tested to production models may be described.

Units tested and the rationale for the selected sample size.	Include serial numbers. Reference may be given to the relevant page of the test report
Intended use and intended environments	Reference should be given to the relevant clause of risk assessment.
Applicable standards and test methods	The name of the standards (with dates) and emissions limits or immunity test levels. Reference may be given to the relevant page of the test report.
Deviations from the basic EMC standards or from this collateral standard	Deviations in tests like amplitude, frequency range (if any) from that prescribed by the standard should be given. Justification is necessary if the device has been tested at severity levels lower than that prescribed by the standard.
Applicability / tests not performed	If any test prescribed by the standard has been skipped, technical justification for the same should be given here. Your description may look like *"Power frequency magnetic field immunity test was not carried out since the device does not use any such component that is susceptible to such fields"*
If the procedure specified by Annex E or an equivalent procedure is used: – a justification for any special environments identified or adjustments made – the adjusted reasonably foreseeable maximum EM DISTURBANCE levels – the resulting final immunity test levels, rounded to the nearest whole number or, if a decimal, to a single significant digit – details of the methods and data sources used in determining the appropriate immunity test levels	If the device is intended to be used in special environments then the device will be required to be tested to higher immunity levels. A justification of the test levels using relevant phenomenon present at that specific location should be given.
Immunity test level for each immunity test and emissions compliance class and group	Severity level to which immunity test was carried out and limits for emissions (including class and group)

	should be specified here. The information is similar to that given in table in EMC assessment section. Reference of appropriate page of the EMC test report can also be given.
Immunity pass/fail criteria	Performance criterion for deciding compliance to immunity tests has to be decided here after carrying out risk analysis. Reference can be given to section of risk analysis.
Environmental conditions as required by the relevant basic EMC standards	Normally, basic standards prescribe conditions like temperature, RH and altitude (of the test lab) should be given here. This information is also given in the test report. Reference of appropriate page of the EMC test report can also be given.
Compliance summary statement	Compliance of the ME EQUIPMENT or ME SYSTEM with each test.
Test data that support the compliance determination for each test performed	Include units of measurement
Device configuration during the test, including a block diagram	Block diagram of the device and all peripherals and auxiliary equipment used.
Device settings and operating modes	List by test.
Device power input voltages and frequencies	Record the device power input voltages and frequencies for each test.
Any connections to the terminal for connection of a potential equalization conductor, if used	Include information on connection to the terminal for connection of a potential equalization conductor used during testing, if any.
Testing of permanently installed large equipment or large system: frequencies, power and modulation of the RF test sources and test distances used.	
Description of any patient-coupled cable termination used	
Description and position of	The layout of excess cable shall be

interconnecting cables.	noted. The length, shielding, ferrites and other construction details should be described. Photographs are also helpful.
Simulators, accessories and auxiliary equipment	Describe simulators, accessories and auxiliary equipment used, including patient physiological and subsystem simulation.
Documentation of any special equipment or system hardware or software needed to perform the tests	Auxiliary equipment may be required in some cases for performance monitoring of the equipment under test. Description of such equipment should be given here.
Test equipment used, including calibration or maintenance dates	Details of the test equipment used like make, model and their calibration due dates should be given. Reference of appropriate page of the EMC test report can also be given.
Test parameters used, e.g. frequencies, phase angles, as applicable	Reference of appropriate page of the EMC test report can also be given.
Dwell time for each immunity test requiring a dwell time	Dwell time for tests like conducted RF susceptibility, radiated susceptibility and power frequency magnetic field should be given. Reference of appropriate page of the EMC test report can also be given.
ESD test points	Photograph or drawing depicting the exact ESD test points with discharge method identified
Measured conducted and radiated emissions	Tabular data of at least the six highest emissions for each test shall be included. Reference of appropriate page of the EMC test report should be given.
Measured harmonics and flicker emissions	Tabular data of at least the six highest emissions for each test shall be included. Reference of appropriate page of the EMC test report should be given.
Equipment or system modifications	Equipment or system modifications needed in order to pass any of the emissions or immunity tests (like change of filter, additional shielding,

	transient suppression device etc. should be given). A statement that they will all be incorporated into production units.
Effects on the equipment or system that were observed during or after the application of the test disturbances, and the duration for which these effects persisted.	All degradation in performance should be documented in detail here. Reference of appropriate page of the EMC test report should be given.
Photographs of each test setup including the equipment or system and all peripherals and auxiliary equipment used.	Photographs should be as detailed as possible to prove that correct testing has been carried out. Reference of appropriate page of the EMC test report should be given.

8.13.1 Risk Management Of Medical Devices

The standard EN 60601 also requires that risk management should be carried out for medical devices for EMC and safety which is on the same lines as done for the LVD. The risk management procedure remains the same as with LVD as discussed earlier and involves process like risk analyses (include hazard identification, risk estimation etc.), risk evaluation, risk reduction and identification of residual. Thereafter a risk management file should be drawn up, which can be a part of the technical documentation.

An example of risk management for electrical safety has already been discussed in the chapter 3. As far as the risk management for the EMI is concerned, the risk is both ways i.e. the device is at risk due to electromagnetic disturbance arising out of the operating environment and the device itself may pose a risk to other equipment in the vicinity by generating electromagnetic disturbance. An example is presented here which will guide the reader through the risk management process. It should be noted that the following is just an example, the reader is advised to carry out his own assessment based on the foreseeable risks in his equipment. We will consider the example of a *patient bed-side monitor* (hereafter referred to as *device*) which is used in ICUs to continuously monitor patient parameters like ECG, heart rate (HR), blood pressure (BP), oxygen saturation (SPO$_2$), body temperature and respiration rate (RR). The ECG data is taken via various leads connected to the patient, the temperature data is taken via the temperature probe attached to the patient, the SPO$_2$ data is taken from a finger clip while that of BP is taken from a cuff. All the data is displayed on the LCD screen of the device. Needless to say, that this device is crucial for monitoring vital body parameters and that any malfunction in the readings may cause the medic to take an action which is unwarranted and which could put the patient's life in danger.

8.13.2 Limits Of The Product And Intended Use

This section should identify the intended use of the device including operational environment, likely misuse, use limits, time limits etc. The standard EN 60601-1-2 identifies three types of operational environment:

–   Home health care environment that includes environment in house, hotels, airports, schools etc.

–   Professional health care environment that covers hospitals, dispensaries, clinics, emergency rooms, patient rooms but NOT near HF surgical equipment and MRI devices.

–   Special environment that includes area near HF surgical equipment, MRI device, military area (near radars) and heavy industrial areas (power plants, automotive manufacturing plants etc.

Your explanation could look like this:

Intended use	This device is designed for use in a professional health care environment. It is for indoor use only and works on 230V supply from normal utility socket. It can be kept on a table or kept in a rack besides the patient's bed.
Misuse of the product	Using the device in environment not intended for its use like in special environment which can lead to a hazard. Using the device beyond its environmental specifications may give rise to temperature hazard.
Use limits	Installation environment: Commercial, indoors only Environmental limitations: Temperature range: 0 deg C to 40 deg C. Humidity range: 10% to 80% humidity Altitude limitations: Ground to 2700m EMI precautions: It should not be used within one meter of HF surgical equipment or MRI equipment or in environment with high electromagnetic fields like

	transmitters, radars, cell phone towers or in environment that can lead to the generation of high static charges, surges or high transients (like near drill machines)
Space limits	*The device manual provides details about space requirements for the product as regards to mounting.*

8.13.3 Environment Of Use

Here the environment in which is supposed to operate should be given here. Your description may look like this:

Home health care environment	*NO*
Professional health care environment	*Yes*
Special environment	*No*

8.13.4 User Groups

The type of users that are supposed to operate the device has to be given here. Your description may look like this:

Qualified staff	*Yes*
Non-professional users	*No*
Apprentices/students	*No*
Children	*No*
Elderly persons	*No*
People with limited physical abilities	*No*
People with limited mental abilities	*No*

8.13.5 Device Lifecycles

The following table illustrates how device life cycles should be identified:

Transport.	*Transport of the product for internal or external relocation.*
Installation	*Assembly, setting, testing, programming, start-up, all operating modes, switching ON, switching OFF, resetting, fault-finding and troubleshooting.*
Operation	*Switching ON, setting, programming, wireless setting, LAN settings, switching OFF, resetting, all operating modes.*
Maintenance	*Cleaning and housekeeping, changing lamp, Fault-finding and troubleshooting (operator intervention).*
Disposal	*By authorized dealers/distributors, authorized collectors of e-waste.*

8.13.6 EMI Hazards During Device Lifecycle

The EMI hazards have to be identified at every stage of the product lifecycle. Your information may look like this:

Life cycle	*EMI Hazards Exist?*
Transport	*No*
Installation	*No*
Operation	*Yes*
Maintenance	*Yes*
Disposal	*No*

8.13.7 Risk Estimation

After having identified the possible hazards in various device lifecycles, the risk posed by the hazards can be estimated on the basis of the severity of the harm and the probability of the harm. The following table can serve as guidance for the severity of the harm. The effect on the patient is not only due to the device malfunction but also due to unwarranted action on part of the medic. Although the severity has been classified as catastrophic, severe, moderate and minor, other intermediate severities may also be applicable for some equipment types. Your description may look like this:

Severity group	People	Equipment / Facility
Catastrophic	Death of patient due to unwarranted application of medication, damage to heart due to unwarranted application of de-fibrillator etc.	Major facility damage like damage to the whole room.
Severe	Disabling injury/illness like lowering or increase in BP due to unwarranted application of medication etc.	Major subsystem loss or facility damage like damage to the whole device
Moderate	Small injury requiring minor medical intervention like cuts, localised scalding, localised haemorrhaging.	Minor subsystem loss or facility damage. Like changing of component or wire
Minor	Short term discomfort like sprain, itch, prick, shock sensation etc. requiring first aid only	No equipment damage.

Likelihood	Expected rate of occurrence
Likely	Once in a month
Possible	More than once per year, but not more than five times a year
Rare	Once per year
Remote	Once in five years

8.13.8 Risk Evaluation

The severity of the harm and the probability of the harm can be combined. The environment here is professional health care environment. Each of the EMI phenomena can then be evaluated. Your description may look like the following table:

Phenomenon / Hazard	Risk Evaluation
Electrostatic discharge.	During routine operation and maintenance. Likelihood: Likely. There are chances that human ESD even can occur when operator touches the equipment. Severity: Severe. The charge on the device can affect the artefact (waveform) on the device which may cause the medic to take unwarranted action. The charge may also appear on ECG leads connected to the patient.
Radiated susceptibility	During routine operation and maintenance. Likelihood: Likely. There are chances that this hazard can occur when operator or visitor or patient operates a cell phone near the equipment. Severity: Minor
Conducted RF susceptibility	During routine operation and maintenance. Likelihood: Likely. There are chances that this hazard can occur when operator or visitor or patient operates a cell phone near the equipment. Severity: Moderate The RF current may be induced on ECG leads connected to the patient. But since the current is very small, the effect may be moderate.
Power frequency magnetic fields	During routine operation and maintenance. Likelihood: Remote. This may occur if the device is operated near high magnetic field generating devices like MRI equipment. But since the professional operating environment excludes such sources nearby, hence the likelihood may be remote. Severity: Minor

Electrical Fast Transients	During routine operation and maintenance. Likelihood: Possible. The transient may be generated if equipment like power drill is operated near the device. Severity: Severe. The transient may cause severe distortion of the ECG waveform on the device which may look like the heart is going into fibrillation (irregular beating) causing the medic to take unwarranted action. The transient voltage may also appear on ECG leads connected to the patient but due to low energy may not cause death but may lead to long term problems.
Combination wave surge	During routine operation and maintenance. Likelihood: Rare. The surges may appear on power mains due to faults. Severity: Catastrophic. The surge voltage may also appear on ECG leads connected to the patient which may cause death.
Damped oscillatory surge	During routine operation and maintenance. Likelihood: Remote. The surges may appear on power mains due to faults. Severity: Severe. The surge voltage may also appear on ECG leads connected to the patient which may cause major health problems.

The risk evaluation regarding hazards arising out of the equipment should also be considered. Your description may look like the following table:

Phenomenon / Hazard	Risk Evaluation
Conducted emissions	During routine operation. Likelihood: Likely. Severity: Minor.

Radiated emissions	During routine operation. Likelihood: Likely. Severity: Minor.
Harmonic Emissions	During routine operation and maintenance. Likelihood: Likely. Severity: Moderate.
Flicker Emissions	During routine operation. Likelihood: Likely. Severity: Minor.

8.13.9 Risk Reduction/Elimination

Your description can look like this:

> The risk is reduced by using inherent design features or by using technical measures. The equipment has been tested to the applicable EMI phenomenon and severity level corresponding to the operational environment. The performance of the device during the test has been monitored and the following deviations from the normal operation is not acceptable:
> - component failures.
> - change in programmable parameters.
> - reset to factory defaults (manufacturer's presets).
> - change of operating mode.
> - Spurious sounding of alarm even if parameters are within limits.
> - Alarm not sounded even when a parameter goes out of limits.
> - Cessation or interruption of any intended operation, even if accompanied by an alarm
> - error of a displayed numerical value sufficiently large to affect diagnosis or treatment;
> - Spurious distortion on ECG waveform that would interfere with monitoring;
> - artefact or distortion in an image in which the artefact would interfere with monitoring;

The procedure for risk reduction is given in the following table:

Phenomenon / Hazard	Risk reduction	Evidence
Electrostatic discharge.	Appropriate design measures at PCB level for high speed have been taken such as multilayer boards, ground planes etc. ESD suppression components have been used for patient interfacing cables and cables leading from front panel to block ESD current from reaching the patient. Front panel display and keypad have metal film sandwiched between two PVC sheets and connected to earth to drain ESD charges. Proper earthing has been ensured so that the ESD currents are drained safely.	The device has been tested to 15 kV air discharge and 8 kV contact discharge. There has been no degradation of performance during and after the test. Further patient leads have been monitored for high voltage and none was found appearing. (Refer clause ____ of report no. ____ dt. _____copy of which is attached in section ____ of this documentation)
Radiated susceptibility	Appropriate shielding measures have been taken to shield the equipment against RF. Further shielding effectiveness has been ensured over penetrations and apertures such as RF gaskets for seams, mesh for ventilation openings, shielded glass for displays etc.	The device has been tested to 10 V/m in frequency range of 80MHz to 2.7GHz. Both AM and pulse modulations have been used. There has been no degradation of performance. (Refer clause ____ of report no. ____ dt. _____copy of which is attached in section ____ of this documentation).

Conducted RF susceptibility	Appropriate line filters have been employed on power mains to block RF currents from entering the equipment.	The device has been tested to 10 V (which is greater than EN 60601-1-2 which recommends 3V) in frequency range of 150kHz to 80MHz There has been no degradation of performance. (Refer clause ____ of report no. ____ dt. _____copy of which is attached in section ____ of this documentation).
Power frequency magnetic fields	The device does not have any such components like CRT displays or Hall effect sensors that are susceptible to such fields.	The device has been tested to 100 A/m (which is greater than EN 60601-1-2 which recommends 30 A/m). There has been no degradation of performance. (Refer clause ____ of report no. ____ dt. _____copy of which is attached in section ____ of this documentation).

Electrical Fast Transients	Appropriate line filters have been employed on power mains to block transients. Transient suppression devices like tranzorbs have been employed at DC inputs of all PCBs. Proper earthing has been ensured so that the transient currents are drained safely.	The device has been tested to 4kV (which is greater than EN 60601-1-2 which recommends 2 kV) with repetition frequency of both 5kHz and 100kHz. There has been no degradation of performance during and after the test. (Refer clause ____ of report no. ____ dt. _____copy of which is attached in section ____ of this documentation).

Combination wave surge	Appropriate line filters have been employed on power mains to block surges. Surge suppression devices like GDTs have been employed at power mains to block surges. Proper earthing has been ensured so that the transient currents are drained safely.	The device has been tested to 4kV CM and 2 kV DM (which is greater than EN 60601-1-2 which recommends 2kV CM and 1kV DM). There has been no degradation of performance during and after the test. (Refer clause ____ of report no. ____ dt. _____copy of which is attached in section ____ of this documentation).

Damped oscillatory surge	Appropriate line filters have been employed on power mains to block surges. Surge suppression devices like GDTs and MOVs have been employed at power mains to block surges. Proper earthing has been ensured so that the transient currents are drained safely.	The device has been tested to 2.5kV as per EN 61000-4-18 as an additional measure. There has been no degradation of performance during and after the test. (Refer clause ____ of report no. ____ dt. ____copy of which is attached in section ____ of this documentation).

Conducted emissions	Appropriate line filters have been employed on power mains to block RF currents from leaving the equipment.	The device complies with Class B limits as per EN 60606-1-2 prescribed for non-industrial equipment with a margin of over ___ dB. (Refer clause ____ of report no. ____ dt. ____copy of which is attached in section ____ of this documentation).

Radiated emissions	Appropriate shielding measures have been taken to shield the equipment against RF. Further shielding effectiveness has been ensured over penetrations and apertures such as RF gaskets for seams, mesh for ventilation openings, shielded glass for displays etc.	The device complies with Class B limits as per EN 60606-1-2 prescribed for non-industrial equipment with a margin of over ___ dB. (Refer clause ____ of report no. ____ dt. ____copy of which is attached in section ____ of this documentation).

Harmonic Emissions	Appropriate harmonic filters have been employed on power mains to block harmonics currents from leaving the equipment	The device complies with limits as per EN 60606-1-2 prescribed for non- industrial equipment with a margin of over ___ dB. (Refer clause ____ of report no. ____ dt. _____copy of which is attached in section ____ of this documentation).
Flicker Emissions	Such emissions arise when the current drawn from the mains varies drastically and is observed in equipment employing components that draw discontinuous current (like heaters, compressors etc). Since the device does not employ such components, flicker cannot be generated.	The device complies with limits as per EN 60606-1-2 prescribed for non-industrial equipment with a margin of over ___ dB. (Refer clause ____ of report no. ____ dt. _____copy of which is attached in section ____ of this documentation).

8.13.10 Residual Risks

Although possible risks are reduced by designing and testing, some residual risks still remain. These are addressed by appropriate warnings/instructions in the manual which can include but not restricted to:

- Cell phones should not be used within 1 meter of the device.

- Tools like drills, cutters, sanders etc. should not be operated within 3 meters of the device.

- The device should not be used in vicinity of cell phone towers.

Medics should ensure that ESD charges are drained by appropriate methods before operating the equipment.

8.13.11 The Essential Requirements (ER) checklist.

The ER checklist (as per Annexure I of the directive) should be reproduced verbatim from the directive and presented as a table with labelled columns, such as ER, applicable applied standard, demonstration of compliance, and location of documentation. Each ER should be addressed and applicability is indicated. If the ER is applicable, reference of the standard or procedure used should be indicated. Explicit document meeting the requirement should be named and its location given. If a particular ER is not applicable, an explanation should be provided. The ER checklist functions as a signpost that provides identification and location of the supporting documentation.

8.13.12 Results of Bench Testing

This should include results of in-vitro/animal studies, simulated use testing and validation of software and the results of special processes (e.g. sterilization validation report(s)). The testing should follow a pre-defined protocol, which should include the parameters to be measured, measuring and test equipment to be used including calibration arrangements, statistical treatment of results and acceptance criteria, together with necessary formal approval of the report

8.13.13 Clinical Evaluation

For class III medical devices, for documenting that the device achieves performance for which it is intended and to determine any undesired side effects (and whether these constitute a risk), a clinical evaluation is required to be carried out. Clinical data includes data from market experience of the same or similar devices (particularly relevant to well established devices), prospective clinical investigations and information from the scientific literature. The scientific literature will often relate to medical devices other than those being assessed. The manufacturer must therefore establish the extent to which the scientific literature is relevant to his device(s). The results from bench testing may be used to establish the extent to which the characteristics of the device(s) being assessed are similar to those of the device(s) covered by the scientific literature, and therefore the relevance of that scientific literature. The manufacturer should make clear where clinical data is being used to demonstrate conformity with each of the applicable essential requirements for the particular device(s) concerned.

The evaluation should be accompanied by a Clinical Evaluation Report (CER). A compliant CER demonstrates that the clinical evaluation process conforms with the one outlined by the MDD. The evaluation process should support strong clinical evidence that your device achieves its intended purpose without exposing users and patients to further risk. A list of possible elements to be included in the report is as follows:

- General information: device and manufacturer name.
- Concise physical and technical device description and intended application.
- Outline of intended therapeutic or diagnostic claims.
- Clinical evaluation and data types.
- Summary of clinical data and review.
- Describe analyses used to assess performance, safety, and relevance/accuracy of product literature.
- Conclusions about safety, performance, and conformity

Further, design considerations and critical components for EMC can be documented as done in section 8.12.9

8.14 SPECIFIC REQUIREMENTS FOR RED

Conformance to the RED requires EMC testing and electrical safety evaluation, the documentation is on similar lines as for the EMC and LVD. However, risk assessment is not required for both. Reference of the testing and conformance to harmonised standard (EN 301 389) series can be made to the reports section of the technical documentation.

8.15   SPECIFIC REQUIREMENTS FOR ROHS

As discussed earlier, to declare conformity with RoHS directive, the approach can be of conformity to the harmonised standard *EN63000 (Technical documentation for the assessment of electrical and electronic products with respect to the restriction of hazardous substances) and* due diligence (with sample testing as per *EN 62321-1 :Determination of certain substances in electrotechnical products* if required). To understand the things in a better way, we will consider the example of a product namely half height turnstile (see fig 8.4). This product is used for access control at locations like a metro railway station. Among other things, the product consists of a three-pronged turnstile, motor, SMPS, display and controller card. It operates on 230V, 50 Hz mains supply.

Fig 8.4: Half Height Turnstile

Your documented procedure for compliance can be like this:

> *To comply with RoHS directive, the harmonised standard EN63000, due diligence and sample testing as per EN 62321-1 approach was followed. Reasonable precautions/steps have been taken to identify and weed out suspect components. Checks/controls have been put at*

point of material supply and production along with a series of appropriate steps to prevent any non-compliance from occurring. Further, it has been ensured that the system of checks is being carried out. The compliance process is consisting of following steps:

Thereafter the procedure as per clause 4.3 of EN 63000 should be followed, which consists of the following steps:

8.15.1 Tasks To Be Undertaken By The Manufacturer (Cl. 4.3.1 Of EN 63000)

The Fig. 8.5 summarises EN 63000 clause requirements. We will consider each of the clauses 4.3.2, 4.3.3 and 4.3.4 in detail and see how the documented information should be.

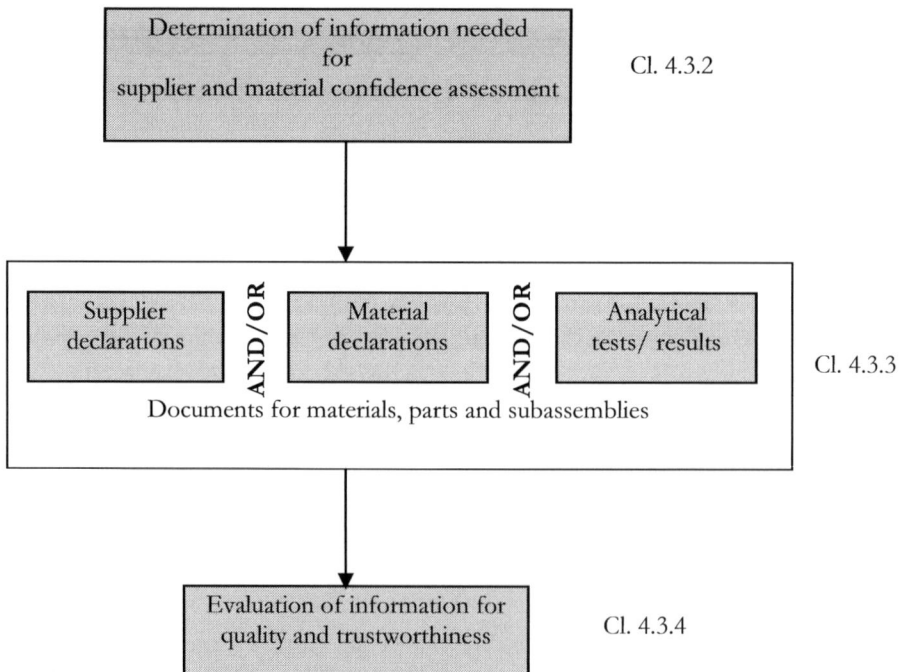

Fig. 8.5: Clause Requirements of EN 63000

8.15.2   Determination Of Information Needed (Cl. 4.3.2)

The clause 4.3.2 states that the types of technical documents that are required for materials, parts and/or sub-assemblies should be based on the manufacturer's assessment of the following:

–   The probability of restricted substances being present in materials, parts or sub-assemblies.

–   The trustworthiness of the supplier.

Materials that are added during the production process (such as solder, paint, adhesives) should also be considered as part of the assessment. EN 63000 also notes that the manufacturer may apply technical judgement when carrying out the materials assessment, as some substances are unlikely to be contained in certain materials (e.g. organic substances in metals).

8.15.2.1 Identification of components by initial screening

The manufacturer should carry out an initial screening of the parts list in order to identify if there are any materials, parts or sub-assemblies in the product which do have or not have any risk of containing RoHS substances above the maximum concentration values. The manufacturer should apply their own technical judgement of whether RoHS substances are not found in certain materials, parts and sub-assemblies. For example, an un-coated stainless-steel screw does not contain any RoHS substances above the maximum concentration values. Similarly, an un-painted wooden case (e.g. for a hi-fi speaker) or a fabric filter (e.g. for a vacuum cleaner) will also not contain any RoHS substances above the maximum concentration values. The manufacturer may decide to include their technical assessment of these parts in the product technical documentation and to not ask suppliers to provide materials declarations for such parts.

8.15.2.2 Assessing the probability of banned substances being present

The manufacturer should use following table as guidance to identify components/materials that have a high probability of containing a RoHS restricted substance. The table includes only six banned substances. The manufacturer should prepare similar table for all ten banned substances. For example, a manufacturer may decide to regard a material, part or sub-assembly as containing high risk materials if the table indicates that there is a high probability of it containing two or more RoHS substances. The manufacturer's description may look like this:

> All those components were identified which, on the virtue of their operation and composition, and the probability of containing the banned substances were evaluated as Low (L), Medium (M) or High (H. A list of components is as follows:

Component/ Material	Restricted substances						Remarks
	Hg	Cd	Pb	Cr(IV)	PBB	PBDE	
Solder	L	L	H	L	N/A	N/A	
Transformer	L	M	H	L	L	M	
Motor	L	M	H	L	L	M	
Wires	L	H	H	L	L	M	
Fuse	L	M	H	L	L	L	
PCB substrate	L	L	L	L	L	L	
Resistors	L	M	H	L	L	L	
Capacitors	L	L	H	L	L	M	
Inductors	L	M	H	L	L	M	
Relay	L	H	H	L	L	L	
Switch	M	H	M	L	L	L	
ICs and BGAs	L	L	H	L	L	L	
Connector	M	L	H	L	L	H	
Line filter	L	L	H	L	L	L	
LCD display	H	L	H	H	L	L	
Glass cover	H	L	H	H	L	L	Pb could be exempt
Metal Frame	L	M	H	H	N/A	N/A	Unpainted
Screw	L	M	M	H	N/A	N/A	Unpainted
Enclosure	L	M	H	H	N/A	N/A	Unpainted

Note: The manufacturer may have more probabilities than L, M or H.

8.14.2.3 Identification of vendors and assessing their trustworthiness

When a manufacturer accepts a RoHS materials declaration from a supplier, the manufacturer should consider his/her relationship with the supplier and take into account the supplier's reputation and the manufacturer's level of confidence in the supplier. Manufacturers should look to establish a level of trust with their suppliers. Some suppliers will readily provide documentation showing RoHS compliance of materials, parts and sub-assemblies that they supply, but others may have lesser levels of awareness and cannot produce the necessary information. In extreme cases, forged documents stating compliance may also be given. Most manufacturers will already have a defined process of supplier qualification as part of their quality management system. This system should be extended to capture supplier qualification information critical to RoHS. This information could be based on audit, past experience etc. The aim should be to determine if a supplier understands the RoHS restrictions and has effective management systems in place to ensure RoHS conformity.

In order to develop confidence and to ascertain to what extent can a supplier supply conforming products, the supplier's own conformity management system may be scrutinized by the manufacturer. The supplier's conformity management system should describe the quality control processes and procedures that are applied to ensure that series production remains in conformity, including how the supplier assesses their own suppliers. The conformity management system may also describe any additional inspections or examinations which the supplier carries out (e.g. XRF screening of incoming materials, parts and sub-assemblies, audits of the supplier's own suppliers etc).

The complexity of the supplier's conformity management systems will depend on the size and nature of the supplier's manufacturing operations which can be of the following types:

  - A supplier who manufactures materials, parts or sub-assemblies at one facility where all of the other products manufactured and/or repaired at this facility are also required to be RoHS compliant. In this case, the conformity risk will reside in materials, parts or sub-assemblies coming from the supplier's own supply chain.

- On the other hand, if the supplier's manufacturing facility also manufactures and/or repairs non-RoHS compliant products then the supplier's conformity management system will need to address conformity risks arising from possible cross-contamination between RoHS compliant and non-RoHS compliant process lines. In this case, the supplier's conformity management system should include procedures for checking contamination as explained in the sections that follow.

- If the supplier sub-contracts some of their manufacturing process (e.g. painting and coating processes) then the supplier's conformity management system should also take account of the quality control systems which are applied at the sub-contractor's facility. A complex sub-assembly or component may involve a number of manufacturing processes which take place at several different facilities and locations. Ensuring that series production remains in conformity requires adequate quality control systems to be in place at all of these facilities.

After using a certain supplier over a sustained period of time, the manufacturer's knowledge and trust in the supplier's conformity management system may increase. Where there may be less confidence in a supplier then an in-person audit (by the manufacturer or an independent auditor) can be a more cost-effective alternative to the manufacturer carrying out independent testing. An inspection of the supplier's manufacturing processes and conformity management system can not only aid understanding from both manufacturer and supplier, but can also build a level of trust. Alternatively, a supplier could provide evidence that they have effective RoHS compliance systems in place by gaining accredited certification to the IECQ Standard *QC 080000: Electrical and Electronic Components and Products Hazardous Substance Process Management System Requirements*. The manufacturer should extend their existing supplier qualification and may use the following criteria to categorise the supplier's trustworthiness for RoHS conformity:

- **Type A**: Supplier has a very good understanding of RoHS, has comprehensive and effective systems in place to ensure RoHS compliance, and incorporates selective analysis of high-risk components /materials that it purchases from its suppliers.

- **Type B:** Supplier has some understanding of RoHS and has a system for ensuring RoHS compliance but may be lacking in some respect, e.g. does not incorporate selective analysis of high-risk components /materials that it purchases from its suppliers.
- **Type C:** Supplier does not understand RoHS requirements or does not have systems to ensure compliance and does not check declarations from its suppliers for incoming components/materials.

The manufacturer may, if required, have more than the above criteria, if required.

18.15.2.4 Material declarations and evidence of compliance

To determine what types of documents are required for supplier parts, the manufacturer should establish an assessment matrix which combines the materials confidence assessment and the supplier confidence assessment. One example of an assessment matrix is given in Table 8.4 wherein the risk levels have been identified as I, II and III.

Table 8.4: Risk levels

		Probability that the material contains RoHS substance		
		Low	Medium	High
Trustworthiness of supplier for RoHS	Type A	Risk Level I	Risk Level II	Risk Level II
	Type B	Risk Level II	Risk Level II	Risk Level III
	Type C	Risk Level II	Risk Level III	Risk Level III

The documents required to reduce or eliminate the risk could be as follows:

Risk Level I

Supplier declaration and/or contractual agreement is required wherein the supplier declares that all materials supplied by him/her will be RoHS compliant.

Risk Level II

Materials declaration from supplier is required for each material, part and/or sub-assembly.

Risk Level III

Materials declaration is required for each material, part and/or sub-assembly which has a *low probability* of containing RoHS substances AND recent analytical test report are required for each material, part and/or sub-assembly that has a *medium or high probability* of containing RoHS substances.

8.15.3 Collecting The Information (Cl. 4.3.3)

This clause states that the manufacturer should collect following documents on materials, parts and/or sub-assemblies.

a) Supplier declarations confirming that the restricted substance content of the material, part, or subassembly is within the permitted levels *(without actually mentioning the value)* and identifying any exemptions that have been applied. The declarations should cover a specific material, part and/or sub-assembly, or a specific range of materials, parts and/or sub-assemblies.

and/or

b) Material declarations providing information on specific substance content *(the actual concentration values)* and identifying any exemptions that have been applied.

and/or

c) Analytical test results /reports using the methods described or referenced in *EN 62321-1: Determination of certain substances in electrotechnical products.*

In order to streamline the process of collecting the information and to ensure that the process is flawless, an assessment matrix can be generated. For that the manufacturer should:

- Generate the bill of materials (BOM) for the product model to produce a list of part numbers for all of the materials, parts and sub-assemblies which are contained in the finished product, and
- Identify and assign the supplier(s) for each of these part numbers.

The manufacturer should then use the assessment matrix to assess this list of part numbers and suppliers in order to produce a table which summarises the types of documents which the suppliers are required to provide for these parts. The manufacturer should include this table in the RoHS technical documentation for the product model to demonstrate how the results of the manufacturer's assessment are used to collect the required types of documents for the materials, parts and/or sub-assemblies in the product. The manufacturer's matrix could be as per the table below

Supplier code	Probability that RoHS substances may be present		Type of documents		
	Part no.	Risk level	Supplier declaration	Materials declaration	Analytical test report
111111	aaaaa	I	Y	N	N
222222	bbbbb	I	Y	N	N
333333	ccccc	I	Y	N	N
444444	ddddd	I	Y	N	N
555555	eeeee	II	--	Y	N
666666	fffff	II	--	Y	N
777777	ggggg	III	--	Y	Y
888888	hhhhh	III	--	Y	Y
999999	iiiii	III	--	Y	Y
121212	jjjjj	II	--	Y	N
343434	kkkkk	I	Y	N	N

Note: For copies of declarations and analytical test reports, please refer to section _____ of this documentation.

8.15.4 Evaluating The Information (Cl. 4.3.3)

Clause 4.3.4 of EN63000 requires the Manufacturer to establish procedures that shall be used to evaluate the documents provided by the supplier in order to determine their quality and trustworthiness. The manufacturer should evaluate, in accordance with these procedures, the source and content of each document received in order to determine whether or not the material, part, or sub-assembly meets the specified substance restrictions. This evaluation will enable the manufacturer to decide whether the documents provide sufficient evidence of compliance to justify their inclusion in the technical documentation.

8.15.4.1 Evaluating the quality and trustworthiness of supplier declarations

The manufacturer should read the supplier's declaration carefully to assess what the supplier is declaring and whether this declaration is of sufficient quality and trustworthiness to be included in the technical documentation for the product model. If a particular document is not considered to be of sufficient quality or trustworthiness, then the manufacturer shall determine what further actions are necessary like requesting additional information from the supplier or undertaking his own substance analysis.

Clause 4.3.4 also requires the manufacturer to evaluate, in accordance with their procedures, the source and content of each document received. Was the document received directly from the supplier or via a third-party broker? If the manufacturer received the document from a third-party broker, how does the manufacturer ensure that they receive updates from the broker when the supplier produces updated compliance documentation for their parts?

To claim compliance to EN 63000, the manufacturer should insist that the content of the supplier's material declaration must:

– Contain an unambiguous statement that all ten RoHS restricted substances are not present above the maximum concentration values, or if an exemption is claimed the statement should specify the particular exemption(s).

– Enable identification of the part number for the list of supplier materials, parts or sub-assemblies which are contained in the manufacturer's product.

– Be signed by an executive officer at the supplier who has authority to sign on behalf of the company.

The following types of declarations are *not* acceptable to the surveillance authorities and cannot be included in the technical documentation:

– Information that is generic in nature i.e. which does not identify the specific material, part and/or sub-assembly, or even a specific range of materials, parts and/or sub-assemblies

– Declarations which do not confirm that the restricted substance content of the material, part, or sub-assembly is within the permitted levels.

– Declarations that contain such statements as *no intentional use* of RoHS substances.

- Declaration is limited to only one substance. It is most frequently observed that for electrical/electronic product the declaration is only regarding the non-use of lead in solder!
- Declarations that contain disclaimers such as *made best efforts* or *to the best of our knowledge* which reduce the trustworthiness of the declaration.
-

8.15.4.2 Evaluating the quality and trustworthiness of analytical test reports

The clause 4.3.4 requires the manufacturer to establish procedures that shall be used to evaluate the analytical test reports provided by suppliers in order to determine their quality and trustworthiness. Manufacturers often do not examine analytical test reports closely enough and there is a lack of knowledge on their interpretation which can result in an over-reliance on the reports provided. The manufacturer should use the following criteria when they assess any test reports which are provided by their suppliers.

How old is the report? All test reports should be dated to indicate when the report was issued and when the test was conducted. The list of materials, parts and sub-assemblies used to manufacture the final product may change over time, so if the test report was written before these changes, it may no longer be relevant to the final product. A manufacturer with a frequently changing product design or production processes may need to update more regularly any test reports that they require for the final product.

Is the test laboratory accredited to ISO/IEC 17025 standard? Test reports from ISO/IEC 17025 accredited laboratory has more credibility and acceptability since a laboratory must apply the appropriate test methods specified in EN 62321 as a part of their accreditation process and requires them to be abreast of scientific and technological advances in relevant areas. Hence all analytical test reports should be from an accredited laboratory only.

Does the test report cover all of the relevant homogenous materials in the part or subassembly that require testing? The test report(s) should identify which homogenous materials within the part of sub-assembly contain high risk materials that require testing, and which materials within the part or sub-assembly do not require testing (e.g. because they do not contain high risk materials). A homogenous material that contains high risk materials may not require testing for all ten of the

RoHS substances. For example, organic substances such as PBBs and PBDEs are not found in metal parts. For electronic components on an assembled printed circuit board, further analysis may be required because it can be difficult to assess compliance at the homogenous material level.

Do the test results confirm compliance? For each RoHS substance that was tested in each relevant homogenous material, the test results should show that the measured concentration of the RoHS substance does not exceed the maximum concentration values, or that the use of the RoHS substance in that homogenous material is covered by a valid exemption. It has been observed that the surveillance authorities are often presented with RoHS test reports where it is clearly evident that the product is non-compliant. This is a clear instance where no checking process is in place and a simple step can prevent the occurrence of larger problems.

8.15.5 Periodic Review Of The Technical Documentation (Cl. 4.3.5)

The clause 4.3.5 of EN63000 states that the manufacturer should:

– perform a periodic review of the documents contained in the technical documentation to ensure that they are still valid;

– ensure that the technical documentation reflects any changes to materials, parts or subassemblies in accordance with clause 4.3.3.

It also includes a note which highlights that article 7(e) of the RoHS directive requires the manufacturer to keep the technical documentation up-to-date with any changes to the product design and manufacturing process. Therefore, the manufacturer needs to have processes in place to:

- Periodically review any product design and manufacturing changes to generate a new, up-to-date list of part numbers and supplier codes for all materials, parts and sub-assemblies which are contained in the finished product, and

- Check that the technical documentation includes materials declarations (and test reports if required) for any new part numbers and supplier codes in this updated list, and that the documents for existing part numbers and supplier codes continue to meet quality and trustworthiness requirements.

> *The RoHS and non-RoHS components are stored in segregated areas which are clearly identified to avoid confusion. Part number format is such that it is easy to distinguish between RoHS and non-RoHS components. For example, all RoHS complaint components begin with the letters **RC** in bold e.g. RoHS compliant wire is designated as "RC-10GHavewellCu" while non-RoHS is designated as "10GFinoflexCu". Appropriate awareness in this regard is created amongst employees through training sessions and meetings. The QS has documented procedure. No. xxxx for quality check so as to ensure that components are not stored in wrong place. The procedure for such checks is documented in the quality manual section _____ .*

8.15.6 Checking Contamination

After identifying compliant components, attention should be given towards developing procedures and checks to avoid contamination / mixing of compliant and non-compliant components during various stages of manufacturing process like storage, retrieval, assembly, production, final inspection, packaging and dispatch.

8.15.6.1 Storage.

Special attention should be given to avoid mixing of RoHS and non-RoHS components during storage. The quality system should have procedures for numbering and documented procedures for quality checks so as to ensure that components are not stored in wrong place. The data generated after these checks should be and stored as evidence. The manufacturer's description may look like this:

8.15.6.2 Retrieval

When components are being issued, it should be ensured that the correct component is issued. The quality system should have procedures for issuing and documented procedures for quality checks so as to ensure that wrong components are not issued. Data generated after these checks should be generated and stored as evidence. The manufacturer's description may look like this:

> *While issuing components, appropriate checks are in place to ensure exact components are being issued. The requisition slip contains a column which has to be ticked to ensure this. Further entries are made in appropriate documents to ensure proper issue of RoHS components. The procedure for such checks is documented in the quality manual section _____ .*

8.15.6.3 Assembly and production

Components can also mix during production process. For example, if there is only one wave soldering machine that is being used for RoHS and non-RoHS soldering, contamination can occur during replacement of solder in the bath. Ideally there should be two distinct assembly lines and appropriate quality check to ensure right components in the right line. The data generated during these checks should be stored as evidence. The manufacturer's description may look like this:

> *Two separate lines exist for RoHS and non-RoHS assembly clearly identified by proper sign boards. Appropriate procedures (no. zzzz) and checks are in place during various manufacturing processes like wave soldering, wiring, final assembly etc to avoid contamination and mixing. The procedure for such checks is documented in the quality manual section _____ .*

8.15.6.4 Final inspection and test.

Mixing of products should also be avoided during final inspection and test. In fact, there have to be procedures for final inspection and test to identify any mixing that has happened during production. Sample testing may be carried out to rule out mixing and appropriate data should be generated and stored as evidence. The manufacturer's description may look like this:

> **Appropriate procedure no yyyy for final inspection and test to identify that RoHS products are checked and certified to be in compliance. The procedure for such checks is documented in the quality manual section _____ .**

8.15.6.5 Packaging and Despatch.

Contamination can occur during packaging and despatch for example when non-RoHS packaging material is used to pack RoHS components. Appropriate quality procedures and checks should be in place to avoid this. The manufacturer's description may look like this:

> *Separate packing material exists which itself is RoHS compliant. Final boxes are appropriately marked indicating whether the packed product in RoHS or non-RoHS. Checks are in place to verify this and is documented in the quality manual section _____ .*

8.16   DESCRIPTION OF QUALITY SYSTEM.

Since the CE marking is based on testing and evaluation of only one representative sample of the product, the quality system of the manufacturer should ensure that all samples coming out of the production line are identical to the sample that was tested/evaluated for compliance. This section should give a description to prove how the manufacturer's quality system ensures this. One can include a detailed description of the quality system, production processes, quality checks, quality procedures regarding design/component/vendor change, inventory management and so on.

All the elements, requirements and provisions adopted by the manufacturer shall be documented in a systematic and orderly manner in the form of written policies, procedures and instructions. That quality system documentation shall permit a consistent interpretation of the quality programmes, plans, manuals and records. It shall, in particular, contain an adequate description of:

(a) the quality objectives and the organisational structure, responsibilities and powers of the management with regard to design and product quality.

(c) the design control and design verification techniques, processes and systematic actions that will be used when designing equipment pertaining to the equipment model covered.

(d) the specific procedures to control variants if the DOC applies to similar models based on testing of one variant

Your description make look like:

> *For every product there exists a list of components and subassemblies. If the purchasing department wishes to change components or subassemblies, a procedure is in place to get appropriate authorisation for this change to occur. For every product there exists a defined method of assembly. If the manufacturing department wishes to change the production method, a procedure is in place to get appropriate authorisation for this change to occur.*

(e) the procedures to assess whether a specific change will affect conformity and the corresponding manufacturing, quality control and quality assurance techniques, processes and systematic actions that will be used to assess the change. For example, if an EMC or critical component for product safety is changed, how is it ensured that the change will not affect EMC or safety conformity

Your description for an EMC critical component change may look like this:

> *The EMC authority will review this change in the light of:*
> 1. *Will there be increased emissions?*
> 2. *Will there be increased susceptibility?*
> 3. *In cases where EMI suppression components like EMI gaskets, power line filters/MOVs are changed, the procedure is to review enclosure shielding & filtering aspects. The relevant (emission/immunity) tests may also be repeated.*

(f) the examinations and tests that will be carried out before, during and after manufacture, and the frequency with which they will be carried out.

(g) the quality records, such as inspection reports and test data, calibration data, reports concerning the qualifications of the personnel, etc.

(h) the means of monitoring the achievement of the required design and product quality and the effective operation of the quality system.

(i) RoHS quality procedures and quality checks to avoid contamination / mixing of compliant and non-compliant components during various stages of manufacturing process like storage, retrieval, assembly, production, final inspection, packaging and dispatch.

8.17   TEST REPORTS AND CERTIFICATES.

This section should include copies of all test reports, certificates, declarations both internal and third party which will help to prove that the manufacturer has taken all efforts to make the product compliant with the requirements of the directive. It may be soft copy form.

–   EMC directive: Test reports (in house or third party) of testing as per applicable EN standard.

–   LVD: Test reports (in house or third party) of safety evaluation and testing as per applicable EN standard.

–   MDD: EMC test reports (in house or third party) of testing as per applicable EN standard (EN 60601-1 series) and test reports (in house or third party) of safety evaluation and testing as per applicable EN standard (EN 60601-1 series).

–   RED: EMC test reports (in house or third party) of testing as per applicable EN standard (EN 301489 series) and test reports (in house or third party) of safety evaluation and testing as per applicable EN standard (EN 301489 series).

CONCLUSION

The technical documentation is an all-important process of CE marking since it documents the entire conformity assessment process. As discussed earlier, this is the first document that is demanded by the surveillance authorities if they suspect that the product is dangerous or if there is any complaint. It is important to note that the documentation contains all the conformity assessment data in order to prove that the manufacturer, on his part, has taken all the steps to ensure product conformity. This document can also serve as a legal defence in case of any unauthorised change in the products that affects conformity.

References

Directive 2014/30/EC relating to electromagnetic compatibility.

Directive 2014/35/EC relating to electrical equipment designed for use within certain voltage limits (i.e. the low voltage directive).

Directive 2014/53/EC regarding placing radio equipment on the community market.

Directive 2015/683/EC regarding protection restricting the use of certain hazardous substances in electrical and electronic equipment (EEE).

Directive 2014/53/EC regarding placing of medical devices on the community market

EN 63000: Technical Documentation for the assessment of electrical and electronic products with respect to the restriction of hazardous substances.

EN 62321-1: Determination of certain substances in electrotechnical products. Introduction and overview.

. . .

ANNEXURE - I

List of harmonised standards under the medical devices directive

EN 60118-13:2005 Electroacoustics - Hearing aids -- Part 13: Electromagnetic compatibility (EMC) IEC 60118-13:2004
EN 60522:1999 Determination of the permanent filtration of X-ray tube assemblies IEC 60522:1999
EN 60580:2000 Medical electrical equipment - Dose area product meters IEC 60580:2000
EN 60601-1:2006 EN 60601-1:2006/A1:2013 Medical electrical equipment - Part 1: General requirements for basic safety and essential performance IEC 60601-1:2005
EN 60601-1-1:2001 Medical electrical equipment - Part 1-1: General requirements for safety - Collateral standard: Safety requirements for medical electrical systems IEC 60601-1-1:2000
EN 60601-1-2:2015 Medical electrical equipment - Part 1-2: General requirements for basic safety and essential performance - Collateral Standard: Electromagnetic disturbances - Requirements and tests IEC 60601-1-2:2014
EN 60601-1-3:2008 /A11:2016 Medical electrical equipment - Part 1-3: General requirements for basic safety and essential performance - Collateral Standard: Radiation protection in diagnostic X-ray equipment IEC 60601-1-3:2008
EN 60601-1-4:1996 /A1:1999 Medical electrical equipment - Part 1-4: General requirements for safety - Collateral standard: Programmable electrical medical systems IEC 60601-1-4:1996
EN 60601-1-6:2010 Medical electrical equipment - Part 1-6: General requirements for basic safety and essential performance - Collateral standard: Usability IEC 60601-1-6:2010
EN 60601-1-8:2007 / A11:2017 Medical electrical equipment - Part 1-8: General requirements for basic safety and essential performance - Collateral Standard: General requirements, tests and guidance for alarm systems in medical electrical equipment and medical electrical systems IEC 60601-1-8:2006
EN 60601-1-10:2008 Medical electrical equipment - Part 1-10: General requirements for basic safety and essential performance - Collateral Standard: Requirements for the development of physiologic closed-loop controllers IEC 60601-1-10:2007

EN 60601-1-11:2010
Medical electrical equipment - Part 1-11: General requirements for basic safety and essential performance - Collateral standard: Requirements for medical electrical equipment and medical electrical systems used in the home healthcare environment IEC 60601-1-11:2010
EN 60601-2-1:1998 /A1:2002
Medical electrical equipment - Part 2-1: Particular requirements for the safety of electron accelerators in the range of 1 MeV to 50 MeV IEC 60601-2-1:1998
EN 60601-2-2:2009
Medical electrical equipment - Part 2-2: Particular requirements for the basic safety and essential performance of high frequency surgical equipment and high frequency surgical accessories IEC 60601-2-2:2009
EN 60601-2-3:1993 /A1:1998
Medical electrical equipment - Part 2: Particular requirements for the safety of short-wave therapy equipment IEC 60601-2-3:1991
EN 60601-2-4:2003
Medical electrical equipment -- Part 2-4: Particular requirements for the safety of cardiac defibrillators IEC 60601-2-4:2002
EN 60601-2-5:2000
Medical electrical equipment - Part 2-5: Particular requirements for the safety of ultrasonic physiotherapy equipment IEC 60601-2-5:2000
EN 60601-2-8:1997 /A1:1997
Medical electrical equipment - Part 2: Particular requirements for the safety of therapeutic X-ray equipment operating in the range 10 kV to 1 MV IEC 60601-2-8:1987
EN 60601-2-10:2000
Medical electrical equipment - Part 2-10: Particular requirements for the safety of nerve and muscle stimulators IEC 60601-2-10:1987
EN 60601-2-10:2000/A1:2001 IEC 60601-2-10:1987/A1:2001
EN 60601-2-11:1997/A1:2004
Medical electrical equipment - Part 2-11: Particular requirements for the safety of gamma beam therapy equipment IEC 60601-2-11:1997
EN 60601-2-12:2006
Medical electrical equipment - Part 2-12: Particular requirements for the safety of lung ventilators - Critical care ventilators IEC 60601-2-12:2001
EN 60601-2-13:2006
Medical electrical equipment - Part 2-13: Particular requirements for the safety and essential

performance of anaesthetic systems IEC 60601-2-13:2003
EN 60601-2-13:2006/A1:2007 IEC 60601-2-13:2003/A1:2006
EN 60601-2-16:1998 Medical electrical equipment - Part 2-16: Particular requirements for the safety of haemodialysis, haemodiafiltration and haemofiltration equipment IEC 60601-2-16:1998
EN 60601-2-16:1998/AC:1999
EN 60601-2-17:2004 Medical electrical equipment - Part 2-17: Particular requirements for the safety of automatically-controlled brachytherapy afterloading equipment IEC 60601-2-17:2004
EN 60601-2-18:1996 Medical electrical equipment - Part 2: Particular requirements for the safety of endoscopic equipment IEC 60601-2-18:1996
EN 60601-2-18:1996/A1:2000 IEC 60601-2-18:1996/A1:2000
EN 60601-2-19:2009 Medical electrical equipment - Part 2-19: Particular requirements for the basic safety and essential performance of infant incubators IEC 60601 IEC 60601-2-19:2009
EN 60601-2-20:2009 Medical electrical equipment - Part 2-20: Particular requirements for the basic safety and essential performance of infant transport incubators IEC 60601 IEC 60601-2-20:2009
EN 60601-2-21:2009 Medical electrical equipment - Part 2-21: Particular requirements for the basic safety and essential performance of infant radiant warmers IEC 60601-2-21:2009
EN 60601-2-22:1996 Medical electrical equipment - Part 2: Particular requirements for the safety of diagnostic and therapeutic laser equipment IEC 60601-2-22:1995
EN 60601-2-23:2000 Medical electrical equipment - Part 2-23: Particular requirements for the safety, including essential performance, of transcutaneous partial pressure monitoring equipment IEC 60601-2-23:1999
EN 60601-2-24:1998 Medical electrical equipment - Part 2-24: Particular requirements for the safety of infusion pumps and controllers

IEC 60601-2-24:1998
EN 60601-2-25:1995 /A1:1999 Medical electrical equipment - Part 2-25: Particular requirements for the safety of electrocardiographs IEC 60601-2-25:1993
EN 60601-2-26:2003 Medical electrical equipment - Part 2-26: Particular requirements for the safety of electroencephalographs IEC 60601-2-26:2002
EN 60601-2-27:2006 Medical electrical equipment - Part 2-27: Particular requirements for the safety, including essential performance, of electrocardiographic monitoring equipment IEC 60601-2-27:2005
EN 60601-2-27:2006/AC:2006
EN 60601-2-28:2010 Medical electrical equipment - Part 2-28: Particular requirements for the basic safety and essential performance of X-ray tube assemblies for medical diagnosis IEC 60601-2-28:2010
EN 60601-2-29:2008 Medical electrical equipment - Part 2-29: Particular requirements for the basic safety and essential performance of radiotherapy simulators IEC 60601-2-29:2008
EN 60601-2-30:2000 Medical electrical equipment -- Part 2-30: Particular requirements for the safety, including essential performance, of automatic cycling non-invasive blood pressure monitoring equipment IEC 60601-2-30:1999
EN 60601-2-33:2010 Medical electrical equipment - Part 2-33: Particular requirements for the basic safety and essential performance of magnetic resonance equipment for medical diagnosis IEC 60601 IEC 60601-2-33:2010
EN 60601-2-33:2010/A1:2015 IEC 60601-2-33:2010/A1:2013
EN 60601-2-33:2010/A2:2015 IEC 60601-2-33:2010/A2:2015
EN 60601-2-33:2010/AC:2016-03
EN 60601-2-33:2010/A12:2016
EN 60601-2-34:2000 Medical electrical equipment - Part 2-34: Particular requirements for the safety, including essential performance, of invasive blood pressure monitoring equipment IEC 60601-2-34:2000

EN 60601-2-36:1997 Medical electrical equipment - Part 2: Particular requirements for the safety of equipment for extracorporeally induced lithotripsy IEC 60601-2-36:1997
EN 60601-2-37:2008 Medical electrical equipment - Part 2-37: Particular requirements for the basic safety and essential performance of ultrasonic medical diagnostic and monitoring equipment IEC 60601-2-37:2007
EN 60601-2-39:2008 Medical electrical equipment - Part 2-39: Particular requirements for basic safety and essential performance of peritoneal dialysis equipment IEC 60601-2-39:2007
EN 60601-2-40:1998 Medical electrical equipment - Part 2-40: Particular requirements for the safety of electromyographs and evoked response equipment IEC 60601-2-40:1998
EN 60601-2-41:2009 Medical electrical equipment - Part 2-41: Particular requirements for basic safety and essential performance of surgical luminaires and luminaires for diagnosis IEC 60601-2-41:2009
EN 60601-2-43:2010 Medical electrical equipment - Part 2-43: Particular requirements for basic safety and essential performance of X-ray equipment for interventional procedures IEC 60601-2-43:2010
EN 60601-2-44:2009 Medical electrical equipment - Part 2-44: Particular requirements for the basic safety and essential performance of X-ray equipment for computed tomography IEC 60601-2-44:2009
EN 60601-2-45:2001 Medical electrical equipment -- Part 2-45: Particular requirements for the safety of mammographic X-ray equipment and mammographic stereotactic devices IEC 60601-2-45:2001
EN 60601-2-46:1998 Medical electrical equipment -- Part 2-46: Particular requirements for the safety of operating tables IEC 60601-2-46:1998
EN 60601-2-47:2001 Medical electrical equipment - Part 2-47: Particular requirements for the safety, including essential performance, of ambulatory electrocardiographic systems IEC 60601-2-47:2001
EN 60601-2-49:2001 Medical electrical equipment - Part 2-49: Particular requirements for the safety of multifunction patient monitoring equipment IEC 60601-2-49:2001
EN 60601-2-50:2009 Medical electrical equipment - Part 2-50: Particular requirements for the basic safety and essential performance of infant phototherapy equipment

IEC 60601-2-50:2009
EN 60601-2-51:2003 Medical electrical equipment - Part 2-51: Particular requirements for safety, including essential performance, of recording and analysing single channel and multichannel electrocardiographs IEC 60601-2-51:2003
EN 60601-2-52:2010 Medical electrical equipment - Part 2-52: Particular requirements for basic safety and essential performance of medical beds (IEC 60601-2-52:2009)
EN 60601-2-52:2010/AC:2011
EN 60601-2-54:2009 Medical electrical equipment - Part 2-54: Particular requirements for the basic safety and essential performance of X-ray equipment for radiography and radioscopy IEC 60601-2-54:2009
EN 62366:2008 Medical devices - Application of usability engineering to medical devices IEC 62366:2007

ANNEXURE – II

List of harmonised standards under article 3.1 (b) of radio equipment directive (RED) relating to EMC

ETSI EN 301 489-1 Electromagnetic Compatibility for radio equipment and services: Common Technical requirements
ETSI EN 301 489-2 Specific conditions for radio paging equipment
ETSI EN 301 489-3 Specific conditions for Short-Range Devices (SRD) operating on frequencies between 9 kHz and 246 GHz
ETSI EN 301 489-4 Specific conditions for fixed radio links and ancillary equipment
ETSI EN 301 489-5 Specific conditions for Private land Mobile Radio (PMR) and ancillary equipment (speech and non-speech)
ETSI EN 301 489-6 Specific conditions for Digital Enhanced Cordless Telecommunications (DECT) equipment
ETSI EN 301 489-7 Specific conditions for mobile and portable radio and ancillary equipment of digital cellular radio telecommunications systems (GSM and DCS)
ETSI EN 301 489-8 Specific conditions for GSM base stations
ETSI EN 301 489-9 Specific conditions for wireless microphones, similar Radio Frequency (RF) audio link equipment, cordless audio and in-ear monitoring devices
ETSI EN 301 489-10 Specific conditions for First (CT1 and CT1+) and Second Generation Cordless Telephone (CT2) equipment
ETSI EN 301 489-11 Specific conditions for terrestrial sound broadcasting service transmitters
ETSI EN 301 489-12 Specific conditions for Very Small Aperture Terminal, Satellite Interactive Earth Stations operated in the frequency ranges between 4 GHz and 30 GHz in the Fixed Satellite Service (FSS)
ETSI EN 301 489-13 Specific conditions for Citizens Band (CB) radio and ancillary equipment (speech and non-speech)

ETSI EN 301 489-14 Specific conditions for analogue and digital terrestrial TV broadcasting service transmitters
ETSI EN 301 489-15 Specific conditions for commercially available amateur radio equipment
ETSI EN 301 489-16 Specific conditions for analogue cellular radio communications equipment, mobile and portable
ETSI EN 301 489-17 Specific conditions for Broadband Data Transmission Systems
ETSI EN 301 489-18 Specific conditions for Terrestrial Trunked Radio (TETRA) equipment
ETSI EN 301 489-19 Specific conditions for Receive Only Mobile Earth Stations (ROMES) operating in the 1,5 GHz band providing data communications
ETSI EN 301 489-20 Specific conditions for Mobile Earth Stations (MES) used in the Mobile Satellite Services (MSS)
ETSI EN 301 489-22 Specific conditions for ground-based VHF aeronautical mobile and fixed radio equipment
ETSI EN 301 489-23 Specific conditions for IMT-2000 CDMA Direct Spread (UTRA) Base Station (BS) radio, repeater and ancillary equipment
ETSI EN 301 489-24 Specific conditions for IMT-2000 CDMA Direct Spread (UTRA) for Mobile and portable (UE) radio and ancillary equipment
ETSI EN 301 489-25 Specific conditions for CDMA 1x spread spectrum Mobile Stations and ancillary equipment
ETSI EN 301 489-26 Specific conditions for CDMA 1x spread spectrum Base Stations, repeaters and ancillary equipment
ETSI EN 301 489-27 Specific conditions for Ultra Low Power Active Medical Implants (ULP-AMI) and related peripheral devices (ULP-AMI-P)
ETSI EN 301 489-28 Specific conditions for wireless digital video links
ETSI EN 301 489-29 Specific conditions for Medical Data Service Devices (MEDS) operating in the 401 MHz to 402 MHz and 405 MHz to 406 MHz bands
ETSI EN 301 489-31

Specific conditions for equipment in the 9 kHz to 315 kHz band for Ultra Low Power Active Medical Implants (ULP-AMI) and related peripheral devices (ULP-AMI-P)
ETSI EN 301 489-32 Specific conditions for Ground and Wall Probing Radar applications
ETSI EN 301 489-33 Specific conditions for Ultra Wide Band (UWB) communications devices
ETSI EN 301 489-34 Specific conditions for External Power Supply (EPS) for mobile phones
ETSI EN 301 489-35 Specific requirements for Low Power Active Medical Implants (LP-AMI) operating in the 2 483,5 MHz to 2 500 MHz bands
ETSI EN 301 489-50 Specific conditions for Cellular Communication Base Station (BS), repeater and ancillary equipment
ETSI EN 301 489-51 Specific conditions for Automotive, Ground based Vehicles and Surveillance Radar Devices using 24,05 GHz to 24,25 GHz, 24,05 GHz to 24,5 GHz, 76 GHz to 77 GHz and 77 GHz to 81 GHz
ETSI EN 301 489-52 Specific conditions for Cellular Communication Mobile and portable (UE) radio and ancillary equipment

ANNEXURE - III

List of harmonised standards under EMC Directive

EN 55011:2009/A1:2010 Industrial, scientific and medical equipment - Radio-frequency disturbance characteristics - Limits and methods of measurement
EN 55012:2007/A1:2009 Vehicles, boats and internal combustion engines - Radio disturbance characteristics - Limits and methods of measurement for the protection of off-board receivers
EN 55014-1:2006/A2:2011 Electromagnetic compatibility - Requirements for household appliances, electric tools and similar apparatus - Part 1: Emission
55014-2:1997/A1:2001, Electromagnetic compatibility - Requirements for household appliances, electric tools and similar apparatus - Part 2: Immunity - Product family standard
EN 55015:2006/A2:2009 Limits and methods of measurement of radio disturbance characteristics of electrical lighting and similar equipment
EN 55015:203 Limits and methods of measurement of radio disturbance characteristics of1 electrical lighting and similar equipment
EN 55022:2010/AC:2011 Information technology equipment - Radio disturbance characteristics - Limits and methods of measurement
EN 55024:2010 Information technology equipment - Immunity characteristics - Limits and methods of measurement
EN 55032:2012/AC:2013 Electromagnetic compatibility of multimedia equipment - Emission requirements
EN 55035:2017 Electromagnetic compatibility of multimedia equipment - Immunity requirements
EN 60204-31:1998 Safety of machinery - Electrical equipment of machines - Part 31: Particular safety and EMC requirements for sewing machines, units and systems
EN 60204-31:2013 Safety of machinery - Electrical equipment of machines - Part 31: Particular safety and EMC requirements for sewing machines, units and systems
EN 50370-1:2005 Electromagnetic compatibility (EMC) - Product family standard for machine tools - Part 1: Emission
EN 50370-2:2005 Electromagnetic compatibility (EMC) - Product family standard for machine tools - Part 2: Immunity

EN 60870-2-1:1996 Telecontrol equipment and systems - Part 2: Operating conditions - Section 1: Power supply and electromagnetic compatibility
EN 60947-1:2007/A2:2014 Low-voltage switchgear and controlgear - Part 1: General rules
EN 60947-2:2006/A2:2013 Low-voltage switchgear and controlgear - Part 2: Circuit-breakers
60947-3:2009, EN 60947-3:2009/A1:2012 Low-voltage switchgear and controlgear - Part 3: Switches, disconnectors, switch-disconnectors and fuse-combination units
EN 60947-4-1:2010/A1:2012 Low-voltage switchgear and controlgear - Part 4-1: Contactors and motor-starters - Electromechanical contactors and motor-starters
EN 61000-3-2:2006/A2:2009 Electromagnetic compatibility (EMC) - Part 3-2: Limits - Limits for harmonic current emissions (equipment input current <= 16 A per phase)
EN 61000-3-2:2014 Electromagnetic compatibility (EMC) - Part 3-2: Limits - Limits for harmonic current emissions (equipment input current â‰¤ 16 A per phase)
EN 61000-3-3:2008 Electromagnetic compatibility (EMC) - Part 3-3: Limits - Limitation of voltage changes, voltage fluctuations and flicker in public low-voltage supply systems, for equipment with rated current <= 16 A per phase and not subject to conditional connection
EN 61000-3-3:2013 Electromagnetic compatibility (EMC) - Part 3-3: Limits - Limitation of voltage changes, voltage fluctuations and flicker in public low-voltage supply systems, for equipment with rated current <= 16 A per phase and not subject to conditional connection
EN 61000-3-11:2000 Electromagnetic compatibility (EMC) - Part 3-11: Limits - Limitation of voltage changes, voltage fluctuations and flicker in public low-voltage supply systems - Equipment with rated current <= 75 A and subject to conditional connection
EN 61000-3-12:2011 Electromagnetic compatibility (EMC) - Part 3-12: Limits - Limits for harmonic currents produced by equipment connected to public low-voltage systems with input current > 16 A and <= 75 A per phase
EN 61000-6-1:2007 Electromagnetic compatibility (EMC) - Part 6-1: Generic standards - Immunity for residential, commercial and light-industrial environments
EN 61000-6-2:2005/AC:2005 Electromagnetic compatibility (EMC) - Part 6-2: Generic standards - Immunity for industrial environments
EN 61000-6-3:2007/A1:2011/AC:2012 Electromagnetic compatibility (EMC) - Part 6-3: Generic standards - Emission standard for residential, commercial and light-industrial environments

EN 61000-6-4:2007/A1:2011 Electromagnetic compatibility (EMC) - Part 6-4: Generic standards - Emission standard for industrial environments
EN 61000-6-5:2015 Electromagnetic compatibility (EMC) - Part 6-5: Generic standards - Immunity for equipment used in power station and substation environment
EN 61000-6-5:2015/AC:2018-01 Electromagnetic compatibility (EMC) - Part 6-5: Generic standards - Immunity for equipment used in power station and substation environment
EN 61131-2:2007 Programmable controllers - Part 2: Equipment requirements and tests
EN 61204-3:2000 Low voltage power supplies, d.c. output - Part 3: Electromagnetic compatibility (EMC)
EN 61326-1:2013 Electrical equipment for measurement, control and laboratory use - EMC requirements - Part 1: General requirements
EN 61326-2-1:2013 Electrical equipment for measurement, control and laboratory use - EMC requirements - Part 2-1: Particular requirements - Test configurations, operational conditions and performance criteria for sensitive test and measurement equipment for EMC unprotected applications
EN 61326-2-2:2013 Electrical equipment for measurement, control and laboratory use - EMC requirements - Part 2-2: Particular requirements - Test configurations, operational conditions and performance criteria for portable test, measuring and monitoring equipment used in low-voltage distribution systems
EN 61326-2-3:2013 Electrical equipment for measurement, control and laboratory use - EMC requirements - Part 2-3: Particular requirements - Test configuration, operational conditions and performance criteria for transducers with integrated or remote signal conditioning
EN 61326-2-4:2013 Electrical equipment for measurement, control and laboratory use - EMC requirements - Part 2-4: Particular requirements - Test configurations, operational conditions and performance criteria for insulation monitoring devices according to IEC 61557-8 and for equipment for insulation fault location according to IEC 61557-9
EN 61326 2 5:2013 Electrical equipment for measurement, control and laboratory use - EMC requirements - Part 2-5: Particular requirements - Test configurations, operational conditions and performance criteria for devices with field bus interfaces according to IEC 61784-1
EN 61547:2009 Equipment for general lighting purposes - EMC immunity requirements
EN 61800-3:2004/A1:2012 Adjustable speed electrical power drive systems - Part 3: EMC requirements and specific test methods

EN 62040-2:2006/AC:2006
Uninterruptible power systems (UPS) - Part 2: Electromagnetic compatibility (EMC) requirements

ANNEXURE - IV

List of harmonised standards under LVD

EN 60065:2002/AC:2007 Audio, video and similar electronic apparatus - Safety requirements
EN 60204-1:2018 Safety of machinery - Electrical equipment of machines - Part 1: General requirements
EN 60335-1:2012/A13:2017 Household and similar electrical appliances - Safety - Part 1: General requirements
EN 60664-1:2007 Insulation coordination for equipment within low-voltage systems - Part 1: Principles, requirements and tests
EN 60947-1:2007/A2:2014 Low-voltage switchgear and controlgear - Part 1: General rules
EN 60950-1:2006/A2:2013 Information technology equipment - Safety - Part 1: General requirements
EN 60974-1:2012 Arc welding equipment - Part 1: Welding power sources
EN 61010-1:2010 Safety requirements for electrical equipment for measurement, control, and laboratory use - Part 1: General requirements
EN62040-1:2008/A1:2013, Uninterruptible power systems (UPS) - Part 1: General and safety requirements for UPS

INDEX

Printed in Great Britain
by Amazon